Conversations in the Rainforest

Conversations in the Rainforest

Culture, Values, and the Environment in Central Africa

Richard B. Peterson

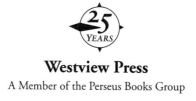

Westview Press
A Member of the Perseus Books Group

All rights reserved. Printed in the United States of America. No part of this publication may be reproduced or transmitted in any form or by any means, electronic or mechanical, including photocopy, recording, or any information storage and retrieval system, without permission in writing from the publisher.

Copyright © 2000 by Westview Press, A Member of the Perseus Books Group

Published in 2000 in the United States of America by Westview Press, 5500 Central Avenue, Boulder, Colorado 80301-2877, and in the United Kingdom by Westview Press, 12 Hid's Copse Road, Cumnor Hill, Oxford OX2 9JJ

Find us on the World Wide Web at www.westviewpress.com

Library of Congress Cataloging-in-Publication Data
Peterson, Richard Brent.
   Conversations in the rainforest: culture, values, and the environment in Central Africa/
Richard Brent Peterson.
     p. cm.
   Includes bibliographical references and index (p.).
   ISBN 0-8133-3709-7 (hc.)
   1. Environmental ethics—Africa, Central. 2. Human ecology—Africa, Central. 3. Environmental protection—Africa, Central. I. Title.

GE42 .P47 2000
179'.1'0967—dc21

                                                                             99-088780

The paper used in this publication meets the requirements of the American National Standard for Permanence of Paper for Printed Library Materials Z39.48–1984.

10   9   8   7   6   5   4   3   2   1

To the farmers and foragers of the Ubangi and the Ituri who shared with me their lives, their knowledge, and their friendship. And in memory of Nsede Papaju, Asani Mbolé, and Jean Bosco.

Contents

Acronyms ix
List of Illustrations xi
Note on Nomenclature xiii
Acknowledgments xv

1 Introduction   1
   Conservation for Whom? 2
   The Limitations of Imported Knowledge, 4
   Thinking About Our Thinking, 5
   The Limitations of Indigenous Technical Knowledge, 6
   Decolonizing the Mind:
      Beyond ITK Toward Afrocentricity, 8
   Afrocentricity and the Environment, 9
   The Tale of the Microhydro and the Mami Wata, 10
   Bridge Blueprints, 13

2 Central African Land Ethics:
   Theoretical Questions and Research Perspectives   21
   Ecology, Values, and Value Transformation, 22
   Environmental Ethics in Global Perspective, 23
   Environmental Ethics in Central Africa, 24
   Universalism and Pluralism: The "One/Many Problem", 26
   Is Nature Normative? Premodern World Views
      and Postmodern Ecological Science, 29
   What Is Africa? Listening to Local People's Voices, 38
   So What? Central African Environmental Ethics
      and Environmental Justice, 43
   Repair and Reconstruction: Signs of Hope
      Grounded in African Soil, 47
   Methods and Research Process, 48
   Summary and Conclusion, 57

3 Narratives on Nature: An Opening Conversation
   with Ubangian Farmers   60
   A Narrative Approach, 62

Loko: Amid Forest and Savanna, 63
Bogofo: An Initial Group Discussion, 68
Conclusion, 82

4 **Parts of a Whole: Nature, Society, and Cosmology in the Ubangi** — 86

Making a Living: Implicit Means of Conservation, 87
Living with Neighbors:
  The Individual-Community Relationship, 106
Living with Meaning in an Interconnected World, 117
Conclusion, 141

5 **Reservations About Nature Reserves: Local Voices on Conservation in the Ituri** — 147

Background: The Ituri Forest, 148
From Mambasa to Epulu, 151
Badengaido: Living Within the RFO, 158
Mbuti Perspectives on the Forest and the RFO, 190
The Bapukele Elders: "The Animals Have Received
  Their Independence and Are Destroying Our Food", 208
The Upshot, 214
Paulin Mboya: "If Local People Don't Support the
  Reserve, It Will Fail", 217
Conclusion, 219

6 **One Step Removed: The Voices of University-Educated Project Staff and Local Academics** — 225

Voices of Contrast, 227
Searching for Common Ground, 241
Conclusion, 250

7 **Conclusion: Lessons for Environmental Practice, Theory, and Ethics** — 253

Loko, the CEUM, and Individual-Community Relations, 254
Epulu, the RFO, and Human-Environment Relations, 259
Conclusion, 275

Appendices — 279
References Cited — 297
Index — 309

Acronyms

ADFL	Alliance of Democratic Forces for the Liberation of Congo-Zaire
CEFRECOF	Centre de Formation et Recherche en Conservation Forestière (Center of Training and Research in Forest Conservation)
CEUM	Communauté Evangelique de l'Ubangi-Mongala (Evangelical Community of the Ubangi-Mongala)
CODIBO	Communauté pour le Développement Intégral de Bodangabo (Community for Integral Development of Bodangabo)
CPCL	comités permanents de consultation locale (standing committees of local consultation)
GIC	Gilman Investment Company
ICCN	Institut Congolais pour la Conservation de la Nature (Congolese Institute for the Conservation of Nature)
IES	Institute for Environmental Studies
ISP	Institut Supérieur de Pedagogie (Institute of Teacher Education)
ITK	Indigenous Technical Knowledge
NGOs	nongovernmental organizations
NYZS	New York Zoological Society
PDC	parish development committee
RDC	regional development committee
RFO	Réserve de Faune à Okapis (Okapi Wildlife Reserve)
WCS	Wildlife Conservation Society

Illustrations

Boxes

2.1	Focus group discussion guide	54
5.1	Notice to employees of CEFRECOF	198

Figures

1.1	Research problem: Connecting divided shores	14
1.2	Democratic Republic of Congo: Political administrative divisions and research bases	17
3.1	Northwestern Congo, Équateur Region: Ubangi and Mongala Subregions	70
5.1	Northeastern Congo, Orientale Region: Ituri Forest and surrounding area	150
B.1	Circles of investigation	290

Photographs

1.1	The major east-west road through the Ituri Forest	2
1.2	A microhydro mill in the Ubangi Subregion	11
3.1	The savanna-forest ecotone near Loko	64
4.1	A Mbandja man gathers caterpillars in the forest near Loko	89
4.2	A group of women farmers plants peanuts in the Ituri	102
5.1	The okapi, a rainforest giraffe endemic to Congo and most abundant in the Ituri	149
5.2	An informational placard about the RFO at the Bronx Zoo	220

5.3	An informational placard about the RFO at the Bronx Zoo	221
6.1	A local villager hired by an Italian commercial logging company wields a chain saw on mahogany logs	235
6.2	Two men produce pit-sawn mahogany lumber in the Ituri	236
7.1	An experimental agroforestry garden at Epulu	269
7.2	Primary forest and plantain gardens intermingled in the eastern Ituri	275

Note on Nomenclature and Congo's Changing Context

It is not uncommon for many changes to take place in a country between the time of one's research there and the time of completing one's writing. In Congo, the scope of these changes goes beyond the ordinary to include a seven-month civil war and the installation of a new president, Laurent Kabila, in 1997. Unfortunately, Kabila has been unable to bring peace and restoration to many parts of the country. Currently, the country is in the throes of a second civil war that has escalated to the point of drawing in the forces and interests of numerous surrounding nations.

Despite these tumultuous changes, many of the stories, perspectives, and conclusions recorded in this book still pertain. The forests, projects, and people I write about remain even though they are going through trying times. Several of the people whose stories I tell have died, and by most accounts current political-economic conditions in the areas of the country in which I worked are worse than they were in 1995. And although the lessons and insights I recently gained have lost some relevance given the mercurial state of affairs in Congo, the research and narratives remain pertinent and valuable.

Knowing which place-names to use in this book has been a difficult decision. In 1995 when this research was carried out and when many of the experiences I record took place, the country was still named Zaire, and longtime dictator Mobutu Sese Seko was still very much in power. Zaire is also the name people used in all of our taped conversations, many of which I quote from directly in the pages that follow. In May 1997, the Alliance of Democratic Forces for the Liberation of Congo-Zaire (ADFL) succeeded in ousting Mobutu and officially renamed Zaire the Democratic Republic of Congo. Official recognition of the country's new name soon followed, and Democratic Republic of Congo reflects a national nomenclature and identity that I wish to respect.

Thus after much deliberation I have decided on the following approach: In most cases I use the name "Congo" except when recounting events or describing situations that are specific to the realities of Zaire in 1995, for example, when I am describing Mobutu's legacy or the social- and political-economic conditions at the time of my research. In such cases I use the phrase "Zaire (Congo)" in the text as well as in the directly quoted narratives; the "State" is often used as a generic reference to the national government. I refer to formerly Zairian authors of secondary sources as "Congolese." In order to protect peoples' identities, I use pseudonyms throughout the book and have changed certain place-names.

Acknowledgments

It's five A.M. The city of Madison is beginning to wake up, and I have yet to go to bed. As the faint fingers of dawn reach up over the horizon and pull the night to a close, so I close this long story with some final words of gratitude. No matter the degree of eloquence I might try to achieve in acknowledging my debts to others, words will always fall short in comparison to the deeds of kindness, direction, and support that I have received during this journey of research and writing.

I am first grateful to the taxpayers of the United States who, whether they know it or not, have supported my research and writing through the National Science Foundation Fellowship Program. Without the Foundation's indirect support and direct financial aid, this journey would have been much more difficult. I am also grateful to Steve Lawry for the kindness of initiating the Lawry Travel Fellowship that helped me get back to Africa.

My journey has been graced by guidance from a number of mentors, all of whom fulfill the ancient and almost holy meaning of that word. Thanks to Cal DeWitt for carefully reading drafts, the ever-present ideas, and quick and helpful feedback. I owe a great deal to Bill Thiesenhusen for the encouragement and inspiration he has given me throughout my years at the University of Wisconsin. Yi-Fu Tuan, Joe Elder, and Aliko Songolo have provided not only valuable assistance but, along with Cal and Bill, served as stellar examples of how to be a part of the academy and not lose one's soul. I am also very grateful to Jan Vansina for sharing with me his thorough knowledge of Central Africa that in numerous ways has deepened my own. It has been an honor to be mentored by all of you.

This book could not have been written without the support and priceless gifts of time, editing, discussion, and feedback offered by numerous friends and colleagues. Thanks to Steve Vavrus for all those breakfast rendezvouses that not only sharpened my thinking but also helped keep me sane. And to Fran Vavrus, I am especially indebted for the excellent insights, encouraging words, and challenging doses of dialectics she brought to this endeavor. Marie Claire Salima eased the pain of transcribing hours and hours of tape, Bob Thornbloom and Ken Satterberg lent their help on deciphering certain difficult Ngbaka phrases, and Maureen Servas beautifully redrew figure 1.1. I am grateful to you all. Thanks to Ajume Hassan Wingo for much enlivening philosophical banter, and to David Giampetroni I owe the world for lavishly sharing with me his skills as a writer and his love as a friend. To the people at Westview—Karl Yambert, Jennifer Chen, Michelle Trader, Cathleen Tetro—

you've been a pleasure to work with! Thanks for your patience, flexibility, and help in getting the word out. And I couldn't have asked for a better copy editor than Jon Howard, whose deft and sensitive alterations polished my prose. Many thanks—whatever errors that remain are my own.

I would likely have had little to do with Africa were it not for my parents, Robert and Ruth Ann Peterson. Their adventure of leaving family and home to work in Congo in 1952 has yielded a gold mine of cultural, intellectual, and spiritual riches for them and for their children over these many years. I am grateful for this (yet another) gift they have given me.

The time I spent "back home" was enriched and facilitated by institutional support from the CEUM (Communauté Evangelique de l'Ubangi-Mongala); CEFRECOF (Centre de Formation et Recherche en Conservation Forestière), and the ICCN (Institut Congolais pour la Conservation de la Nature), all of which went out of their way to help—from lending motorcycles, to granting research clearance, to purchasing supplies. To Mbio ya Kotake, Sebutu, Andabo, Paul and Sheryl Noren, Roy and Aleta Danforth, and many others at Loko—many thanks for the welcoming hospitality, enlivening discussions of people and forests, and logistical assistance. Thanks for the same to Robert Mwinyihali, Richard Tshombe, Kambale Kisuki, Leonard Mubalama, John and Terese Hart, and numerous other staff and workers at Epulu. I am also grateful to Pères Marcel Henrix and Honoré Vinck for their kind hospitality at both Bokuda Moke and Bamanya, for many delightful and enlightening discussions, and for pointing me to obscure and helpful references. For the pleasure and fruitfulness of my stay and research at Centre Aequatoria in Bamanya, I also owe a great deal to librarians Guillaume Essalo and Charles Lonkama. But it is the farmers and foragers of the Ubangi and the Ituri who shared with me the intimacies of their lives, struggles, and joys, their words, food, homes, and communities, to whom I owe an even greater debt. *Merci mingi. Buku oyo ezali mpo na bino.*

Finally, I want to express my deep gratitude to my life partner, Debra Rothenberg, for collaborating in meeting the challenges of fieldwork, for exploring with me the wonders of the natural world, and for sharing in the excitement of intellectual and emotional discovery. Whatever merit this book may have is due in large part to her fine mind and heart. And to my son, Benjamin, I am grateful for bringing many a smile to my face, many a sense of wonder to my soul. You have both not only graciously borne with my absorption in this work but, in more ways than I can name, helped to make it better. Thank you!

Richard B. Peterson

1

Introduction

The bells of Kisangani's cathedral toll six A.M. Swinging my legs over the bed I rise, remembering that I had decided not to take the boat. The dilapidated commercial tug, which was to have taken me back down the Congo River to the Democratic Republic of Congo's Équateur administrative region from where I would head back to America, had left at four A.M. Instead I would travel by Land Rover back to Congo's Ituri Forest, extending what was to have been but a two-week visit to a fledgling conservation project in March 1986. I was quite unaware of the consequences this decision would have on my life.

Later that day, field ecologist Roy Radcliffe and I drive the 460 kilometers to Epulu, a small town on the Epulu River with a long history as a base for foreign researchers.[1] Working with the Radcliffes and their local Congolese team, over the next several months I become absorbed in the exciting challenge of getting a wildlife research and conservation project off the ground. We plane lumber, collect bricks from crumbling Belgian colonial state houses, build and repair houses, replace roofs, and then set off deep into the forest with Mbuti elders Moke and Masisi in search of the proper place to set up a study site. We settle on the confluence of the Afarama and Edoro Rivers as the area in which to initiate the first field-based study of Congo's elusive okapi (*Okapia johnstoni*—a rainforest giraffe). Soon a little ring of wattle-and-daub houses with *mangongo*-leaf roofs (*Marantaceae*—a broad-leaved understory plant) begins to take shape, marking the spot where I will spend much of the next three years.

Now, more than ten years later, I swing my legs over the bed and head downstairs to my basement office in Madison, Wisconsin, to continue my work on a manuscript entitled "The Community of Life: Human Values and the Environment in Central Africa." It is a long way from Kisangani to Madison, a long way from tracking okapi through mud and rain to writing about human values and the environment. Yet across these distances of space and time, persistent questions trouble me, questions that will require bridges from there and then to the here and now. Writing this book is a way to both build and cross those bridges.

Conservation for Whom?

My questions have mostly been about people, especially those living in or near areas of Central Africa's rainforest hallmarked for conservation. Initially these questions centered around the dilemma of how to meet local people's needs while keeping these biologically rich areas of forest intact. Traveling the trans-Africa highway during those first years in the Ituri, I realized that the dilemma in that forest was rapidly becoming more complex as new immigrants continued to arrive from the densely populated highlands to the east, clearing larger and more intensive farms from forests along the road. My specific question became how and why these immigrant farmers moved to the frontier, obtained land, and used land and forest differently as compared to the Ituri's indigenous farmers. The question led me to cross over from field research with Ituri farmers and foragers to study at the University of Wisconsin–Madison's Institute for Environmental Studies (IES).

The major east-west road through the Ituri Forest. The Route d'Ituri has served as a conduit for immigrants from the bordering eastern highlands who bring with them an agricultural style both more intensive and extensive than that of the Ituri's indigenous farmers.

Introduction

At Wisconsin I wrote a master's thesis on the various environmental and social impacts immigrants' farming practices have on the forest and on the Ituri's indigenous peoples, most notably the hunter-gatherers known collectively in the Ituri as the Mbuti. But I also probed the root causes of the environmental degradation I saw. This in turn led me to look beyond the forest to events and structural changes taking place in immigrants' homelands. Specifically, I emphasized how land tenure systems that promote a growing unequal distribution of homeland impel people to migrate to the rainforest. Contrary to the beliefs of some conservationists and to popular images of slash-and-burn cultivators destroying the rainforest, I found that immigrant farmers are rarely the fundamental cause of ecological destruction. Rather their environmental behavior is often strongly influenced by social and political-economic processes beyond their control. This was a fairly straightforward conclusion, yet it left me with a question that became the title of the final chapter in my thesis: "Conservation for Whom?"

As I spent time with the Ituri's farmers and foragers and continued to learn how conservationists view their work of protecting biodiversity, I became more convinced of the fundamental differences between the world views and values of these two groups of people, both of whom are deeply involved in Central Africa's forests. The conflicts I saw and continue to see emerging between conservationists and local forest people, as well as the failure of many environmental sustainability projects to really take hold among local communities, seem to be supremely rooted in these often conflicting ways of looking at and valuing the forest. For the most part, conservationists have justified and promoted the protection of Central Africa's rainforests based primarily on values and perceptions inherent in Western conservation science and culture. They seek to preserve biodiversity, sustain a scientific laboratory, stem global warming, and protect an aesthetic wonder. In contrast, people who live in these areas more often view the forest as the source of their life, both materially and culturally—a source of daily sustenance and of meaning, a resource to be cared for in order that it remain to provide for the generations that follow.

It also became clear how Western scientific ways of learning about the forest—in all its human and nonhuman biological richness—so often blind conservationists to local knowledge, to local ways of learning about and caring for the forest. The very epistemologies that guide much of conservation research often bias researchers against carefully listening to what local people have to say and against looking within indigenous culture for locally comprehensible values and practices to serve as the foundation for conservation projects. To put it broadly, conservation science often biases its practitioners against asking what might comprise the components of a local land ethic.

I began to ponder the types of approaches that would allow researchers to experience more deeply what local people encounter in their daily lives vis-à-vis the forest. I began to ask whether local tradition—within local people's long-standing ways of understanding the human-environment relationship—could serve as a better foundation than researchers' imported epistemologies for the work of building ecological and social sustainability.

The Limitations of Imported Knowledge

One of the consequences of the choice I made that early morning in Kisangani was the beginning of an association with field ecologists and conservationists in Africa through which I have learned a great deal. They are some of the most fascinating people I have met: bright, often brilliant, driven, brave, adventurous, caring, deeply concerned about the loss of natural areas. Sometimes they are also concerned about the people who live within those areas, for many of them have come to realize that to preserve biodiversity and to protect wilderness they must consult and seek authentic participation with the people who live in and around those areas.

This is a very positive step, yet few conservationists have acquired training in "paying attention to the people" who live with the animals and plants and in the ecosystems they are keen to preserve. Mostly trained in the natural sciences, conservationists will often approach the complex study of human-land relations with the same methods they use in their own disciplines of biology, botany, and ecology. Eager to get hard quantitative data, conservation researchers may quantify forest resource use, map human settlement patterns, or determine the amount of time people allocate to different activities. Though this knowledge may be useful, it largely serves the interests and agendas of outside agencies and often bypasses the needs, views, attitudes, beliefs, and interests of local people. Capturing quantitative social data to help determine externally based conservation policy is often given priority over creating rapport and carefully listening to the people who have lived in these wilderness areas since long before foreign scientists arrived. Consequently, resources for building ecological sustainability that exist within local culture often go unnoticed. In the end, despite exhibiting some degree of increased sensitivity to local people, conservationists often remain limited in their ability to truly incorporate local people, their lives, their interests, their needs, their realities, their understandings of the natural world into the important work of sustaining biodiversity.

Over the ten years that I have been involved with conservation and environmental projects in Central Africa, these problems have continued to trouble me. We often fail in our work—whether that is oriented toward conservation of the environment or improving the quality of life for human beings within the environment—when we do not pay sufficient attention to local culture and instead found our projects upon imported meanings, concepts, and practices. Building on foreign rather than local cultural foundations laid down over generations of habitation, we often find that our goals and means make few connections with the ways and understandings of local people. In the end, more often than not our work fails or, like a car running on the wrong kind of fuel, simply sputters along. If we can speak of building structures for environmental sustainability, our structures—like the pine-framed, oceanfront condominiums continuously wiped out by local weather conditions—are often built from materials that simply don't stand up under local cultural conditions.

Thinking About Our Thinking

Earlier I mentioned "bridges" across space and time—from Congo to an American research university, between then and now. What I'm talking about is building a bridge between theory and practice, science and values, quantity and quality, a bridge that information and data must cross if they are to become practical wisdom.

The question of how to increase the cultural viability of ecological and social sustainability projects within Central Africa's forest region serves as one pillar of that bridge. It is based on the conviction that rather than relying solely on imported materials from outside local rainforest cultures we would do well to start building environmental projects with local materials (local ecological values, perceptions, ethics, and epistemologies). But finding those materials, both locally and across the region, is not easy. It takes time—lots of time—talking with local people who live close to the land, reading people who still retain aspects of local culture but who now write about land and people's relations to land from a distance. These have been my tasks in this project: journeying back to Congo to listen to and record what local land-users have to say about their relationship to the natural environments they depend upon; talking with university-trained Africans, both scholars and practitioners, about their views of what a land ethic rooted in the forest cultures of Central Africa might look like; and reading the work of various Africans writing about land and culture. This work is in part a search for local materials, local ways of understanding and relating to nature, with which to build a more firm foundation for ecological and social sustainability within Central Africa's forest region.

But it takes more than one pillar to support a bridge. There are other questions, perhaps, one might say, even broader questions about the environmental problems facing not just Africa, not just conservationists and local people, but the entire planet. The dominant social science paradigm encourages one to focus more narrowly on a specialized field of inquiry, moving up a pyramid of learning to a pinnacle of expertise. But I keep finding myself looking more and more broadly, moving down and outward in my journey of wondering what we are doing wrong to the natural world and what types of changes it will take to truly move us away from destruction to repair.

Having worked at the level of conservation policy, I feel drawn to a deeper level where the fundamental environmental questions have more to do with ethics, spirituality, and value transformation than science and technique. Science and technology can serve as vital instruments for meeting the environmental challenges facing us, but this requires that the heads, hearts, and hands that use them and the values that undergird their use operate more in accord with what the natural world and what certain traditions within the human cultural world have been trying to teach us about how to live well on the planet. It has become clear that the gravity of our environmental crisis requires that we think about our thinking, reexamine our values, open ourselves to other ways of knowing besides our Western scientific epistemology, acknowledging and abiding by limits to what we can know and what we can do. It demands nothing short of *metanoia*—conversion or transformation—not only of our minds but also of our hearts, values, and lifestyles.

In short, I am convinced that any purposeful application of ecological knowledge will be successful to the extent that it is guided by an understanding of the ways human cultures, certain individuals, and specific communities—past and present—create and practice viable environmental ethics. The grand diversity of human culture across this planet offers us a rich and varied array of creative ethical, religious, and philosophical traditions upon which to draw for transformation to a more ecological way of living. Searching those traditions for the ecological and ethical wisdom they offer us as common inhabitants of one earth serves as the second pillar of the bridge.

But how do we tie these two pillars together—the first pillar growing out of more practical questions stemming from my actual field experience on the ground in Central Africa; the second growing out of more theoretical gropings for ethical and spiritual resources for transformation to sustainable livelihoods? I began to see, yet again, that the practical and the theoretical can never really be separated. Some of the same questions holding up my first pillar applied equally to the second. Lessons learned from asking how environmental sustainability projects in Central Africa can be more directly founded on local culture also provide material for the question of how to build a world ethic for living sustainably out of the world's diverse cultural traditions. It has become obvious to me that the theoreticians are looking at some ethical, religious, and philosophical traditions much more closely than others and that African traditions—for one reason or another—are not achieving the attention they deserve.

This two-pillar bridge, then, has brought me to focus on the traditions and thought systems of the various human groups who have lived in the rainforests of Central Africa, to see what they have to offer for creating ecological sustainability both practically on the ground and theoretically as applied to the current debate on global ecological ethics. My primary goal is to increase environmental theoreticians' and practitioners' understandings of the particular ethical and cultural contributions to ecological sustainability found within indigenous Central African communities.

The Limitations of Indigenous Technical Knowledge

One response to the conceptual framework I've outlined is that the work it prescribes has already been done, that the bridge has already been built, that all the questions I ask have already been considered and answered to a degree. One important attempt has indeed been made through the theory and practice of Indigenous Technical Knowledge (ITK, sometimes called Indigenous Knowledge Systems; see Brokenshaw et al. 1980). Applied to many domains, ITK approaches are based on the belief that local people often, if not always, know better than outside experts. Listening to them and building on their local technical expertise can greatly enhance the viability of any environmental project by increasing the level of connection and understanding such projects generate among local people. In this way, approaches that incorporate ITK have definitely improved upon earlier exogenously defined project approaches that still abound despite the influence of ITK.

Yet ITK is not a panacea, nor is it free from generating serious questions. The most obvious question stems from its technical bias. When ITK remains only at the level of technique—exchanging the foreigner's expertise for that of local people—it can easily degenerate into simply another technological fix. Although ITK offers fixes of an alternative variety, it remains predisposed to treating complex and multifaceted problems and solutions as primarily, if not solely, technical matters. In that way ITK approaches can often ignore or fail to seriously consider the many contextual variables—social, cultural, political, and economic—that are often key factors in generating and resolving the dilemmas ITK addresses.

Problems also arise when those professing an ITK approach extract ITK out of its local context and use it for purposes other than serving the needs of those who developed that knowledge. Often researchers will justify such extraction by citing the need to extend the application of ITK to other milieus or to preserve ITK from possible extinction. However, in so doing they transform ITK into an abstraction antithetical to its purported nature of being grounded and fully integrated into the particularities of local environments. As an abstraction, ITK can become more a basis for new institutions, academic course offerings, and careers than a means of solving local people's problems.[2] If knowledge is power, then control of that knowledge indeed becomes a critical issue. With the greater means of control available to them, ITK researchers have sometimes gained more power from ITK than have those within and among whom such knowledge has originated.

Finally, although ITK certainly represents a step in the right direction, it remains locked within the objectivist, empirical bias characterizing most all of Western science. This seriously limits its ability to improve on exogenous approaches to solving environmental problems or to conserving biodiversity since, like them, it rarely engages and attempts to understand the metaphysical, spiritual, and religious dimensions of people's experience and of their relation to land. The realm of land's *meaning*—what land symbolizes and how it is understood in the broader metaphysical context of people's lives—often goes ignored by ITK practitioners attempting to solve objective technical problems generated by human interaction with the land. Rarely do such realms gain any attention from biologically trained conservationists seeking to preserve biodiversity. Yet by not trying to get at people's individual and collective fundamental understandings of nature, of human nature, and of human beings' relation to nature—realms that often require respect for the nonverifiable, the subjective, the numinous—ITK approaches, and certainly Western conservationist approaches, like so many other exogenously founded approaches to conservation and sustainable development, end up building solutions on shallow and poorly rooted foundations. I would like to dig those foundations deeper.[3]

Decolonizing the Mind:
Beyond ITK Toward Afrocentricity

Let me step back from the local focus and examine issues of land and people on a broader scale. A brief discussion of how the concept of Afrocentricity relates to my topic may be helpful.

Approaches such as ITK have certainly lessened the degree of disjuncture between local and exogenous ways of relating to the environment and solving environmental problems in Africa. But we need to go farther. We need to bring metaphysical dimensions of land and human beings' relation to land into the quest for solutions to the real-life environmental challenges many people in Africa currently face.

But how can we understand the meaning of land or people's relationship to land without first trying to understand their views on their own human identity? Such questions are difficult for any of us to answer; they are made especially difficult when one's identity has been subjected to the dehumanizations of colonialism. Yet the determination of many Africans to examine and articulate their own identities in the wake of colonialism's attempts to define them has provided many sources to draw from to enhance our understanding. One is the rather extensive body of work by African writers, philosophers, theologians, and political leaders organized around the theme of Afrocentricity. Among other queries, this literature asks, "What does it mean to be African? What is Africa? How do we define ourselves and our ways of thinking of the world so as to reclaim our authentic voices silenced by so many years of being defined and having our world defined by others?" In the words of Camara Laye, these works represent an attempt to "decolonize the mind," which must characterize any search for understanding the meaning of land and of people's relation to land.

Attempts by Africans to decolonize the mind are not new. They have existed ever since Africa came in contact with oppression at the hands of the West under the slave trade; on through Negritude[4] and African socialism, the political writings of Senghor, Nyerere, and Nkrumah holding up the dignity and humanity of African peoples amid colonial conquest and exploitation; on through many of the struggles for independence waged across the continent. They continue today in the latest versions of emancipation being crafted in South Africa, as well as in the numerous struggles for freedom from neocolonial leaders who try to define Africa and African peoples in ways that enhance only their own positions of power and wealth.

Contemporary African philosophers, writers, and theologians have not failed to keep pace alongside these political articulations and actions on behalf of emancipation and self-definition. Indeed, they have often carried their theorizing directly into the realm of concrete political action. They have moved beyond simply critiquing the West and the liberal tradition, beyond postmodern deconstructions of Africa itself, to bring their important work of reflection, research, and imagination directly to bear on the real-life situations of contemporary African nations and peoples.

Afrocentricity and the Environment

Do such expressions of African identity address questions of land and people's relations to land and the real-life environmental crises Africa currently faces? Yes and no. Very few philosophers and theologians have drawn directly on the rich diversity of African thought to address current environmental problems. The works of Malawian theologian Harvey Sindima, Congolese scholar Mutombo Mpanya, Tanzanian C.K. Omari, and various South African and Zimbabwean theologians and communities of faith do indicate, however, just how rich such an approach can be. In addition, various African scientists and conservationists have begun confronting the Western approaches to conservation under which they have been trained with approaches based on a diversity of African understandings of nature and the human-nature relationship. And one must not ignore the writings of many African novelists and poets in which the theme of land—both its ruin and its restoration—plays a central role.

In addition to my work at the local level, I would like to join with these beginnings by bringing work on African thought, religion, and identity[5] to bear specifically on the question of the environment. This is seldom crossed but extremely important terrain, for it is Africa's environment that will ultimately be the source of its holistic development and healing from the wounds of colonialism. It is also an environment increasingly ravaged by neocolonial operations, from large-scale logging schemes to extractive mining industries to massive hydroelectric dams. Most attempts to repair or avert such ravages have not been based on Afrocentric understandings of land and have often served to exacerbate rather than improve environmental conditions. How can work to sustain African environments be centered on African ideas, concepts, and understandings of land and people's relation to land rather than continue on with the game of dependency on inappropriate Western models?

These are the questions that motivate my work. To try to answer them, and to isolate specific Central African contributions to land ethics, will require a better grasp of Central African metaphysical, spiritual, and religious understandings of nature and humanity. To be useful such knowledge must also be applied to the questions of technique and of practice—the questions ITK jumps to directly. Are such "African land ethics" being put into practice in the contemporary real-life situations of rural African peoples? If so, how? How can such Afrocentric environmental practices and ethics alter theories and policies brought from outside to address Africa's environmental problems? Where do Western and Afrocentric approaches complement each other, and where does each need to be corrected?

Research along these lines, as I have mentioned, requires a review of contemporary African writing on African thought. But it also requires going to and talking with Africans themselves, especially those whose livelihoods remain tied directly to the land. That is what this book is about: listening and giving expression to the voices of

African farmers, hunters, fishers, and foragers who live on and with the land daily. I have chosen not to reduce their voices to percentages or statistics but to simply share the words they used in responding to the following question: "What does it mean for you to live well on the land?" Of course, this question propelled our conversations into often long and involved talk of the meaning of land, of the ancestors and their ways of taking care of the land, of the different forces causing the ruin of the land, of the Western organizations acting under a different approach to how land is to be cared for and the difficulties such different approaches hold for them.

I make no claim that these voices are representative, only that they are sincere expressions of people living, like many of their neighbors, under increasingly difficult environmental conditions. I believe it is as important to listen to these voices as it is to describe—through vast and often impersonal statistical techniques—voice patterns across large decontextualized spaces.

The Tale of the Microhydro and the Mami Wata

A story told to me by one of the Congolese farmers with whom I worked in 1995 illustrates, more clearly than any of the reasons given above, how important it is to look beyond what meets the eye, to look deeper or more broadly than our epistemic biases normally allow in trying to understand and work with any aspect of Africa's environment.

Tata Fulani is a man in his early sixties, born and raised in the Bodangabo region of northwestern Congo. He is an enterprising man with lots of ingenuity, characteristics that have helped him in his long-standing work as a farmer and mechanic, trades he has plied throughout the Ubangi. He has recently returned to his home village and focused his energy on his own farm and a small enterprise centered around the microhydro mill he has built at a small waterfall on the Ebale River. He has faced many difficulties stemming not only from the technical challenges of converting falling water to mill power but also, more significantly, from the repercussions such actions hold when one undertakes them within the context of natural and social worlds charged with spiritual power, expectations, prohibitions, and dangers. Unlike what we have come to believe in the West, here in Central Africa the natural, social, and spiritual worlds are inextricably linked, and one's actions hold consequences on every front. One afternoon while we were sitting on the banks of the Ebale River, Tata Fulani recounted his own experiences with microhydros and Mami wata[6]:

"I really can't give you an explanation for these things because I myself am their victim and it's really a great sorrow.... It happens with a lot of things. For instance, if you're a farmer and you have a good harvest, and you sell some of it and acquire some wealth, may no one in your family happen to die. If they[7] die, people will say they died because you killed them in order that their spirit work for you, and that's

Introduction

A microhydro mill in the Ubangi Subregion. The dam at the left edge of the photo diverts river water through a penstock to the turbine house, where it powers a small (six-twelve inches in diameter), locally made turbine.

how you've grown wealthy.... You can never get a real explanation for it, there is no clear evidence. You ask someone how they know it happens and they will reply only, 'You know all about it, you just don't want to believe it.' But in fact you really don't know where it comes from. And they will begin to see you, ... well take me, I am seen here throughout this zone, from here to the limits of the region, it is said that I am a person who kills people....

"Now with regard to microhydros, people will tell you how important it is to realize that the Mami wata also lives in the river, and that you must first give her a person in order that she agree for your turbine to be built and succeed. Mine, my microhydro, when we were building this site, we brought my father here. It came time to mount the waterwheel. We finished the task and upon my return, I learned that my father was dying. It was a horrible coincidence and I was so discouraged. Here was my father, and people began to say that I had killed him for this work here, the microhydro. And my father, I love him, and he loves me. And in the way he died in my arms, ... now I am without a father, a father who had given me good advice, I am without.... And people when they hear of it, they show me no sympathy, no grief; they see me as a bad person because I have killed my father. So it goes....

"I think it stems partly from jealousy but it's also because people are afraid, they are afraid of death. They are afraid; for example they say, 'If I buy a truck, people will kill me because I have a lot of things.' Or another thinks, 'Eh, if I build a house with a tin roof, people will kill me because they will be jealous of my things and they will kill me.' You see, it is very complicated. Some people run away from taking any action in order that others not kill them. Others are full of fear in that they see someone doing something and say he or she wants to kill people. You see ... it is not a good thing. But people believe in it....

"Another guy came here one day and said to me, 'I've come to ask you to go to the Mami wata here in this river to ask her to give me some money so that I can marry.'

"I exclaimed, 'What? Does a Mami wata live here?'

"He said, 'Yes she lives here. Why are you incredulous? Why are you asking me that? You know that you speak to the Mami wata here a lot.'

'Me?'

'Why do you deny it?'

'Eh, *bandeko*!! I don't know the Mami wata, nor do I speak with her; what would I say to her if I did and for what purpose?'

"And so I began to teach this guy *ti-i-i*.... He never fully agreed with me but he did return home. Later my son came and I told him about it and he said, 'Father, you let him go for nothing. He did you wrong and you had every right to bring him to the State.'

"And I replied, 'My son, I bring him to the State and what do I get?'

'No, no, no' ..., he replied. 'But if that's the case, let's take it to his own people.'

"I agreed but asked him how were we going to do it? My son was smart and went to this guy and told him to come, that Father wants to go ahead with the thing he had asked about. The man came. I told him, 'Okay this thing you are asking for, I've decided that it's best if you write it all out in a letter and then give it to me. Write all the names of the people you would have killed in order that I can then take it to the Mami wata. Write everything in such manner. And when you are ready to write it, you need to write it while you're sitting in the outhouse.'

"He obeyed me—he wrote everything, everything. He brought me the letter, gave it to me and I told him that after two days, I would send word for him to come to receive the answer that has arrived. He went away. Two days later my son went to tell him that his answer had come. They came, but now we took the young man to tell his parents of the problem he had brought to us and all that had ensued. We told them, 'We don't practice these things but neither do we want to make a big affair out of it. We only want to let you know that he is not a good child. What we are doing here on my farm is not what he thinks it is.'

"And so they took the letter and read it in front of all his people. Those whose names he'd written down to be killed were sitting right there and heard everything, everything. It became a huge *likambo* (problem). His parents and elders were furious with him and put him in a major prison. But I said to them, 'The real reason for all of this is only because he is lazy. If he cuts a garden and sells the produce he will get some money easily. You don't get money from the wind, you don't get money by sitting, you only get it from the sweat of your brow. If you take him before the State, they will only levy a fine and you yourselves will have to pay it because of his laziness. The truth is you did not raise him well and now the laziness with which you raised him has brought him to this type of affair and it is your shame. To draw attention to this and make it a big deal will only add to your shame.' They agreed and said to him it's all over and they let him be. Look at what happens to a person like that....

"These days it's hard for someone to start a project like this. Take the experience of Olomi: ... The local chief died and the people said he died because Olomi had built a microhydro. It is not good."

* * *

This story reminds us that before we conserve forests, before we establish projects for building environmental and social sustainability, before we gather ITK, we must first take very seriously people's beliefs about the environment and about spiritual forces that affect their relationship to the environment. Local cosmologies that affect how people perceive land and the symbolic realms within which people give meaning to land provide the deep foundations that are needed for building durable structures of sustainability. Without understanding these symbolic realms, we often miss important realities that affect the failure or success of our work.

Bridge Blueprints

The garage and technical services of the small mission station in Congo where I grew up jokingly calls itself Fiasco Engineering. Their logo is a beautiful bridge; its roadway arches gracefully from each shore but then mysteriously fails to connect in the middle, one side a good several meters below the other. An engineer scratches his head underneath the words, "FIASCO ENGINEERING: GETS IT RIGHT THE FIRST TIME."

In attempting to build my own bridge linking the shores of practice and theory, the local and the global, ecology and religion, the natural and the supernatural, I often feel like that engineer. It is hard to make connections between shores with a long history of separation, each with inhabitants who still protest against any bridge-building. It is well nigh impossible to get it right the first time. Yet if such challenges continue to thwart even tentative attempts at making connections, then the two divided shores will continue producing incongruities such as those I have alluded to. Conservationists, trained primarily in the natural sciences, will continue to bypass the cultural, religious, and practical resources for ecological sustainability within local communities. Meanwhile, ethicists, theologians, and an increasing number of scientists who recognize the important roles culture, religion, and ethics can play in promoting conservation will continue failing to ground their discourse in specific

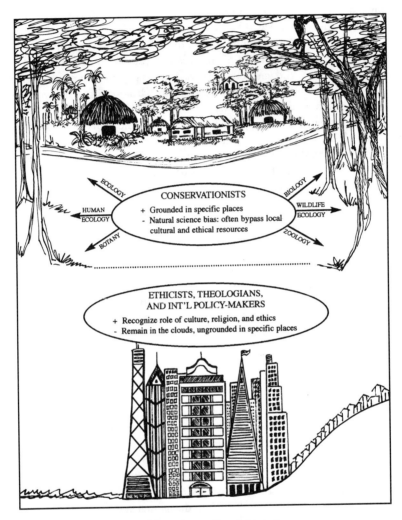

FIGURE 1.1 Research problem: Connecting divided shores.

places where culture, conservation, and ecology interact. Furthermore, their discourse will continue to favor certain cultural resources over others (see Figure 1.1).

The incongruity between these two related yet disparate situations provokes a research problem from which I have derived three questions that form the basis of this work:

1. What are some of the indigenous, locally comprehensible cultural components of Central African land ethics?

2. How can actual projects working for ecological and social sustainability in the region be based more firmly on these culturally grounded foundations?
3. What and how can Central African land ethics—drawn from within specific communities living close to the land, as well as from those who reflect on the land from a distance—contribute to the international agenda of shaping global ethics for sustainable living?

In seeking answers to these questions, I hope to contribute both to local practical actions for sustainable livelihoods grounded on Central African soils, as well as to ongoing work within international policy arenas where global ecological ethics are gaining increasing recognition. Within the latter, this work helps to fill the hole in the debate left by the underrepresentation of African cultural perspectives on what constitutes "right" relationship between humans, land, and God.[8]

In Chapter 2, I set these questions within the broader theoretical context they demand. I first ask what it might take to move people to adopt a more ecologically balanced livelihood. How can and does sincere transformation actually take place? I then inquire into the meaning of environmental ethics within both global and Central African contexts, an inquiry that provokes some thorny theoretical problems:

- the "one/many problem," or the relationship between universalism and pluralism in the quest for environmental ethics;
- the nature-culture relationship and the ways in which it has been observed by premodern world views and by postmodern ecological science;
- the questions of "What is Africa?" and "Whose discourse on Africa should one listen to?";
- the problem of the discrepancy between environmental ethics and environmental realities especially in the African context; and
- the problem of connecting ethics and action—are there any concrete signs of hope grounded on African soils?

After offering this theoretical context, I describe my research process and the particular methods I have used to learn about Central African land ethics (see Appendix B for suggested additional avenues of exploration into my topic). I conclude the chapter with a summary of the key points of reference underlying this work.

Chapter 3 contains an initial account of what local people who still make their living directly from the land have to say regarding what it means to live well on the earth. I begin the chapter by describing and defending the narrative approach that I have chosen for writing about the people with whom I conducted my research, as well as for writing about my own experience of doing research. This includes describing my journey to Loko, one of the two environmental projects in northern Congo that serve as case studies in this work. Loko is a nonprofit, mission–affiliated project involved in public health and environmental sustainability work in the

Ubangi and Mongala Subregions of northwestern Congo (see Figure 1.2). It is supported by and affiliated with the Evangelical Covenant Church of America and the CEUM, a Congolese Protestant church involved in ministries of education, health care, literacy, and sustainable development. I am quite familiar with the region surrounding Loko, having been born and raised there.

After I set the stage in Chapter 3, I center on an opening conversation I held with a group of farmers living in the village of Bogofo, near Loko. My method is to let these Ubangian[9] farmers speak for themselves regarding the cultural, ethical, and practical resources for ecological ethics from within their own cultural systems, past and present. But I also include my own voice, at times recounting my side of our exchange, at times analyzing what these farmers tell me, bringing their words and my interpretations into conversation with other voices in other literatures. I conclude the chapter by drawing out from this single but incredibly rich conversation several key points about Central Africans' relationships to the land, which I continue to examine in the subsequent chapter.

Chapter 4 remains situated in the Ubangi and adds both complexity and contradiction to the themes of Chapter 3. Its material comes from my conversations with individual land-users most often recorded out on the land and preceded or followed by our working the land together. The chapter is organized around three relational themes: the relationship between humans and land (in its broadest sense, the natural realm), humans and each other (the social realm), and humans and God (the cosmological realm). However, its primary goal is to show how thoroughly these relationships are intertwined, how none of them can be viewed in isolation. Thus Chapter 4 highlights the interconnectedness between these three key relationships as illustrated in the narratives of the local land-users with whom I worked, as well as in the writings of other African observers of land and culture. I also discuss some of the implications such interconnectedness holds for contemporary environmental actions, especially in light of the ways in which the bonds holding these relationships together have been and continue to be frayed by a variety of forces. I conclude the chapter by pointing to several actions that provide some hope amid the dramatic changes Central African people and their traditions have undergone and continue to go through.

In Chapter 5, I journey to the Ituri Forest, my second area of research (see Figure 1.2). The chapter's case study is based within the communities of and near Epulu, a small forest town and center of the newly created Réserve de Faune à Okapis (RFO), a 13,000-square-kilometer multiuse conservation area. Epulu is also the home of CEFRECOF, a tropical forest training and research center involved in botanical, zoological, and socioecological research and the training of Congolese scientists. CEFRECOF is jointly administered by the ICCN and NYZS/The Wildlife Conservation Society. I am also familiar with the Ituri area, having lived in Epulu from 1987 to 1989 while working for the Wildlife Conservation Society.[10] The Ituri is also the region where I conducted graduate research on human immigration into the rainforest.

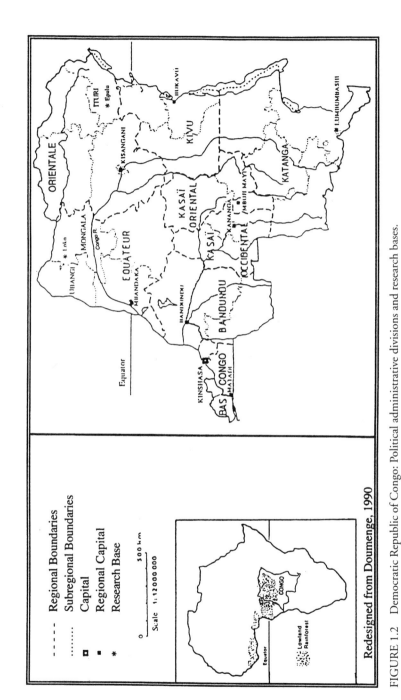

FIGURE 1.2 Democratic Republic of Congo: Political administrative divisions and research bases.

In Chapter 5 I focus on the relationships between local communities and the RFO. The chapter's primary purpose is to record the sentiments of local people living in and near the RFO in regard to the reserve and the way in which it is affecting their lives. Such a case study of local communities and conservation provides insights into how Central African perspectives on the relationship between people and land differs significantly from those assumed by the theory and practice of conservation science. The chapter concludes with a summary of these differences, preparing the ground for the discussion of their implications for the various conservation projects working in the Ituri that I take up in the final chapter.

Prior to that, however, in Chapter 6 I offer up additional voices to this examination of Central African land ethics by discussing the major themes that emerged from two focus groups I conducted with the university-trained staff of both the Loko and Epulu projects. Our focus group discussions centered on the environmental challenges the staff face in their work and what, if anything, they feel their ancestral traditions have to offer in the way of solutions. The chapter also includes the voices of two professors teaching at local institutes of higher education. It is important to hear the voices of project staff and local academics, in addition to those of local land-users, since their perspectives also shape the ways in which social and environmental sustainability can take root within local communities. The chapter points out the differences as well as the similarities between the views of the university-trained and those of local land-users in order to clarify areas needing negotiation and areas of common ground on which to further the work already begun.

Finally, in Chapter 7 I conclude the book by returning to the two case study projects at Loko and Epulu and applying what I have learned to what I perceive are the paramount problems facing each project. In the case of Loko, I focus on the relevance of Central African understandings of the individual-community relationship for Loko's work of promoting sustainability. This is followed by exploring the implications Central African understandings of the human-environment relationship hold for the conservation efforts taking place within the Ituri's Réserve de Faune à Okapis. I conclude the chapter by discussing what Central African ways of thinking about land and people can contribute to environmental theory and the growing debate over global environmental ethics.

I turn now to the task of situating this inquiry into the nature of Central African land ethics within a broader theoretical context.

Notes

1. Epulu used to be called Camp Putnam after Patrick Putnam, a Harvard anthropology student who ended up settling along the beautiful river in the 1930s, operating a sort of zoo and lodge for trans-Africa travelers (see Mark 1995). Later the Belgian colonial government set up a station for capturing okapi under the directorship of Captain Jean de Medina. In the late 1950s, the anthropologist Colin Turnbull, who was to bring knowledge of the Ituri to many through his classic *The Forest People*, used Epulu as the base for his work with the Mbuti.

Introduction 19

2. One example is CIKARD, the Center for Indigenous Knowledge for Agriculture and Rural Development based at Iowa State University; however, one can find evidence of the institutionalization of indigenous knowledge systems throughout the academic realm, from course offerings to conferences to computer networks. The following quote, taken from one of CIKARD's numerous web pages, illustrates my point: "[CIKARD]... focuses its activities on preserving and using the local knowledge of farmers and other rural people around the globe.... Its goal is to collect indigenous knowledge and make it available to development professionals and scientists" (URL: http://www.physics.iastate.edu//./cikard/CIKIntro.html). Another CIKARD web page provides links to the web sites of fourteen other such centers around the world—nine in Africa, three in Europe, one in Canada, and one other besides CIKARD in the United States (http://www.physics.iastate.edu//cikard/worldtext.html).

3. Let me clarify that in speaking of the local, or African and the nonlocal, or Western, I do not mean to imply that the former, by the fact of its being local and African, is automatically superior or closer to the "truth"; I am speaking more to the matter of "cultural fit" and do not mean that the local is in any sense more pure or less adulterated than the nonlocal. My research has made it very clear to me that local farmers' views on "tradition" or on land and land use are filtered through their numerous and complex experiences of both the local and the nonlocal. For instance, many of the local people and all of the project staff and local academics I worked with have been heavily influenced by the perspectives and beliefs of Western Christianity and Western education. In sum, although I often tend to be more critical of the nonlocal West than of the local African, I am aware of the dangers of romanticizing the African and the local and of treating them in any essentialist fashion.

4. Negritude began as a student movement in Paris in the 1930s. It soon mushroomed into a forceful and widespread political and philosophical movement critiquing colonialism and affirming Black identity. It is exemplified in the writings and actions of Léopold Senghor, Aimé Césaire, Léon Damas, and Jean-Paul Sartre, among others, and it played a key role in bringing independence to the continent.

5. I primarily consult works written by Africans not because their Africanness in and of itself brings them closer than non-African Africanists to the "truth" but because by virtue of being born and growing up within an African culture they have experienced Africa very differently than have non-African Africanists. Neither do I presume that their perspectives are free of Western influence; in fact the interaction of their African heritage and the Western influences on their lives in many cases adds to their ability to reflect on these issues.

6. Mami wata are water spirits, belief in whom is found throughout much of West and Central Africa; see Drewal 1988.

7. In both Lingala and Swahili, the third person singular pronouns (*ye* and *yeye* respectively) are gender-neutral. I have tried to match this gender inclusivity in English by using the plural "they" except where such use changes the intended meaning. In such cases, I have either alternated use of masculine and feminine pronouns or included both. When narratives explicitly refer to one gender or the other, I have translated the pronoun as "he" or "she" accordingly.

8. In this book, I use the word "God" mostly to avoid the stylistic awkwardness of such terms as "Supernatural" or "Ultimate Reality." In using the term I do not imply a preference for any one understanding of "God" or even for a necessarily theistic world view.

9. Although Loko and Bogofo are located in the Mongala and not the Ubangi Subregion, here and throughout the remainder of the book I will use the term "Ubangi" to refer in general to the area of northwestern Congo situated beneath the bend of the Ubangi River. This includes

all of the Ubangi Subregion and the Mongala Subregion north of the Congo River. Such usage of the term has precedents (see Vansina 1990, 66, 116–117, 257).

10. I served as project manager for Projet Okapi, a Wildlife Conservation Society project, the principal focus of which was to conduct the first ecological study of the rare rainforest giraffe in the wild.

2

Central African Land Ethics: Theoretical Questions and Research Perspectives

Before we listen to the voices of the Central African farmers and foragers, project staff, and local academics with whom I worked, it is necessary to situate the paths of inquiry we traveled together on a "map" of the theoretical territory that surrounds them. Opening up that map, I have encountered a varied and complex terrain, a landscape with uncharted regions, little-traveled domains, and lots of question marks scribbled here and there. But I have also discovered lots of notes, guideposts, and directions left by other travelers who have ventured here before me. Without those aids, I would have likely folded up this map and looked for an alternative terrain that afforded easier passage. Instead I have ventured a course across difficult territory, a course, I must underline, that is only one of many possible alternatives.

My route begins with and winds through a number of theoretical questions, and the ensuing discussion will illuminate how my path of inquiry relates to and is informed by broader fields of investigation. First I examine how people manifest their values in relation to the earth (i.e., their "environmental ethics"). I ask What are the various means through which those values are revealed? and also By what means might those values be transformed? I then offer both global and Central African perspectives on this manifestation and transformation of people's environmental values through the diverse media of cultural expression.

This discussion inevitably traverses tricky terrain that includes the following question: How does the universal need for improving our relationship to this one earth—home to all peoples—jibe with the incredibly diverse, culturally grounded possible means of doing so? I contend that universalism and pluralism are not mutually exclusive; neither are the views of the powerful culture of science and the views of other cultural traditions in regard to how people do and "should" relate to the earth. It is important that I situate my task within these broader dilemmas, these various facets of the "one/many problem," in order to show its relevancy across time and space. The present and the past, the one and the many—all have a part to play in helping people achieve better ways of living with the earth.

After crossing through this somewhat abstract territory, my course heads directly into Central Africa, leading me to ask these difficult questions: Which Africa and whose Africa am I traveling in? Is this journey of discovery worthwhile, for even if there are Central African land ethics, have they made any difference? How does one reconcile Central Africa's contemporary environmental realities with whatever environmental ideals may be found there? Are there *any* signs of hope, signs that people are achieving sustainable livelihoods even amid excruciating difficulties?

After exploring those questions, I finally reach the local places and paths of inquiry I traveled with local land-users, project staff, and local academics. I conclude with a brief discussion of the methods we used to travel these paths together and a summary of the key points of reference that have influenced my course.

Ecology, Values, and Value Transformation

It is not the ecologists, engineers, economists, or earth scientists who will save spaceship earth, but the poets, priests, artists, and philosophers.

—**Lawrence S. Hamilton**

Contrary to the statement above, safeguarding the earth demands the contributions of both the sciences and the humanities; it demands, as I've suggested, a bridge connecting the two. Such statements, however, do reflect a growing willingness to acknowledge that at its roots the environmental crisis is a crisis about values, about what we as humans desire and believe at the inner core of our beings. Professor of ethics J. Ron Engel articulates the problem clearly:

> Contemporary threats to biodiversity are of such an order of magnitude that it is difficult to conceive how more and better management, knowledge, education, political participation, or economic incentives will suffice. The grip of the modern development world view is so strong that only a fundamental shift in what people believe to be of ultimate concern will be powerful enough to motivate them to search for a more ethical relationship to the diversity of life, and effect the change in heart and in social behavior required.
>
> Historically, religious myth, symbol, and ritual have served as the primary vehicles for ... motivating personal and collective transformation (Engel 1993, 193).

Let me be clear from the outset: This work in no way denies the need for good science; for sound economics in service of fostering community and ecological well-being; for appropriate engineering of technologies geared to sustainable lifeways; and for more education. All of those are extremely important, but they do not provide the context for this work. Its theoretical context is instead the exploration of fundamental human values vis-à-vis the natural world and how those values can be transformed—of the questions "How does conversion to ecologically sound livelihood happen?" and "What motivates people to change from paths of destruction to paths of restoration?"

I begin by supporting Engel's hypothesis that meaningful transformation happens primarily through the means of religious myth (or, more broadly, story), symbol, and ritual, means largely misunderstood, demeaned, and destroyed by the ideas and forces of the modern period. Now in a postmodern[1] age that recognizes modernity's biases and damages, these seeds of transformation may be finding new soil. Evidence of this reappraisal of the symbolic and religious, especially in regard to our need for transformation and new ecological ethics, abounds, not only within the humanities but also within the sciences and international policymaking.[2]

Personal accounts of transformation also note the key role played by symbolic action and meaning. Whether through hearing patterns of truth in a story's profound metaphors or re-encountering the transcendent nature of existence through ritual, people speak of experiencing a profound "turning around" whereby "the scales [fall] from our eyes, and we perceive our place and the places of others in a different way than we did before" (Sallie McFague, in symposium dialogue recorded in Rockefeller and Elder 1992, 178). Old Testament scholar Walter Brueggemann, in his book *Finally Comes the Poet* (1989), articulates the recognition common to many that the "places of resistance" within us that keep us from change can finally be reached only by story, metaphor, and symbol, not by new rules and laws. During a recent conference entitled "Institute for Ecology, Justice, and Faith," restoration ecologist Bill Jordon spoke convincingly that one reason the modern environmental movement is failing is its "puritan relationship to ritual." Currently he is working to create ways in which ritual and restoration can be combined so as to transform restorationists as well as degraded natural areas, to combine technical ecological efficacy with transformative human meaning. Finally, Barry Lopez has articulated as well as anyone the transformative power of story in bringing our "interior landscapes" into line with the truths we discover in the "exterior landscape":

> A story draws on relationships in the exterior landscape and projects them onto the interior landscape. The purpose of storytelling is to achieve harmony between the two landscapes, to use all the elements of story—syntax, mood, figures of speech—in a harmonious way to reproduce the harmony of the land in the individual's interior. Inherent in story is the power to reorder a state of psychological confusion through contact with the pervasive truth of those relationships we call "the land" (Lopez 1988, 68).

This book rests on the belief that human stories, rituals, and symbols are key not only in revealing ecological ethics but in transforming them as well; that in order to build bridges in the world we must first build bridges within ourselves.

Environmental Ethics in Global Perspective

If Engel's hypothesis and these witnesses' observations are correct, then it behooves us to draw upon the vast diversity of the world's cultural and philosophical traditions for resources to aid us in the arduous task of transforming values and bringing human behavior more into line with ecological imperatives. Several writers have voiced the vital need for just this sort of work:

> But we need to allow "every voice to be lifted," to gather together in mutual interaction and transformation the many cultural heritages of humanity, some that have been unjustly dominant and yet do not lack precious resources, and others that have been deeply silenced and rightly claim space to flower again (Ruether 1992, 11).

> Diverse conservation values and ethics—differently grounded and articulated, but pragmatically convergent—should be developed out of indigenous world views and asserted as a matter of local autonomy and pride (J. Baird Callicott, quoted in Engel 1993, 189).

> One task on the agenda ... is to reconceptualize our inherited moral ideas so that they can do justice to the full complexity of interactions within and between biological and social communities.
> This task is much too large for any single ethicist or combination of ethicists working in isolation in different parts of the world, or for any one school of thought or particular cultural, scientific, philosophical, or religious perspective. It requires that ethicists and scientists from different religious, philosophical, and cultural backgrounds have the opportunity to work in close and sustained collaboration with one another (Engel 1990, 19).

However, all of these cultural resources for building ecological ethics must also be informed by sound scientific knowledge about how nature works. Myth, symbol, and ritual that contradict relationships in the natural world can be more destructive than constructive of sound ecological ethics. All three of the authors above thus concur with other colleagues in that every tradition will need to be transformed or reconstructed so as to comply with twentieth-century revelations and realities. Those include, within postmodern physics, the account of a common creation story, as well as ecology's insights on biotic relationships; but they also include the technological possibility of nuclear catastrophe and the untold ecological devastation that has resulted from historically unprecedented growth in population and material consumption. Such revelations and realities were hardly imaginable when many traditions were born. As ecofeminist theologian Rosemary Radford Ruether puts it: "There is no ready-made ecological spirituality and ethic in past traditions.... The radical nature of this new face of ecological devastation means that all past human traditions are inadequate in the face of it. Whatever useful elements may exist ... must be reinterpreted to make them usable in the face of both new scientific knowledge and the destructive power of the technology it has made possible" (1992, 206).

Environmental Ethics in Central Africa

Ruether also reminds us that some of these traditions have received less attention than others; some have even been deliberately silenced. Indeed, most writing within the field has until recently focused on the bases for ecological ethics found within Western traditions. Perhaps that is appropriate since most of the authors are Westerners. Significant problems arise, however, when Western traditions are posited as ultimately superior to all other cultural traditions for serving as the basis for universal ethical principles. Such tendencies fall error to what African philosopher Tsenay Serequeberhan calls the "metaphysical delusion" that "European humanity is properly speaking isomorphic with the humanity of the human *as such.*" Under such a delusion, "European cultural-historical prejudgments are passed off as transcendental wisdom!" (1991, 7, italics in original).

As part of the effort to counter such tendencies, I would like to explore cultural resources found outside the Western tradition, specifically the contributions found within African ethical and environmental thought and practice. Much work has been done on the contributions to ecological ethics offered by Asian, Australian, and Native American cultural traditions, but African contributions have received far less attention,[3] perhaps for the same reason why so much of Africa has been ignored: the faulty assumption that it has been more a source of problems (including environmental ones) than a source of solutions. This work, then, seeks to nullify those incorrect assumptions and correct that imbalance by collecting and analyzing a variety of African cultural resources for their eco-ethical principles, exploring how they might promote a change in value orientation toward the natural world. Such principles, I hope to demonstrate, have great relevance for all peoples seeking to restore an ecologically sound relationship to the earth.

However, Africa is alive and pulsing with diversity,[4] and any continentwide generalization becomes immediately suspect. Yet within the immense number of cultural and philosophical variations found throughout the continent some distinct patterns emerge. Such a view—which illustrates the false dichotomy between generality and particularity—is held by Ghanaian philosopher Kwame Gyekye. Gyekye writes that evidence of common features found across African cultural systems justifies the "assertion that ethnic pluralism does *not* necessarily or invariably produce absolute verticalism with respect to cultures, allowing no room for shoulder rubbing of any kind" (1987, 194, italics in original). Gyekye goes on to recount evidence from numerous ethnographical and philosophical studies from throughout the continent in support of such commonalities.[5] To speak of general patterns emerging out of the immense diversity of Africans' perceptions and relations to the natural world is not to deny the particularity and pluralism vital to any search for globally applicable environmental ethics. It is rather to illustrate the dialectical[6] relationship between universalism and pluralism, particularity and generality. I have more to say about this relationship below.

Nevertheless, it remains less problematic to speak of commonalities on a regional rather than a continental scale. One simple reason is the ecological and geographical variety of the African continent. If it is true, as Robin Horton argues, that African theoretical systems are closely adjusted to the "prevailing facts of personality, social organization, and *ecology*" (1967, 172, emphasis added), then it will be hard to make any accurate generalizations between rainforest-dwelling peoples, for instance, and their Sahelian pastoralist neighbors to the north. For this reason, among others, Africanist historian Jan Vansina contends that there did exist a common tradition for the forested region of equatorial Africa (1990), but to extend the scope of such a tradition to, or even to speak of the same singularity within, East Africa or southern Africa is much more problematic (Vansina, personal communication). Therefore, although I may draw from literatures that do not set such clear regional limits, my research focuses most extensively on the forested region of Central Africa.

Prior to bringing the Central African context of this work into focus, it may be helpful to see how its subject matter projects against more universal backdrops. No attempt to shape global environmental ethics out of the world's diverse cultural traditions can ignore the relationship between universalism and pluralism, the problem of the one and the many. Nor can it ignore the question of how the views of science, particularly the science of postmodern ecology, compare to the views of other cultural traditions on the relationship between humans and the environment.

Universalism and Pluralism: The "One/Many Problem"

We live in a postmodern age, one that champions particularity over generality, pluralism over universalism. Such preferences serve as important correctives to the ways in which knowledge of both natural and social realities has been socially constructed so as to allow dominant groups to maintain power over others. Any talk of a universal or world ethic for sustainable living is therefore immediately suspect by many as being some new type of colonialism. Ecotheologian Sallie McFague summarizes the concern:

> There is a danger of a new universalism. We are at a time when women and a number of other people whose voices have not been heard—the so-called underside of history— are insisting upon different experiences, different voices, social forms different from the oppression under which they have lived. The danger is that we could move into a new type of the universalism which in the past has been a mask, usually, for white male European/American views.... So when asked about a universal ethic, I in turn ask: is this some kind of a watering-down of differences? Is it a refusal to entertain now the very different voices that are coming to us and that have not been allowed to be part of the discourse in the past? ... It's terribly important that the voices that have not been heard now be heard. This includes the voice of the creation (part of symposium dialogue recorded in Rockefeller and Elder 1989, 184).

Such concerns are valid and demand full attention. Any attempt to achieve eco-ethical unity at the expense of marginalizing less powerful voices is unethical on its face. Yet in light of the world's panoply of cultural traditions bearing a multitude of different perspectives, how can we ever reach common ground for confronting our global ecological crisis together? The task is made even more daunting by advocates for cultural relativism arguing that only those "on the inside" of a tradition can rightfully judge its worth. The danger of becoming stuck between scylla and charybdis, between the ethical imperialism of universalism and the ethical gridlock of cultural relativism, is as real as the ecological crisis itself. It is ironic that postmodern correctives, when pushed to the extreme, end up creating dualistic binds not unlike those they meant to remedy. Fortunately, our choices are not limited to a rock and a hard place: There is a middle passage.

Freya Matthews, an Australian ecofeminist philosopher, does a good job of charting out that middle passage. She is fully in favor of pluralism when it comes to crafting a global ecological ethic but finds this no warrant for relativism:

> To insist that we must draw on as wide a range of cultures as possible in piecing together a complex, multi-perspectival ethics is not the same as saying that we must simply accept the rightness of every culture's distinctive attitude to its own environment.... If we allow that the West has got things wrong ... then the possibility has to be allowed that other cultures may have got certain things wrong as well (1994, 5, 6).

Echoing Matthews' concerns about relativism, geographer Yi-Fu Tuan has remarked how relativism disguised as open-mindedness, or the "politically correct" belief that only insiders hold the truth, is actually an extreme close-mindedness that cares so little about the other's point of view that it leaves it unexamined, thus bypassing a rich opportunity for discussion, collective evaluation, and mutual learning (1994b).

Furthermore, distinctions between insiders and outsiders are increasingly becoming less discrete as the entire globe becomes both more interconnected and endangered. In light of such globalization, there remains a grave need to take cultural particularity seriously and respectfully but not to assume, as does cultural relativism, that all practices or all beliefs of all cultures are equally legitimate. Rather globalization, global ecocrisis, and our dire need to benefit from cross-cultural learning require that we sift and winnow all traditions *both from within and from without* in order to draw out from each the most ecologically and socially sound ideas. Sewing these ideas together into a colorful eco-ethical patchwork quilt that is whole yet diverse requires good, sharp needles held in respectful and discerning hands.

J. Ron Engel supports such a dialectic between universalism and pluralism. Arguing that any global eco-ethical agreements must sprout from the grassroots, from the ethical approaches of particular places and peoples, he feels it "makes all the sense in the world to speak at one and the same time of 'a world ethic for living sustainably' and of 'world ethics for living sustainably'" (1994a, 5). The principle of "unity in diversity," a key idea in the science of ecology and a characteristic of all healthy ecosystems, can also guide us through the perils of both universalism and relativism in our search for an ecological ethic and ethics.

Thus it is becoming clear that universalism versus pluralism is really a false dichotomy, a product of dualistic thinking that has great difficulty accepting and embracing complexity and nuance. In place of dichotomy, we need a dialectical exchange of cultural information, wisdom, and knowledge so as to mutually and reciprocally revise cultures Western and non-Western, "modern" and "traditional." Local traditions need to be reexamined in light of postmodern science while historically dominant traditions, such as those of the West, have much to gain from the perspectives and ethics local traditions engender. Matthews sees great hope in this process since it points us toward our commonality as citizens of one earth that needs us to work together rather than in isolation and opposition. As she puts it:

> Representatives of no one regional culture then are in a position to determine the elements of a global ethic of Nature. But as representatives of the global community, we are all in a position to contribute to this debate. We may hope for some consensus on the grounds that, although we all bring different interests and identities to the debate, we can also all recognize that there is something much, much larger than these differences at stake: the preservation of the earth, of which we are all natives and to which all our traditions owe their origins. In other words, we are no longer, any of us, only Aborigines, or Europeans, or Africans. There is also a new, as yet shadowy, dimension of our identity—a transnational, transcultural ecological dimension—which is at this very moment in process of formation. It is this exciting new aspect of our identity which will, more than anything else, I think, create the possibility of a fertile though pluralistic ethical consensus rather than a sterile relativistic stalemate (1994, 7).

In addition to enhancing our sense of unity, the growth of this ecological dimension of our identity may also grant us a new appreciation for diversity, especially for the way in which the earth's biodiversity and our human cultural diversity are related. The term "biodiversity" is most commonly used to refer to the earth's immense number of plant and animal species. Rarely does it incorporate human beings and the cultural diversity they encompass. Yet cultural diversity and biodiversity are integrally related. In fact, the former is but one facet of the latter, for human beings and their cultures are also representations of the diversity of life. Furthermore, the particularity of biodiversity often promotes cultural diversity, as differences in the landscape lead to differences in the ways in which humans live and gain meaning from the land. In turn, diverse human cultures and the specialized knowledge held by their human constituents about their particular environments (a product of culture) are often key in maintaining and promoting biodiversity. Biodiversity and cultural diversity together constitute crucial components of a healthy ecosystem. Both in turn end up being destroyed by systems and ethics that decrease diversity, systems and ethics of homogenization, from monocultured cornfields to monocultured consumerism. Resistance to such forces requires human culture, since nonhuman biodiversity can't speak and act out against homogenization. Where cultural diversity persists through acts of resistance against homogenization on the part of cultural actors, biodiversity is often found persisting alongside.

Cultural and biological diversity, then, are complementary—both serve and need each other. It follows that to achieve one important goal of environmental ethics—the conservation of biological diversity—requires that we maintain and learn from cultural diversity, especially the diverse ways in which different cultures have resisted the forces that seek to destroy biodiversity in all its human and nonhuman expressions. "Unity in diversity," as a solution to the "one/many problem," applies not just to culture nor just to the workings of the natural world, but to both in integral relation to each other.

The principle of "unity in diversity" has even more to teach us. The expressions of biodiversity do not exist in splendid isolation from each other but are integrated into ecosystems that in turn are integrated into the global biosphere. The ecological health of a natural system often tends to decrease when any one part fails to carry out its particular function in relation to the whole. From this aspect of ecosystems, J. Baird Callicott draws a lesson for grappling with the "one/many problem" as it pertains to cultural diversity and global environmental ethics: "Cultural fragmentation—inwardness, isolation, mutual hostility, intolerance—is no less destructive of human and biotic communities than is cultural homogenization. An analogue to ecosystemic integration is needed to complement cultural diversity" (1994, 13). In the most controversial part of his recent book, Callicott goes on to argue that the postmodern scientific world view with its "evolutionary-ecological environmental ethic" can serve as "a tie ... [to] bind the many cultural worlds into one systemic whole" and as a norm against which to compare each tradition's degree of "green-ness" (1994, 13). Such a claim opens the way to the next theoretical problem confronting my research: How do the eco-ethical teachings of the world's cultural traditions relate to the scientific view of postmodern ecology regarding the relationship between nature and culture?

Is Nature Normative? Premodern World Views[7] and Postmodern Ecological Science

Whether we choose to learn from the past or not, the past is our most reliable instructor in reality.

—**Donald Worster**

Sankofa.
—An Akan word meaning "to draw on the past to prepare people for the future"

Even though many Western Christians and certainly most academic theologians would find the primal cosmos of the African ... "primitive" or even "superstitious," the real situation is very different. We may be the ones who are behind the times. Perhaps modern liberal Western theology ... has been vainly striving to reconcile religion to an allegedly scientific world view which is actually becoming more outdated every day. Paradoxically, the traditional African sensibility, to which the indigenous Christian churches respond so inventively, may be more in tune with the "quantum world" than Western theology is.

—**Harvey Cox**

In a presentation to the Institute for Ecology, Justice, and Faith (Chicago, March 15–18, 1995), Callicott honestly acknowledged the privilege his argument grants to science. He went on to defend his claim that postmodern ecological science can serve as a unifier and norm for the "many cultural worlds" with five justifications: (1) for better or worse, science is the current, most widespread world view; (2) mechanistic science, correctly linked to ecosystem degradation, is obsolete and giving way to the more environmentally friendly perspectives of the postmodern scientific revolution championed by the new physics, ecology, and systems theory; (3) postmodern science and premodern world views can work complementarily to create a new world view through the articulation of an "ecologically correct" environmental ethic in local grammars; (4) such work will serve to validate and enhance respect for non-Western cultural traditions; and (5) science operates under a strong tradition of self-criticism.

It is not my task here to judge the validity of each of these justifications. Instead I use them to point to the realization being made by many scientists, philosophers, and environmental ethicists that premodern world views and postmodern science share much in common in regard to the principles they posit for guiding the relationship between human beings and nature. One way of exploring these commonalities is to briefly review how premodern traditions and the science of ecology have understood the relationship between nature and culture and, more specifically, how each has answered the question of whether nature is normative, that is, whether nature itself provides lessons on how we should treat it and each other.

Premodern World Views and the Nature-Culture Relationship

Most premodern traditions, including the foundational elements of today's major religions, deeply reflect humans' intimate contact with nature necessitated by their direct dependence on the natural world to fulfill their needs. Both knowledge and ethics paid keen attention to the natural environments that directly sustained human communities. Learning the "way nature works" was imperative if one wanted to stay alive. Setting up codes of conduct to govern individuals' and the community's relationships to nature served to sustain both the environment and the human community. If nature's ways were violated, humans paid the price. Thus commonly the creators of premodern traditions derived human "oughts" from natural "is's" based upon their intimate knowledge of the natural world.

These perceptions of nature's workings and the ethics derived from them have varied tremendously over time (Glacken 1967) and culture (Tuan 1986), yet three themes seem to persist. First, many premodern world views recognize and respect that nature holds real limits. Through observing limits in the natural world, cultures also place limits on their adherents' behavior so as to avoid calamity. One cannot kill certain animals, hunt at certain times of the year, enter certain sacred forests, or farm any one area too long or too hard. Placing limits on the amount of personal and group material acquisitions helps ease social tensions, but it also lightens humans' impact on the land. Across time and culture, such traditions have acted and continue to act as constraints on human behavior vis-à-vis the natural world.

In contrast, the modern Western world view and experience provide dramatic examples of what happens when humans ignore such constraints. Beginning with the Enlightenment, moderns tried, and continue to try, to supersede nature's limits and control its processes. Evidence of the disregard for constraint and the drive to control nature abound—from the modern period's desanctification of nature as a means to legitimate its rampant exploitation, to the belief in modern technology as human's inexhaustible resource and ultimate mechanism of control. The severity of our current environmental crisis bears witness to the price we are paying for modernity's arrogance.

Nevertheless, we must also acknowledge that such calamitous denials of nature's limits, although supremely exemplified by the modern period, are not limited to it. Even though premodern peoples may have recognized limits and codified them into ethical systems, limits were not always respected. Indeed, the earth is littered with the ruins of civilizations (Babylonian, Greek, Roman) that overstepped nature's bounds and culture's sanctions. It may be that the science of ecology is replicating in a modern and postmodern age the warnings of premodern traditions, respected or not, that we need heed nature's limits. It is yet unclear whether we will do better at abiding by an ecological world view than did civilizations now turned to dust. What is clear is that although ecology is extremely young in the West an ecological world view goes back a long time.

A second characteristic of premodern world views is their discrete recognition of the essential relatedness and unity of all parts of the universe. Interdependence and holism are key concepts of premodern world views. Evidence of such an understanding of the cosmos can be found in numerous cultures and religions across the globe. For many Native Americans this unity is expressed in the belief that all of creation is sacred—"every part of this soil … every hillside, every valley, every plain and grove" (Chief Seattle 1854)—and that "all beings [are] bound together in an unending cycle of mutual dependence" (Cornell 1992, 42; see also Momaday 1994 and Nelson 1983). Similarly, the Hindu doctrine of "unity of life" recognizes that "God, humans, and all living creatures have the same essence which is consciousness" (Pappu 1994, 296). Many African traditions understand the universe as being "full of sacred life," possessing a common "vital force" that permeates not only the animate but also the inanimate worlds (Sindima 1990, Asante 1985). But such understandings also permeated much of Western thought prior to the advent of modern mechanistic world views that separated the whole into parts and removed sacredness from the natural realm. Carolyn Merchant, in her fascinating study *The Death of Nature*, uncovers in detail the Renaissance and pre-Renaissance European world views that conceived of the cosmos "as a living animal. A vast organism, everywhere quick and vital, its body, soul, and spirit were held tightly together" (1980, 104).

Thus John Muir was not being original when he said, "When we try to pick out anything by itself, we find it hitched to everything else in the universe" (Muir 1911,

157). He was only one of the first ecologists to articulate a corrective to modern reductionist and analytical natural science that premoderns had known for centuries. The proposition that interdependence is one of ecology's fundamental precepts is common knowledge and does not need elaboration; the proposition that ecology is reflecting and enhancing certain aspects of these premodern knowledges deserves greater emphasis.

Related to interdependence, a final characteristic of premodern world views is the idea of moral-natural holism. A more historical term for it is "covenant"—the idea that the human and natural worlds are connected not only ecologically but also morally. Under covenant, natural chaos and natural harmony are inextricably linked to social chaos and social harmony. Morality encompasses both the spatial-natural sphere and the human sphere in relation to each other. The quality of these relations is the key factor in determining whether nature and people experience harmony or chaos. Thus in many traditions "sin" constitutes being in wrong or distorted relationships (Ruether 1992, 142). Conversely, moral customs, or ethics, are judged, as Yi-Fu Tuan points out, according to whether or not they promote good relationships (1989, 14). And relationships here are inclusive, encompassing those between people and people, people and land, and people and God. All three facets of relationship are interconnected and affect each other. Since the principle of covenant is such a profound one, it may help to explore more deeply, using the ancient Hebrews' exemplary understanding of the concept.

What is the meaning of "covenant"? Drawing on the work of McCoy (1991), Brueggemann (1977), and Anderson (1984), J. Ron Engel has described seven key components of the covenant relationship as understood by the people of ancient Israel: (1) constituted unconditional commitment on the part of God to the people of God and the earth; (2) based on shared purposes, chiefly, justice and harmony; (3) applied to each member of a community and to the community as a whole; (4) endured over time and was not fulfilled in the short-term; (5) included human apostasy and violence in a continual pattern of repentance, ever-present forgiveness, and renewal; (6) always offered to humans as a matter of choice rather than being forced upon them; and it was (7) anchored in the assurance that God would remain faithful no matter what (1994c). I would add two points that apply to years past as well as today: (8) breaking covenants always bears consequences; and (9) although the process of life is teleological, the future is open and progress is not inevitable (McCoy 1991).

Bearing strong similarities to ideas within other premodern world views, covenant for the ancient Hebrews meant that they and the natural world were vitally and directly connected. As Rosemary Ruether describes it, in covenantal relation, "the social and the natural languages are so intimately integrated that it is something of a distortion even to speak of them as two different relations, rather than as parts of a single socio-natural covenant that binds the creation into one community" (1975, 188). Within that community, human actions directly affect the natural world in such a way that the effects in turn become a form of judgment or blessing on those actions. Nature, in other words, becomes a mirror reflecting the degree to which human beings are abiding in right relation to God and their fellow creatures. It becomes a sign of human beings' ethical state of health, not because nature is meant to serve human beings as a sign but because nature itself is also part and parcel of the covenant between God and creation (Ruether 1978, 1131).

Thus for the ancient Hebrews to commit social injustice was to break the one socionatural covenant and suffer the consequences as they were played out within the natural realm. In the words of the prophet Isaiah, "The earth dries up and withers, the whole world withers and grows sick; … earth itself is desecrated by the feet of those who live in it, because they have broken the laws and violated the eternal covenant" (Isaiah 24: 4–5 NEB). Similarly, to commit injustice against the creation was also to break the covenant with God and with other people since people's well-being was intimately connected to that of creation.

But covenant also held a positive side: the promise of restoration in response to social repentance and transformation. Just as natural chaos signified ethical chaos within ancient Hebrew society, so natural harmony grew out of the restoration of right relationships. Transformation that mended humans' broken relations to God and to each other was met by the restoration and flourishing of the entire creation. "Before you mountains and hills shall break into cries of joy, and all the trees of the wild shall clap their hands; pine-trees shall shoot up in place of camel-thorn, myrtles instead of briars" (Isaiah 55: 12–13 NEB). And it was their faith that served to keep the ancient Hebrews' natural and social worlds intact. As they understood it, covenantal faith was rock-solid faith that God would remain faithful no matter what, even in spite of their faithlessness. People bore the consequences of their faithlessness, but that did not change the faithfulness of God. Rather the consequences served to call them back to repentance and transformation.

From this overview, we begin to see how covenant, a complex symbol of right relation between human beings, God, and the earth, can serve as a particularly powerful metaphor for crafting an environmental ethic. Covenant means that humanity and the earth sink or swim, fall or flourish, together. It inextricably links ecology and justice and reveals that the way to heal the earth is not simply to develop ecological consciousness or to conserve nature but to change the social systems of injustice and domination that are at the root of ecological destruction. Simply cutting household consumption will not do, for that leaves the systemic structure of mass consumption and exploitation untouched (Ruether 1975, 200). It is on the basis of covenant that Ruether states the hard realities of the ecojustice agenda: "The environmental crisis is basically insoluble as long as a system of social domination remains intact" (1981, 59).

But the covenant metaphor can also give people the courage and commitment to keep going "no matter what the 'trends' may be. Whether we are immediately 'winning' or 'losing' cannot shake our rooted understandings of what biophilic life is and should be" (Ruether 1992, 273). It also reminds people of the cyclical nature of renewal and restoration. It does not point to some paradise of restoration in some far distant future and otherworldly realm. Instead it offers people the promise of a continual process of being made whole again and again, here and now on this earth, their true home; not ever in the same way, but in a spiraling historical project of renewal under changing circumstances (Ruether 1981, 69). Covenant as metaphor brings ecology, justice, and faith together into one seamless whole.[8]

Although what I have outlined above draws most heavily on the ancient Hebraic understanding of the covenantal relationship, many other premodern traditions hold very similar themes. For example, in many precolonial African societies, the living members of the community, the ancestors, and the land are all complexly intertwined such that actions taken within one domain hold ramifications throughout the whole. The painstakingly researched novels of Nigerian writer Chinua Achebe illustrate such natural and moral interconnections well. Achebe's most famous novel, *Things Fall Apart* (1958), clearly illustrates how in Igbo tradition, as in that of many other African societies, the way people behaved toward each other had very real consequences for how the land behaved toward them. When Achebe's protagonist, Okonkwo, accidentally kills his clansman, he commits not only a societal offense but an offense against the earth. To keep the land from being ruined, Okonkwo must go into exile in order to cleanse the earth. If Okonkwo refuses his exile, *Ani*, the Earth Goddess, will spoil the land for the entire community. On another occasion, Okonkwo's best friend chastens him for taking part in the Oracle-pronounced murder of his adopted slave son, Ikemefuna: "What you have done will not please the Earth," Obierika tells him. "It is the kind of action for which the goddess wipes out whole families" (1958, 46). Thus, similar to the workings of covenant, in many African traditions the breaking of the social order affects not just the individual committing the offense but the entire social and natural communities.

It is this final characteristic of premodern traditions, the principle of covenant, that I find missing in postmodern ecology. Including it in the work of crafting a global ecological ethic represents one way in which postmodern science can benefit from traditional world views in working to cocreate with them a new ecological world view robust enough to guide us into the twenty-first century.

Ecological Science and the Nature-Culture Relationship

Among the sciences, ecology has come closest to bridging the gap (or, one might say, had the hardest time denying that a bridge exists) between the "is" and the "ought" as the two play themselves out in nature and culture. However, a review of ecology's relatively short life reveals that it has never, even in its postmodern phase, been unambivalent in its assessment of whether nature is normative (Worster 1994). It has consistently oscillated between the heavy influence of the amoral mechanistic-reductionist world view (yielding the approaches of bioeconomics; analytical, quantitative microlevel studies; and ecosystem energetics) and more holistic and organismic perspectives. If the latter do not entirely build a bridge between "is" and "ought," they at least refrain from erecting an absolute wall between the two.

At times, ecologists have been explicit about nature's perceived normativeness. In the 1940s, the so-called Ecology Group founded at the University of Chicago by Warder C. Allee stated openly that its goal was "to extract from nature a set of holistic values to apply to mankind" (Worster 1994, 327). One of the group's colleagues, Ralph Gerard of the University of Chicago's bioscience faculty, went so far as to contend that "a pronounced pattern or observed direction in nature provides man with all the guidance he needs for 'shouldness'" (Worster 1994, 336). In subsequent years, the influences of economics and mechanistic physics significantly weakened this organismic and ethical strain within ecology, but it has never died out completely. Revived during the 1970s age of ecology through such theories as the Gaia Hypothesis (James Lovelock's suggestion that that earth is a single living organism), the strain within postmodern ecology now faces new challenges and opportunities.

What postmodern ecology brings to the debate on nature's normativeness is that any moral lessons to be gained from nature are complicated by the fact that nature itself appears to be far less ordered, stable, and equilibrious than previously thought. Although postmodern ecologists have not completely discarded the idea of natural order, they have certainly found that order much harder to locate. Under the influence of such intellectual movements as Chaos and Systems Theories, postmodern ecology has highlighted the role that irregularity, discontinuity, and nonequilibrium play in determining the way the natural world works. Coupled with this, postmodern ecologists have emphasized the position of the observer and issues of scale as key determinants in our understanding of what nature appears to be doing. Like other postmodern scientists, they realize that natural observers can never get outside of nature and therefore, by their presence, they influence what they see. To postmodern ecologists, what you see in and of nature depends on how and from where you look. In short, as ecologist Cal DeWitt so elegantly states it, nature is less like a clock or even an organism and much more like a Calder mobile with its changing yet ordered arrangements that look very different depending on where one is standing (personal communication).

Such revelations have held diverse implications for ecological ethics, some troubling, others hopeful. Some ecologists find in nature's newly discovered irregularity a justification for the revival of social Darwinism. For others it has meant a retreat from the field and practical action to their labs, in order to squeeze some order out of nature's unpredictability with the aid of sophisticated computer models (Worster 1994, 413, 414). Another group of ecologists, exemplified by Daniel Botkin, interprets the idea that "disturbance is natural" to mean that a more conciliatory approach to economic development and human aspirations for wealth and power is in order. As to whether nature is normative, members of this group "den[y] that there [is] any firm guide to behavior to be found in nature, past or present, or even much reason for limiting human wants or rejecting progress" (Worster 1994, 416). Instead the ecological ethic stemming from their "disturbance-impressed" ecology runs in the opposite direction: Rather than depend on any patterns in nature to guide our behavior, our human knowledge and action must guide nature. In even stronger terms, order is not out there in nature for us to step to; we are the ones who must make it and design, out of nature's cacophonies, a melodious tune by which to make our home, the earth, a comfortable one. To quote Daniel Botkin, "Nature in the 21st century ... will be a nature that we make" (Worster 1994, 415). Chilling.

In contrast, another group of ecologists, led by E.O. Wilson, finds that nature's disorder, unpredictability, and mystery make it all the more important to conserve. The new field of conservation biology has mushroomed nearly overnight and is replete with top-notch ecologists, such as Michael Soulé, who are putting their expertise into practical and ethical action on behalf of conserving biodiversity. Environmental historian Donald Worster summarizes well the upshot of fin de siècle ecology's seemingly contradictory ethical implications:

> By the last decade of the century the science of ecology, after so many intense, complicated theoretical debates, found itself in a more uncertain state of mind about its implications for modern technological civilization than it had been in the two or three decades following World War Two. Yet, surprisingly, it also found itself regrouping around a new conservation ideal that was, if not exactly required by new theories, at least was not contradicted by them. Apparently, moral ideals have a way of unexpectedly precipitating themselves out of the flux of events, the uncertainties of theory. As one set of environmental perceptions and values faded away another began to take its place. Nature, ecologists began to argue, is wild and unpredictable. Nature is in deep, important ways quite disorderly. Nature is a seething, teeming spectacle of diversity. Nature, for all its strange and disturbing ways, its continuing capacity to elude our understanding, still needs our love, our respect, and our help (1994, 419).

Well, then: What of the commonalities between premodern world views and postmodern ecology? Is their any basis for Callicott's statements that "we may confidently say that there are interesting similarities between the ideas of the new science and non-Western traditions of thought"; that "historically, the reconstructed postmodern world view will be Western. Substantively, it will not. It will be more Buddhist than Platonic, more Kayapoan than Cartesian" (1994, 192)? Above, I have underscored some of the persistent themes found in premodern or, in Callicott's terms, "non-Western," world views regarding the nature-culture relationship. I have noted some similar themes within the history of ecology in the West. Postmodern ecology certainly has changed the playing field, but not to the point of precluding any and all perception of natural patterns. From the science of ecology—modern and postmodern—Donald Worster distills three general conclusions about nature's patterns and the implications such patterns hold for ecological ethics:

1. Nature's "is's" do not in and of themselves constitute human "oughts."
    What they do provide are numerous "models of successful adaptation." If they so choose, humans can draw from these models certain lessons to apply to their own value systems and behavior. They may also try to replicate these models in order to achieve a way of living more successfully

adapted to the surrounding natural environment. For example, Wes Jackson and his colleagues at the Land Institute are currently trying to develop a perennial polyculture that protects soils and preserves biodiversity based on the model of Kansas's natural prairies (Jackson and Bender 1984). One condition that model has dictated for their experiments is the observation of limits. Unlike the expansive, heavily machined, monocultured wheat fields of its neighbors, the Land Institute's fields, in order to be successfully adapted to the natural prairies, are limited both in expanse and in the amount of technological intervention they bear.

Furthermore, the observation of limits that marks polyculture's replication of nature extends to the simplified lifestyle of the human community for whom the Land Institute is home. Since humans are also part of nature, should we not also consider such human communities living in sustainable relationship with the rest of creation (and there are numerous other examples besides the Land Institute) as "natural models" from which to learn? Worster notes a common feature among such human communities: They have all created rules to govern their relationship to the rest of nature. Similar to how many premodern communities acknowledged limits, such rules or ethics exemplify the same sort of acknowledgement by present-day communities living in a postmodern age and influenced in part by the insights of postmodern ecology.

2. In accordance with premodern teachings, and in spite of the evidence for nature's disorderliness and individualistic strivings, interdependency remains the name of nature's game.
3. Change in nature is real but it is also various. Change itself changes. It is sometimes slow, fast, cyclical, or linear, sometimes working against us, sometimes for us, but none of these patterns of change can be taken as normative. Rather the ethical implication of this feature of nature is to challenge us "to protect certain rates of change going within the biological world from incompatible changes going on within our economy and technology." It leads us to follow "a pattern of behavior based on the idea that preserving a diversity of change ought to stand high in our system of values, that promoting the coexistence of many beings and many kinds of change is a rational thing to do" (Worster 1994, 432).

Is there any cognate within premodern world views to this third characteristic of nature as illumined by postmodern ecology? None comes easily to mind, although a deeper search (which I cannot conduct in this space) may reveal much more. In any case, nature's changing rate of change does not negate nature's two patterns of *interdependency* and *limits*, observed by both postmodern ecology and premodern world views. It also provides another example, as in the case of covenant, of the cross-fertilization and enrichment that can come from bringing pre- and postmodern traditions into dialogue and mutual revision.

What, then, can we conclude about the relationship between pre- and postmodern world views with regard to their understandings of nature? Callicott's phrasing of the relationship is helpful: He describes it as a process of "translating ecologically sound principles into local cultural grammars." Rather than summarize what he means, I choose to quote his explanation in full because it articulates very nicely one way in which pre- and postmodern thought can work dialectically to inspire transformation toward more ecologically sustainable livelihoods. Callicott writes:

> Thus, detailed cross-cultural comparison of traditional concepts of the nature of nature, human nature, and the relationship between people and nature with the ideas emerging in ecology and the new physics should be mutually reinforcing. Traditional environmental ethics can be revived and, just as important, validated by their affinity with the most exciting new ideas in contemporary science, while the abstract and arcane concepts of nature, human nature, and the relationship between people and nature implied in ecology and the new physics can be expressed in the rich vocabulary of metaphor, simile, and analogy developed in the traditional sacred and philosophical literature of the world's diverse cultures.
>
> One might therefore envision a single cross-cultural environmental ethic based on ecology and the new physics and expressed in the cognitive lingua franca of contemporary science. One might also envision the revival of a multiplicity of traditional cultural environmental ethics, resonant with such an international, scientifically grounded environmental ethic and helping to articulate it. Thus we may have one world view and one associated environmental ethic corresponding to the contemporary reality that we inhabit one planet, that we are one species, and that our deepening environmental crisis is worldwide and common. And we may also have a plurality of revived traditional world views and associated environmental ethics corresponding to the historical reality that we are many peoples inhabiting many diverse bioregions apprehended through many and diverse cultural lenses. But this one and these many are not at odds. Each of the many world views and associated environmental ethics can be a facet of an emerging global environmental consciousness, expressed in the vernacular of a particular and local cultural tradition (1994, 12).

Thus we see again that universalism versus pluralism is a false dichotomy. Based on Callicott's analysis, we might also find that Central African rituals, stories, teachings, and practices involving humans and nature—the subject of this book—may have more to offer our global quest for achieving sustainable livelihoods, and be more in line with an ecological understanding of the way the world works, than the dualistic, materialist, secular, empiricist, and economic theories on which much of the dominant social paradigm is based and on which mainstream culture at large continues to operate. It is to a discussion of the African context of this work that I now turn.

What Is Africa?
Listening to Local People's Voices

Who is speaking about [African gnosis]? Who has the right and the credentials to produce it, describe it, comment upon it, or at least present opinions about it? ...

I would prefer a wider authority: intellectuals' discourse as a critical library and, if I could, the experience of rejected forms of wisdom which are not part of the structures of political power and scientific knowledge.

—V.Y. Mudimbe

Africa as a continent is a reified European geographical abstraction (perhaps derived from "the four parts of the world" in late antiquity?). It does not therefore correspond to anything "out there."

—Jan Vansina

Inquiries and insights like these underscore the complexity and limitations involved in any attempt to speak about Central African land ethics, one dimension of African gnosis. Mudimbe's questions, with which he begins his critical examination *The Invention of Africa* (1988), leave me both daunted and inspired. Daunted, because they point to the difficulties inherent in saying anything meaningfully close to "truth" when power, consciously and subconsciously, plays such a strong role in shaping knowledge. This is especially true in cases where, from a position of relative power, one is seeking to understand cultures of which one is not a member. At the same time, I find the work of Mudimbe, a Congolese philosopher, novelist, and professor at Duke and Stanford Universities, extremely inspiring. If not clearly outlining how, it at least convincingly argues the need for, and begins to point the way toward, a method of exploring African gnosis that truly tries to listen to the marginalized voices of local people rather than only to the voices of power.

A key purpose of Mudimbe's work has been to reveal how Africa and knowledge about Africa are most accurately thought of as a collage of inventions by various epistemes, rather than as indigenously created texts. Among those epistemes, the most powerful and most heard has been that of the West. As he explains:

> The fact of the matter is that, until now, Western interpreters as well as African analysts have been using categories and conceptual systems which depend on a Western epistemological order. Even in the most explicitly "Afrocentric" descriptions, models of analysis explicitly or implicitly, knowingly or unknowingly, refer to the same order. Does this mean that African *Weltanschauungen* and African traditional systems of thought are unthinkable and cannot be made explicit within the framework of their own rationality? My own claim is that thus far the ways in which they have been evaluated and the means used to explain them relate to theories and methods whose constraints, rules, and systems of operation suppose a non-African epistemological locus (1988, x).

In *The Invention of Africa,* Mudimbe, drawing heavily on the work of his mentor, Michel Foucault, outlines the evolution of a series of different African inventions. He states his objective as "rather than simply accept the authority of qualified representatives of African cultures, I would like to study the theme of the foundations of discourse about Africa" (xi). The various discourses about Africa he goes on to "deconstruct" include those of Western anthropologists, missionaries, Western and African apologists of Negritude,[9] early African political leaders such as Kwame Nkrumah and Julius Nyerere, Caribbean proto pan-Africanist E.W. Blyden, and a host of African and Africanist philosophers and theologians. In regard to this work, Mudimbe's ideas in response to the question "What is Africa?" have greatly influenced my research approach.

Mudimbe's insightful analysis leads me to fundamentally question all pictures of Africa that I have viewed, as well as any picture of Africa I might wish to paint. One response to such work—a response the deconstructionist school often inspires—is to give up even trying. After finishing the penultimate chapter of *The Invention of Africa*, in which Mudimbe completes his deconstructive task, I scribbled my response in the margin: "So, if the thing you're after is unattainable, are any attempts to reach it, all being tainted with various subjectivities and motives, legitimate to carry out?" A valid question, perhaps, but when answered negatively it represents too facile a capitulation to postmodern nihilism. The Congolese with whom I grew up have a saying in Lingala: *Kotuna ezali mabe te*, which means "It is not bad to ask." As long as one tries to ask carefully, mindful of and openly stating one's biases, then there is still merit in asking and attempting to answer a question such as what might constitute Central African land ethics. Not asking such sorts of questions too easily reflects recession into one of several abysses: cynicism, universalism, or extreme relativism, some of the dangers of which I have outlined above.

Mudimbe himself does not give up on the endeavor to answer the question "What is Africa?" In fact he calls for African philosophers to continue the task of indigenizing responses to the question by identifying, applying, and problematizing "the basic African epistemic frameworks and principles in analyses of discourses," doing so "in a manner that will make oral traditions useful to the present" (Masolo 1994, 243). But if he raises the banner of African philosophy's task, Mudimbe holds back (understandably, given his methodological preference) from prescribing how exactly one might go about it. He does propose staying within the primary empirical methods of anthropology—but a new anthropology, one that is aware of its own biases, history, and reasons for undertaking what it purports to undertake, one that is philosophically astute and self-critical. Mudimbe summons his readers to see "African *gnosis* and ... its best sign, anthropology, as both a challenge and a promise" and suggests that

> this *gnosis* makes more sense if seen as a result of two processes: first, a permanent reevaluation of the limits of anthropology as knowledge in order to transform it into a more credible *anthropou-logos*, that is a discourse on a human being; and second, an examination of its own historicity. What this *gnosis* attests to is thus ... a dramatic but ordinary question about its own being: what is it and how can it remain a pertinent question mark" (1988, 186)?

The type of anthropology Mudimbe encourages believes neither in fundamental opposition between science and traditional thought, history and anthropology, the

Same and the Other, nor that one simply evolves into the other. Instead it views both sides of such dyads as equally valid discourses that complement rather than demean each other.

Still, this is not reassuring to young Africans who may applaud Mudimbe's critique of how the West has seen them but who long for more concreteness in putting it into practice so as to transform their own identities and the identity of Africa. Kenyan philosopher D.A. Masolo points to the growing perception of Mudimbe as prophet on the part of many Africans who are becoming conscious not only of how the West has misconstrued Africa but also of how their own self-understandings have been distorted by allegiance to Western epistemological categories. Although they accept Mudimbe's challenge to develop an indigenous African epistemological method for understanding who they are and what Africa means, many Africans may demand more guidelines on how to arrive at such methods. They may also show greater concern for the implications these new ways of understanding themselves and Africa hold for concrete political action, as Masolo insightfully points out in response to the question many would ask of Mudimbe:

> How can basic African epistemological principles be integrated into already alienated modes of thinking? This demand will place Mudimbe in a position similar to that once occupied by Kwame Nkrumah and Julius Nyerere in regard to their proposals for a new political and economic order in postcolonial Africa. What Mudimbe does with *gnosis*, that is, with the creation of new order of knowledge, cannot be fully isolated from the ideology of empowerment on which Africa has focused over the past three decades (1994, 189).

The one guideline Mudimbe seems ready to offer the quest for indigenously African epistemological methods is that in some way the voices of local people (he uses the term "subjects"), those "that structuralism too easily pretends to have killed" (Mudimbe 1988, 23), must be heard. Indeed, he harbors some hope in the fact that the passion of the "subject" remains alive and well, that "he or she has gone from the situation in which he or she was perceived as a simple functional object to the freedom of thinking of himself or herself as the starting point of an absolute discourse" (Mudimbe 1988, 200). Local African people do not need Foucault or any other Western intellectual critic to empower them. Rather their subjective discourses emanate out of the rich material within their own conceptual, perceptual, and spatial realities and the concrete dynamics of their everyday lives. The challenge for African intellectuals and sympathetic non-Africans, then, is simply to listen to the voices of local people rather than ride roughshod over them with their own theoretical and methodological grids.

Meeting this challenge is a real problem for the intellectual—African and Africanist—because of the way they themselves are held captive within the Western epistemological training they have already received. Masolo, quoting Foucault, puts it nicely:

> Through and through, then, Africa as we know it (through its Western representation) is a social construct. To produce this construct, the African intellectual has been successfully used. Pulled away from the masses, he has been made both a victim and an

agent of his own alienation. Foucault writes—and Mudimbe also believes—that the African and Africanist intellectuals who have participated in this game "discovered that the masses no longer need to gain knowledge: They know perfectly well, without illusion; they know far better than he and they are certainly capable of expressing themselves. But there exists a system of power which blocks, prohibits, and invalidates this discourse and this knowledge, a power not only found in the manifest authority of censorship, but one that profoundly and subtly penetrates an entire social network. Intellectuals are themselves agents of this system of power—the idea of their responsibility for 'consciousness' and discourse forms part of the system" (Masolo 1994, 183 citing Foucault 1977, 207).

Masolo also makes the important observation that Mudimbe himself is an African intellectual captive of the Western episteme, a fact that fills Mudimbe's work with a certain ambiguity. In Masolo's words:

> Mudimbe's project is doubtlessly a noble one—to recapture *authenticity*. But this desire is not without contradictions. The other must regain himself or herself in "the freedom of thinking of himself or herself as the starting point of an absolute discourse," free from the epistemological codes of Western philosophy. Yet in this same project Mudimbe invests heavily in the positions of Lévi-Strauss and Foucault as the openings that lead to the *bricolage*, the framework of authentic African rationality. This said, pertinent questions remain in Mudimbe's project: which and where is the African epistemological locus? and how does one reach it without recourse to some established method(s), like Mudimbe himself uses, to state its necessity (1994, 187)?

In other words, there is no way for the intellectual (or anyone, for that matter) to crawl out of their own skin. All of our knowledges are conditioned by our languages and the society, culture, and age within which we live.

The questions and gropings of African philosophers like Mudimbe and Masolo present every seeker of knowledge (which includes this writer) with two methodological challenges. As stated above, one is to uncover and state clearly (rather than sweep under the rug) the nature of the biases and conditionings to which we have each been subject. Recognizing that we all hold biases, what our particular biases are, and that bias exists as a matter of degree is a crucial step in achieving any understanding. Once we acknowledge our degree of bias, we can work to minimize it by looking at any particular phenomenon from as many different perspectives as possible. Eric Hoffer, the stevedore-philosopher, reminds us of what happens when we settle for an education that encourages us to reject certain perspectives out of hand: "If it is true, as Bergson says, that 'the true human nature from which we turn away is the human nature we discover in the depths of our being,' then the intellectual is in a hell of a fix. There is no getting away from it: education does not educate and gentle the heart" (1969, 40).

It is perhaps this gentling of the heart that we intellectuals most need in confronting and learning from local African people. Only then can we meet the second challenge, which is to struggle to understand any complex topic, such as Central African land ethics, through a complementarity of lenses that sees both science and traditional thought as being equally valid and worthy ways of perceiving events. Rural sociologist Jack Kloppenburg speaks of the challenge in terms "not ... of choosing between scientific knowledge or local knowledge, but of creating conditions in which these separate realities can inform each other" (1991, 540). Such an approach, incorporating the views of both subjects and objects, contributes to more, rather than less, "objectivity." True objectivity comes from bringing into focus as many views of an issue as are possible, seeing value and distortion in each, so that when they are taken together we have a more whole and "objective" picture rather than a more singular and distorted one. Ecotheologian Sallie McFague describes this approach in her assessment of science from a feminist perspective:

> The feminist criticism accepts both the political and the empirical character of science; its criticism aims at a greater, not a lesser, objectivity for science by broadening the base of who participates in setting scientific agendas so that science might be emancipatory, liberating, beneficial for more people—and for the planet that supports us all.
>
> The important issue here is situated or embodied knowledge resulting in a stronger objectivity, and therefore the possibility of humane scientific projects. The 'view from the body' is always a view from somewhere versus the view from above, from nowhere: the former admits to its partiality and accepts responsibility for its perspectives, while the latter believes itself universal and transcendent, thus denying its embodiment and limitations as well as the concrete, special insights that can arise only from particularity (1993, 95).

Using research methodologies that I detail below, I seek to listen to and learn from those "whose knowledge has been developed by the integration of hand, brain, and heart in caring labor" (Kloppenburg 1991, 531); to hear and record the knowledges of local African people, knowledges embodied in the particularities of social and physical place. In carrying out that task, I admit that these knowledges, like all knowledges, are partial, but I also hope to bring to the fore certain distinctive insights from within the particular and subjective Central African communities with whom I worked. "What constitutes Central African land ethics?"—like the larger query "What is Africa?"—will always remain a question mark yet a question to which local people as well as external observers must be able to contribute. This research seeks to enhance the possibility for such contributions to be made.

So What? Central African
Environmental Ethics and Environmental Justice

Discrepancy between stated ideal and reality is a worrisome fact of our daily experience.
—Yi-Fu Tuan

The West is now everywhere, within the West and outside: in structures and minds.
—Ashis Nandy

Questions easily posed to an attempt such as mine include: "So what? What difference have ethics made? If indigenous cultural resources for building sustainable livelihoods have existed in Central Africa, why do we hear so many horror stories about Central African people barely able to make ends meet on impoverished soils, degraded lands, and denuded forests?" Discrepancy between ideals and realities, ethics and actions, theories and practices is a widespread phenomenon that in and of itself merits volumes of study and explanation (see Tuan 1968). Even when limited to the discrepancy between environmental ethics and environmental realities in Central Africa, such a question remains beyond the immediate scope of this book. However, one response I will offer here is that they can be answered only within the framework of Central Africa's particular history of colonial exploitation, as well as the present-day context of a global economy more interested in Central Africa's natural resources than the welfare of its people and environment.

Numerous studies have documented the changes wrought by colonial and precolonial conquests of Central Africa (Vansina 1990, Jewsiewicki 1983, Rodney 1982, Harms 1981). In the penultimate chapter of *Paths in the Rainforests* (1990), his groundbreaking history of the region, Vansina gives a harrowing account of just how horrific these changes were:

> The violence and utter destructiveness of such colonial wars is often still misunderstood. Routinely village after village was burned down and people fled, sometimes for years, into deep forests where they built only the flimsiest of shelters and depended largely on gathering for food. During the fighting and immediately afterward, casualties among Africans were high, but even more died later from the combined effects of malnutrition, overwork, and epidemics such as smallpox, measles, dysentery, and above all sleeping sickness. In some districts the conquest lasted for years.... The despair engendered by the conquest is evident in many a document similar to the following entry in a diary. This incident took place on the banks of the Aruwimi [a tributary of the Congo River] in 1894, in a district where countless unrecorded tragedies similar to this one recurred, as war raged back and forth for 20 years.
>
> "22 June 1895— ... A man from Baumaneh scouring the forest and loudly shouting for his wife and child who had lost their way there, comes too close to our camp and is shot by a bullet of one of our sentries. They bring us his head. Never have I seen such an expression of despair, of terror" (1990, 244, citing account recorded in P. Salmon 1969–1970, 244).

Although Vansina's history focuses on Central Africa's political traditions, it is clear from his research that such traditions also encompassed individuals' and communities' relations to the natural world. Hence, his assessment of the changes such traditions have experienced over time holds relevance for this study of Central African land ethics.

Traditions, according to Vansina, are processes. They have to change in order to remain alive, and life for traditions means undergoing a constant readjustment between cognitive meanings and physical perceptions. Thus it was not change per se that destroyed the common equatorial tradition he argues existed in the region;

rather it was change occurring outside the autonomy of the tradition's followers. Once the autonomy of Central African peoples was stripped away through colonial structures of oppression—including forced labor; taxes; forced cultivation of cash crops; creation of foreign institutions such as "customary law" and chiefdoms; forced resettlement along roads; destruction of traditional institutions such as initiation rituals, the veneration of the ancestors, and territorial cults; privatization and individualization of land tenure; the list goes on—the tradition that had existed in the region for hundreds of years began to die. Vansina notes that "a tradition is harmed when it loses its ability to innovate efficiently. If the situation perdures, it will die. A tradition dies when its carriers abandon its fundamental principles to adopt those of another tradition" (1990, 260).

Eritrean philosopher Tsenay Serequeberhan echoes Vansina's assessment in noting that the destructiveness of the colonial legacy effectively put the brake on the innovative dynamics crucial to the life and perseverance of African traditions. A tradition dies when it can no longer create. According to Serequeberhan, African traditions have been dealt a serious blow through a simultaneous two-pronged attack: (1) the assimilation and/or conversion of Africans to a European philosophy and way of life; and (2) the suppression and deprecation—as barbaric, savage, and nonhuman—of African cultural traditions by a Western metaphysic manifested both politically and spiritually (1991, 8). Together the dual nature of this attack led Africans into allegiances to other world views, Serequeberhan asserts, while leaving them bereft of any sound tradition to fall back on when assimilation failed.

In short, both these scholars posit that as a result of the colonial experience Central African peoples no longer possessed the liberty to choose, create, and innovate within their traditional cognitive realms as a means to provide new meanings to the momentous changes occurring in their physical, socioeconomic, and political worlds. Coupled with changes in daily physical life were the devastating ecological changes wrought by the destructive colonial extraction of natural resources on a massive scale. Thus it is important to underline that in Central Africa the current discrepancies between eco-ethical ideals and ecological realities are rooted in colonial practices of destruction, both of the Central African environment itself and of the tradition that gave birth, through constant innovation, to eco-ethical ideals and applied them to ecological practice.

Vansina's rather bleak diagnosis is that Central African peoples, even with independence and political autonomy, still lack "the guidance of a basic new common tradition" and remain "bereft of a common mind and purpose" (1990, 248). Indeed, independence has in no way freed Central African countries and cultures from the stranglehold of Western economic, political, and cultural power. Neocolonial alliances between corporate economic interests and the personal interests of Central African political leaders have continued, ramifying the destructiveness and penetration of a global capitalism bent on profit maximization, natural resource extraction, cheap land and labor, and new markets. The very structure of financing what is euphemistically called "development" through lending institutions such as the International Monetary Fund and the World Bank encourages this process (Escobar 1995, George and Sabelli 1994).

Such penetration not only destroys local economies and landscapes but also—when aligned with an effective advertising campaign to create needs, whet desire, and homogenize artistic expression—aggravates the cultural degradation of African young people, a process that was begun during the colonial era. Today many young urban Central Africans are likely to know more about Rambo and Chuck Norris than their own mythic heroes. Few young urbanites partake in any extensive initiation rituals whereby the essence of the tradition is passed down. In short, like numerous other locales, Central Africa has not escaped the influence of monolithic consumerist culture propagated by Western capitalism. In addition to the ecological, cultural, economic, and political influences of colonialism, such current and ongoing postcolonial influences must also be seriously considered in any attempt to explain the discrepancies between environmental ethics and environmental realities in Central Africa.

It is easy to speak of and describe Africa's environmental crisis. It is harder to articulate and analyze the ties and complex weavings that link the crisis to past and present forces of economic, political, and cultural imperialism. Yet the story of Central African environmental ethics remains incomplete without including the distinctive episodes of injustice—political, cultural, and economic—that have been part of Central Africa's environmental and social history.

Put another way, as Westerners we cannot simply jet off to Central Africa to extract from "Central African thought" useful lessons and guidelines to apply to our Western-defined global environmental crisis. Instead we desperately need to ground an ethical response to the environmental crisis much more solidly in the context of historically created relations and conditions of imperialism and subjugation—and the alternative points of view such relations evoke on the part of many non-Western peoples. Solutions and ethical responses to the global environmental crisis born out of a justified anger and intellectual and practical resistance to Western imperialism must be given voice and power to redirect change in ways that redress past injustices (see Athanasiou 1992, Engel 1994b, Tabaziba 1994, Sarathy 1994).

In this sense, many Central African people would likely be quick to support what Wangari Maathai, feminist environmentalist and founder of Kenya's Greenbelt Movement, named as the key tasks for building sustainable societies in her address to the official plenary of the Rio Summit in 1992. She did not call for population control, increased economic growth, or more acreage converted to national parks. For Maathai and many other Africans who have experienced "how the other half dies" (George 1977) an environmental ethic calls for:

- eliminating poverty
- fair and environmentally sound trade
- reversal of the net flow of resources from South to North
- clear recognition of the responsibilities of business and industry
- changes in wasteful patterns of consumption

- internalization of the environmental and social costs of natural resource use
- equitable access to environmentally sound technology and its benefits
- redirection of military expenditures to environmental and social goals
- democratization of local, national, and international political institutions and decisionmaking structures (cited in Athanasiou 1992, 59).

In short, many people in Central Africa, similar to others in the South, inextricably link environmental ethics with their concerns for environmental justice so that what has been wrong can be made right, what has been put down can be raised up, those who have been misled can gain new direction, and those who have been denied a voice can retrieve the power of speech.

Repair and Reconstruction: Signs of Hope Grounded in African Soil

The alternative is, in a sense, always there. From this perspective, there is not surplus of meaning at the local level but meanings that have to be read with new senses, tools, and theories.... The subaltern do in fact speak, even if the audibility of their voices in the circles where "the West" is reflected upon and theorized is tenuous at best.

—**Arturo Escobar**

From the foregoing analysis, one might conclude that any Central African traditions that give guidance for living on the earth are dead and buried with no hope of resurrection. Certainly the devastation of traditions at the hands of colonial and neocolonial forces was and is severe, and thus such a pessimistic conclusion would not be unreasonable. But it would also represent a capitulation to despair and a demeaning of Central African peoples, since it would too easily dismiss their capacities for resistance, transformation, and creativity.

The fact is that despite past and present injustices and the harrowing wreckage they leave—a wreckage that prevails in Western perceptions of Africa—people and communities have not responded powerlessly. As Arturo Escobar so importantly documents, at the local level of so-called Third World communities "the concepts of development and modernity are resisted, hybridized with local forms, transformed" (1995, 51). Not all has been smashed and laid to waste. Underneath the noise of the headlines, behind the scenes that flash across our TV screens, there are voices and signs of hope.

This work documents only a few of these contemporary signs of hope rising out of the social and ecological turmoil wrought by colonialism and neocolonialism. There are many other signs of hope, many other African communities working culturally and politically to create eco-ethical alternatives to environmental destruction by drawing on past and present. Some of their stories have been told elsewhere (Daneel 1994, 1991; Cox 1995, 1994; Brecher, Childs, and Cutler 1993; Lewis and Carter 1993; Western and Wright 1993, Ruether 1995); others remain to be told.

But it is by telling the stories of communities of people who, through combining the best of tradition and modernity, are reconstructing socially and ecologically sustainable livelihoods that horrific headlines go a little less unquestioned, sordid stereotypes begin to dissolve, and awareness of a different Africa starts to grow in small ways. In spite of the legacies of history, the extreme sufferings from poverty, and the structural violence of the world economy, hope exists in Africans' resistance and creative action for ethical, social, and environmental well-being.

I contend with others such as Rosemary Ruether that only when Northern people connect with the real-world stories of marginalized peoples; when they see and hear first-hand the harsh and dreadful realities of poverty in the South and the social causes of environmental degradation operating there; when they stand humbled and amazed by people's indigenous and hopeful acts of resistance to these processes; only then will Northern, if not global, environmental ethics come to mean anything of substance (Ruether 1995).

My course through these five thorny theoretical problems has now brought me to the local places in which I conducted this research. I turn now to describing the specific methods I used in traveling these local paths of inquiry in partnership with local land-users, project staff, and local academics.

Methods and Research Process

Through my research process, I worked hard to meet three primary objectives:

1. Relying principally on the voices of local land-users, my initial objective, and the one I treat most extensively, has been to record and communicate cultural, ethical, and practical resources for ecological sustainability from within the indigenous cultures surrounding two conservation–sustainable livelihood projects in the forest biome of northern Congo.
2. My second objective has been to supplement these local voices with, and relate them to, the voices of university-educated project staff and local academics speaking in regard to Central African land ethics.
3. My third and final objective has been to apply local environmental perceptions, knowledges, and practices to the concrete problems facing each case study project and to discuss the lessons Central African understandings of nature hold for Western environmental theory and ethics.

To meet these objectives, I have used a variety of methods, which I outline below.

Conversation, Lived Experience, and Radical Empiricism

> Now if it is true that linear perspective and literacy prevent coevalness, then there is a good case for trying to understand the world through bodily participation and

> *through senses other than sight. Let us not forget the taste of Proust's petite madeleine, nor music, nor dance, nor the sharing of food, the smell of bodies, the touch of hands.*
> —Michael Jackson

> *First hand knowledge is the basis of the intellectual life. The second handedness of the learned world is the secret of its mediocrity.*
> —Alfred North Whitehead

As I have hinted, using interactive research methods to gain some sense of how Central African people understand their relationship to the natural world provides no grounds for claiming a detached "objectivity." However, interacting with and taking part in the lived experience of Central African people can enlarge one's capacity and availability to hear, understand, and record in their own terms what people are saying about their relationship to the forests, lands, and waters that surround them.

The methods of "interacting" and "taking part in lived experience" hold a variety of meanings for me, some of which I could foresee prior to returning to Congo, many of which came unexpectedly and spontaneously while I was there. One thing is certain: Via these methods I became deeply involved in what Robert Chambers has called "learning by sitting, asking, listening" (1983, 202), spending time with people in that ancient method of conversation. Examination of the origins of the word "conversation" reveals that during earlier times conversing constituted a method of interaction much more profound and substantive than is implied by its modern meaning: to talk informally with another or others, to chat.[10] In using conversation as a methodology I tried to recover some, certainly not all, of these former meanings.

In regard to the format of my conversations with local land-users, project staff, and local academics, I have been guided by the pragmatist tradition as represented in the works of John Dewey (1929, 1980), Richard Rorty (1979, 1982), and anthropologist Michael Jackson (1989, 1995). Drawing on the work of Rorty, Jackson spells out some of the implications pragmatism holds for research methodology:

> The pragmatist regards open-ended, ongoing conversation *with* others as more "edifying" than the task of completing a systematic explanation *of* others. In ethnography, this means abandoning induction and *actively* debating and exchanging points of view with our informants. It means placing our ideas on a par with theirs, testing them not against predetermined standards of rationality but against the immediate exigencies of life (1989, 14, italics in original).

In other words, the creation of knowledge is an interactive endeavor that takes place through a two-way exchange between the researcher as someone from outside the local culture trying to understand it and those within local culture who are trying to understand her or him. In this way, open-ended conversation with others can provide more opportunities for establishing "a ground where we shall both be intelligible without any reduction or rash transposition" (Merleau-Ponty 1964, 122).

Moreover, the open-ended nature of conversation with others should not be interpreted as equating simply to rambling. My conversations with local Congolese land-users, project staff, and academics certainly focused on issues of land: how it is used, how it ought to be used; what is good use, what isn't; land's role in making meaning, meaning's role in making land more than just soil or terrain; taboos and traditional teachings about land; land-based symbols and the role such symbols play in ritual; current destruction of land and its causes; the relationship between land and ancestors; and so on.

Thus open-endedness can be equated to trying one's hardest to refrain from continually controlling the categories, pace, and direction of conversation with people. I was not always as observant of this as I would have liked, but when we were able to interactively set the course of our dialogue together we managed to avoid a lot of the stiltedness and artificial ambience that often accompanies interviewing (the more methodologically correct word for "conversation"). When it came to talking about oral traditions, our conversations constituted more a listening on my part and a telling on theirs. Proverbs, riddles, songs, tales, stories about land, and stories in general are not discussed (at least initially) as much as given and received. Finally, our conversations also included space for people to tell personal and village life-histories as a means of expressing in their own terms important events, places, and experiences.

But verbalization is only one medium through which to participate in lived experience. The best research and the richest experience come through opening all of one's capacities and senses to the wonders of learning and culture. Michael Jackson's work has profoundly influenced me for the way in which it so wholeheartedly exemplifies such an approach. As an anthropologist, Jackson has lived and learned with the Kuranko people of Sierra Leone and, most recently, with the Warlpiri people of central Australia. He engagingly and persuasively articulates and practices a method of anthropology that moves beyond literacy and verbality. The following passage served as a guiding light for my own experience of fieldwork:

> Ethnographic knowledge that is constructed out of verbal statements or likens experience to a text which can be "read," deciphered, or translated, is severely restricted, if for no other reason than that the people with whom anthropologists live and work are usually nonliterate. To desist from taking notes, to listen, watch, smell, touch, dance, learn to cook, make mats, light a fire, farm—such practical and social skills should be as constitutive of our understanding as verbal statements and espoused beliefs. Knowledge belongs to the world of our social existence, not just to the world of academe. We must come to it through participation as well as observation and not dismiss lived experience—the actual relationships that mediate our understanding of, and sustain us in, another culture, the oppression of illness or solitude, the frustrations of a foreign language, the tedium of unpalatable food—as "interference" or "noise" to be filtered out in the process of creating an objective report of our profession (1989, 9).

Jackson's method of lived experience, which he draws from William James's notion of radical empiricism, posits four necessary recognitions that push it beyond traditional, or positivist, empiricism. Those recognitions also guided me in conducting this research. First, experience is temporal. In addition to regular patterns, it holds idiosyncrasies, exceptionalities, fleeting moments, and "things refractory to order" that merit attention. Second, directly felt prereflective experience is as valid for learning as is experience that is reflected upon. For example, "the smell of eucalyptus after rain and biochemical explanations of this same experience should not be ranked in terms of a distinction between subjective appearance and objective reality, for the illumination is as meaningful as the analyses that result from our reflections on it." Third, the order and patterns we conceive in our analyses of research "data" do not so much represent what is really out there as much as our wishful thinking for order to exist. We must "refuse to reduce lived experience to mathematical or mechanical models and then claim that these models are evidence or representations of the essential character of experience." Finally, we need to recognize the ways in which our obsession with finding order and pattern is linked to our peculiarly Western history of needing to control nature and people through instrumental rationality. Instead of judging an idea only against our standards of logic, science, and reason, we also need to see how it corresponds to the "practical, ethical, emotional, and aesthetic demands of life" (Jackson 1989, 12, 13).

In addition to the learning that has come through conversation, this sort of lived experience has also been my teacher, most pointedly during my fieldwork but also during the many years I previously spent in Congo. Part of that lived experience consisted of doing things with people. As I had done in my master's research in the Ituri (Peterson 1991), I again bought a machete and "hired" myself out to farmers to spend the day working with them planting seeds, clearing weeds, and harvesting crops. On other occasions, we would go into the forest to check fish traps or gather wild fruit. Not only did this prove a way to establish trust and rapport and offer some reciprocity (albeit small in comparison) for the valuable knowledge I gained; it also enabled me to learn bodily what it means to earn one's living from the land. While working together we often talked, sometimes we just worked, often we laughed and joked. Afterward we would invariably sit down in some shaded corner and eat together, perhaps returning to the village to share a steaming bowl of rice and *mpondu* (manioc greens). Our conversation would continue, but the meal itself, the smells and touches, the details of lighting a fire, were also valuable ways of coming to know more about what it means to be a Central African rainforest farmer and forager.

Thus I would venture with Jackson, Dewey, James, and Whitehead, that lived experience—the primary research methodology I have employed—is one of the best ways to gain understanding of any subject. This does not mean that one enters into a culture tabula rasa, claiming to have no preconceived theoretical vantage points. Rather, as mentioned above, it requires one to acknowledge one's view as a view from somewhere, then bring that view into dialogue (both in word and deed) with other views so as to try to gain, through this multiplicity of lenses, a greater understanding of the subject at hand.

Here the subject is what Central African peoples and cultures have to say regarding what constitutes the right relationship (i.e., conducive to social and environmental well-being) between humans and the earth. My view is from the somewhere of being part African—having been born and lived half of my life in Congo—and part white, middle-class, Protestant-raised American. My view has definitely been affected by my being thrown into the betwixt-and-betweenness of these two cultures—American and Central African. I have come to this research, therefore, as one who is highly skeptical of yet entrenched within Western epistemological categories such as reason, objectivity, and a host of dualisms (spirit/body, fact/value, feeling/thinking, subject/object, nature/culture). At the same time, I hold a certain facility with and appreciation of Central African lived, thought, and felt experience, in many ways different from, and in some ways bearing similarities to, my Western heritage. But facility and appreciation in no way infer an inside knowledge of the sort possessed by those with whom I worked. My African experience is also muted by my white skin, my missionary-kid upbringing, and my periodic returning to Mputu (Europe or America)—the land of privilege.

In some ways, this experience with a bit of both worlds has made my research easier. Without a doubt it has colored it with a mixed hue of ambiguity. That ambiguity is rooted, on the one hand, in the desire to present aspects of Central African land ethics that to the greatest degree possible are true to and stem directly from the voices of Central African peoples; on the other hand, it is rooted in the knowledge that I can't entirely pull off that presentation, because an "environmental ethic" is on its face a reified relic of my Westernization. Thus the ambiguity stems from the awareness that even though I was able to enter into lived experience to some degree, at the end of my time I left and returned to the more abstract world of the university and the relatively easy life of Madison, Wisconsin, no longer seeing and experiencing (albeit in muted form) the suffering and constraints posed by the exigencies of life in Congo.

Fully embracing rather than denying these ambiguities, I present in this book what I have learned through this research approach of "Africa lived more than studied." It is an approach that "gives incompleteness and precariousness the same footing as the finished and the fixed," an approach that tries "not to subjugate lived experience to the tyranny of reason or the consolation of order but to cultivate that quality which Keats called negative capability, the capability of 'being in uncertainties, Mysteries, doubts, without any irritable reaching after fact & reason'" (Jackson 1989, 16 citing Keats 1958, 193).

Focus Groups

At times, during my research process, conversing with people in a group helped me gain other insights in addition to those yielded through individual conversations. This was especially true during my research with the university-trained staff of each project, but on several occasions I also met and held long talks with groups of local land-users. As a means of facilitating and recording these group conversations, I conducted them in a manner based on the methodology of focus groups.

Focus groups have become a widely used research method for gaining knowledge from a group of people who share one or more common characteristics. "A focus group can be defined as a carefully planned discussion designed to obtain perceptions on a defined area of interest in a permissive, nonthreatening environment," according to Richard Krueger, one of several focus group researchers who have written extensively on the method (1988, 18). Although other researchers have pointed to the inconsistencies inherent in a conversational method that is "carefully planned" while meant to create a "permissive, nonthreatening environment" (Agar and MacDonald 1995), focus groups can provide many benefits to more conventional means of social research.

> Focus groups have advantages over quantitative and other qualitative methods.... Unlike quantitative methods, they emphasize participants' perspectives and allow the researcher to explore the nuances and complexities of participants' attitudes and experiences.... Unlike in-depth individual interviews, they permit researchers to observe social interaction between participants around specific topics.... Unlike naturalistic observational methods, they provide a mechanism by which the researcher can structure the content of such interaction.... Focus groups are neither as strong as participant observation in their ability to observe phenomena in naturalistic settings, nor as strong as in-depth individual interviews in providing a rich understanding of participants' knowledge, but they are better at *combining* these two goals than either of the two techniques alone. Therein, focus groups provide a body of material that differs in form and content from that provided by other research methods (Hughes and DuMont 1993, 776).

Important factors that contribute to a focus group's success include providing the participants with a discussion guide that has enough structure to guarantee that the topic of interest gets discussed, but not so much structure as to inhibit a fluid and synergistic exchange between the participants. The same can be said for the researcher's role in moderating the focus group: "Too much moderator control prevents the group interaction that is the goal; too little control, and the topics might never be discussed" (Agar and MacDonald 1995, 78). According to Alan Johnson, "The moderator's aim is *synergy*—a vibrant group discussion in which individual participants create not a series of controlled and contrived bilateral exchanges with the moderator but rich and meaningful multilateral conversations between themselves" (1996, 522). To achieve synergy requires that the moderator be skilled in building rapport within the group; managing group dynamics so that everyone has a chance to speak and some not speak too much; creating a "safe" atmosphere that encourages people to share their thoughts and opinions openly; using language with which participants can feel comfortable; exhibiting an authentic concern and sensitivity for participants' points of view; maintaining a neutral position and refraining from influencing the course of the conversation toward the moderator's biases; and enabling the conversation to continue to flow smoothly. Finally, focus group success depends on such technical matters as choosing a comfortable and convenient locale and ensuring the sound quality of the recording by using high-quality equipment (Hughes and DuMont 1993, 779).

Choosing participants for a focus group is based not on representativeness but in response to a question: "Who do you want to hear from?" (Krueger 1988, 28). Generally, this means that participants share features that qualify them as a group in one way or another. In each of the two focus groups I conducted with project staff, all participants worked for the project or were closely affiliated with the project (e.g., as university research interns or, in one case, as an exemplary individual putting the project's goals into practice on his own land). These criteria supplied the basis for focus groups comprising eight staff people at Loko, ten at Epulu. In both cases all participants were men, reflecting the gender distribution of each project's managerial staff. Three of the eight participants in the Loko focus group were American missionary staff; the remaining five participants at Loko and all ten at Epulu were Congolese. Several days prior to each focus group I circulated the following discussion guide to staff members in order to give them an opportunity to think about our topic of discussion beforehand (see Box 2.1). In both cases, we met in comfortable conference rooms used regularly by staff to conduct meetings and seminars.

Box 2.1   Focus Group Discussion Guide

MEMO: Regarding a meeting of the Staff of the Agricultural Department of CEUM Loko/Center for Conservation and Tourism at Epulu with Mr. Richard Peterson, Anthropologist, University of Wisconsin, U.S.A.

CEUM Loko, Zaire/CEFRECOF Epulu, Zaire

July 28, 1995 9:30 AM/September 1, 1995 3:30 PM

Introduction

It is indeed a great pleasure to have this opportunity to meet with all of you. As most of you know, I am in the process of completing my research for a doctorate in the Institute for Environmental Studies at the University of Wisconsin–Madison where I will write a dissertation in the domain of environmental ethics. Specifically, my dissertation is on the indigenous cultural resources for environmental ethics found within certain of the ethnic groups living in the extensive forests of Central Africa. By discovering and taking advantage of certain indigenous concepts, guides, prohibitions, beliefs, or limits that have contributed to relatively stable relations between people and the forest (in other words, environmental ethics), we may be more successful in creating environmental projects that are truly supported and understood by local communities. The motive for our meeting is simply to allow us to discuss this theme, and to allow me to hear your thoughts, you who are directly engaged in environmental work with local communities.

Objective

The purpose of this memo is to give you some time before our meeting to think about our topic. Our general topic is "Indigenous Cultural Resources for Environmental Ethics in Central Africa." To begin, it will be necessary for us to try to define our terms and the concepts we will discuss. Thus, the first question will be: "How would you define a way of living that is ecologically and socio-economically sustainable?"

After we have heard our definitions, we can begin with some additional questions and points of discussion. For example:

- Among your cultural traditions (or the traditions of the people with whom you work) pertaining to the natural world, can one find certain practices, principles, or beliefs that aided, continue to aid, or would be able to aid in developing a sustainable relationship between people and nature?
- Were traditional means of utilizing natural resources (cutting fields, hunting, fishing, etc.) truly ecologically sustainable?
- If so, to what would you attribute this: ethics or traditional rules? A very low population density? Use of limited technologies? Or something else?
- If an environmental ethic did exist, what were its components? What were some of the traditional restrictions or prohibitions?
- Among such ethics, are there some that are still functioning today? If yes, why? If no, why not?
- If all traces of traditional environmental ethics are dead, is there any way to revive or restore them in order that people may restrain themselves from the overuse of nature in a manner they understand?
- To complete our discussion, I would like to know, "What are the greatest difficulties or obstacles that hinder the effectiveness of your work?"

Method

In order to facilitate our meeting, I suggest that we use a method known in English as "Focus Groups." Focus groups permit the diverse thoughts, insights, and points of view people hold in regard to a subject to be heard. If they go well, Focus Groups also provide a means for members of the group to learn from each other through the exchange of ideas and the coalescing of different points of view. Normally, Focus Groups are facilitated by a moderator (often the researcher). However, rather than act as a chair, this person's task is simply to ensure that the discussion stays on track and that everyone has a chance to participate. Thus, in contrast to questionnaires or other methods chosen by the researcher alone, Focus Groups allow new questions, additional points of discussion, and the details of the subject being discussed to arise interactively during the course of the meeting.

In conclusion, I look forward to the opportunity to meet and discuss these issues with you. I wish to thank each of you in advance for your help and I hope that our discussion will also be of benefit to you.

In the case of local land-users, focus group participants shared the common feature of being residents of the same village who earned their primary livelihood directly from the land. Chapter 3 focuses entirely on the focus group conducted with land-users in the village of Bogofo, near Loko, and provides the details of its setting and the criteria for choosing its seven male participants. Later, in the Ituri, I conducted a less formal focus group with land-users in the village of Bapukele, near Epulu, which I discuss in Chapter 5. The village meeting shelter where we gathered is more open and fluid than a conference room. Thus the twelve male elders with whom the focus group began were joined at times by curious children and, toward the latter part of the discussion, by a prominent woman from the village. As I had done during in-depth conversations with individuals, I received the participants' permission prior to tape-recording each session. My conversations with focus groups and individuals find their form in the following pages, through hours upon hours of transcription from audiotape. Except for one focus group and one individual conversation transcribed by a Congolese friend in Madison, all other exchanges have been transcribed and translated by this writer, as I am fluent in both Lingala and Swahili. I have taken some liberty in making the literal Lingala, Swahili, or French wording of a sentence more accessible to English-speakers; I have also inserted discursive notes in places where meaning was especially difficult to convey in English. On more than one occasion I consulted with native speakers to verify meanings of particular words and phrases. I lightly edited several passages to improve readability and grammar but without changing the speaker's meaning as I have interpreted it.[11]

Narrative and Interpretation

Finally, I must address the problems presented by narrative and interpretation: Throughout this material interpretations of what people told me play a critical role alongside what people actually said on the tapes. I appreciate the honesty of, and am in full accord with, a now famous adage attributable to anthropologist Clifford Geertz: What we write about our research experiences with other people is "really our own constructions of other people's constructions of what they and their compatriots are up to" (1973, 9). Interpretation is inevitable and indispensable, and trying to avoid it does not necessarily translate into greater objectivity; accepting it as an inevitable subjectivity may even enhance a researcher's ability to empathize with those he or she is writing about.

Eduardo Galeano, the Uruguayan historian, recounts the counsel he received regarding his struggle with his own "subjective interpretation" amid doing what he felt was supposed to be "objective research":

> I had been writing *Memory of Fire* for a long time, and the more I wrote the more I entered into the stories I was telling. I was already having trouble distinguishing past from present: what had happened was happening, happening all around me, and writing was my way of striking out and embracing. However, history books supposedly are not subjective.

I mentioned this to José Coronel Urtecho: in this book I'm writing, however you look at it, backwards or forwards, in the light or against it, my loves and quarrels can be seen at a glance.

And on the banks of the San Juan River, the old poet told me that there is no … reason to pay attention to the fanatics of objectivity:

"*Don't worry,*" he said to me. "*That's how it should be. Those who make objectivity a religion are liars. They are scared of human pain. They don't want to be objective, it's a lie: they want to be objects, so as not to suffer*" (1991, 120, italics in original).

At the beginning of Chapter 3, I recount my personal struggle over how to write about the people who so profoundly shared their lives and experiences with me. The defense of my choice to write using a narrative approach more aptly fits in the beginning of that chapter, where it directly precedes the actual narratives of local people—the heart and soul of this book. Thus I will conclude this accounting of my research methods with this: What follows is my own interpretation of reality, not reality itself. My hope is that readers of this book will discern an honest attempt to reflect participants' own knowledge and experiences.

Summary and Conclusion

In Chapter 1, I described my academic journey using the metaphor of building a two-pillar bridge. To conclude Chapter 2, I provide the key points of reference developed in this chapter, which summarize the framework of the bridge I build in this book:

- We currently face a global environmental crisis that is, at its roots, a crisis about values, about what we as human beings desire and believe at the inner core of our beings, a crisis in our understanding of what we hold to be of ultimate concern.
- The transformation of values happens more through the realm of the symbolic—story, myth, ritual—and through lived experience than through laws, facts, ethics, better management, technique, or scientific investigation—all of which, however, remain very important.
- Every culture has resources to contribute to this possibility of value transformation and conversion to ecologically sound livelihoods, but some cultures have received more attention than others while other cultural voices have been silent and silenced. African contributions to global ecological ethics have received much less attention than contributions from other cultures.
- Africa is alive and bursting with diversity; nevertheless out of this immense variation some general patterns do emerge. This is especially true of the forested region of Central Africa, the cultural and ecological area on which my research focuses.
- The evolutionary ecological environmental ethic of postmodern science and premodern teachings regarding nature and people's relation to nature

(such as those from Central Africa) share much in common, and each holds much to contribute to the other. A helpful way of thinking about this dialectical relationship is that of discovering local cultural translations of ecologically sound principles.
- In trying to illuminate some of the contributions to ecological ethics from Central African cultures, the voices of African subjects—especially people most intimately tied to the land and forests, but also marginalized university-trained practitioners and academics—must be heard.
- Local people's narratives, as well as their oral traditions (a culture's collective body of narratives), provide windows into fundamental metaphysical, epistemological, and ethical understandings of the natural and social worlds. However, these narratives need to be allowed to speak for themselves rather than be subjected to external epistemological grids under the false assumption that only then will they yield knowledge.
- Explanation of the discrepancies between Central Africa's eco-ethical ideals and its environmental realities must begin by setting such a comparison within the context of Central Africa's particular colonial history and its present-day experience of exploitation through the structures and dynamics of a global political economy. What such contextualization teaches us is that Central African environmental ethics are inextricably linked with African concerns for environmental and social justice.
- African eco-ethical traditions are not dead and buried with no hope of resurrection. Hope lives on in the living incarnations of African environmental ethics—communities of people working both culturally and politically to create eco-ethical alternatives to environmental destruction by drawing on both the past and the present.

I turn now to discuss evidence of Central African land ethics from the ground, from the voices, narratives, and practices of local Congolese land-users. I begin this discussion in Chapter 3 with the voices of local farmers living near Loko, the first case study project, situated in the Ubangi area of northwestern Congo.

Notes

1. I use the term "postmodern" throughout this chapter to connote the trends, schools of thought, and movements rejecting, questioning, and otherwise responding to various tenets of modern (post–Middle Ages) thought, including reductionism, dualism, universalism, and secularism.

2. See Engel 1993, 184 for evidence among natural scientists; Shipton 1994 for evidence among social scientists; and Callewaert 1994 for evidence within international policy making.

3. A review of the entire twenty volumes of *Environmental Ethics*, one of the primary journals in the field, yields only one article on African contributions to environmental ethics (Burnett and Kang'ethe 1994), the conclusions of which are controvertible. J. Baird Callicott's recent work (1995) exploring diverse cultural contributions to ecological ethics has only a

short section on Africa that is admittedly the weakest part of the book (Callicott, personal communication).

4. Within the country of Congo (about the size of the United States east of the Mississippi River) alone, there are more than 250 different ethnic groups and/or languages.

5. See especially Gyekye 1987, 189–212 for an overview of such studies and the common features to which they point.

6. In this work, I use the term "dialectical" not in its formal Hegelian sense but in a more informal sense of connoting "both/and" rather than "either/or" thinking. "Both/and" thinking consists of delving into the creative tension inherent in synthesizing what are seemingly opposite characteristics, propositions, and processes.

7. Overholt and Callicott define "world view" as "a set of conceptual presuppositions, both conscious and unconscious, articulate and inarticulate, shared by members of a culture" (1982, 1).

8. I do not limit the meaning of "faith" here to its religious definitions or to "faith in God." Rather I mean having confidence or trust in someone or something.

9. See Chapter 1, note 4.

10. The etymology of the word is revealing and speaks to the depth of interaction the method engenders. "Converse" comes from the Middle English *converse(n)*, which is descended from the Latin word *conversari* (to associate with). "Associate," in turn, comes from the Latin *associat(us)* (joined to or united with). Former meanings of converse, now considered obsolete, include "to have sexual intercourse with" and "to commune spiritually with" or, in its noun form, "spiritual communion." One obsolete meaning of conversation is "close familiarity; intimate acquaintance, as from constant use or close study." The archaic meaning of "conversation" is "behavior or manner of living" (Webster's 1989).

11. In the case of unknown or unclear words, I have consulted the following dictionaries: Lumana Pashi and Alan Turnbull, *Lingala-English Dictionary* (Kensington, MD: Dunwoody Press, 1994); Inter-Territorial Language Committee for the East African Dependencies, *A Standard Swahili-English Dictionary* (Nairobi: Oxford University Press, 1939); Institute of Kiswahili Research, University of Dar es Salaam, *Kamusi ya Kiswahili Sanifu* (Dar es Salaam: Oxford University Press, 1981); Kenneth Urwin, *Langenscheidt's Compact French Dictionary* (New York: Langenscheidt, 1968); and Denis Girard, *The New Cassell's French Dictionary* (New York: Funk and Wagnalls, 1962).

3

Narratives on Nature: An Opening Conversation with Ubangian Farmers

The question arises time and time again: how can one keep faith with the people who adopted one into their world and transformed one's understanding? How can one reconcile the different conceptions of knowledge that obtain in the field and in academe?... I began to see how events could be readily recast as narratives. I hesitated to go beyond this point. Narrative did not stray too far from the lived experience of fieldwork, and I wanted to avoid the radical reconstruction and systematization that mark academic writing.

— **Michael Jackson**

The aim ... can be put like this: to approach certain lives, not to pin them down, not to confine them with labels, not to limit them with heavily intellectualized speculations but again to approach, to describe, to transmit as directly and sensibly as possible what has been seen, heard, grasped, felt by an observer who is also being constantly observed.... The aim, once again then, is to approach, then describe what there is that seems to matter.

— **Robert Coles**

Most of the people whose voices are heard in the pages that follow gain their primary sustenance directly from the natural world. Most of them are farmers, and some hunt, fish, and trap in addition to farming. Others, like the Mbuti near Epulu, continue to practice various modes of hunting and gathering. This dependency on the land does not make them independent of the need for cash to buy manufactured goods and pay for children's schooling. To earn cash most people market a portion of their crops, game, and fish, and many have supplementary means to provide for themselves and their families in addition to working the land—blacksmithing, woodworking, petty trading, pastoring the local congregation, fixing things, brewing corn whiskey, panning gold. A few people had worked for the government, and some now work for one or the other of the two environmental projects that served as my case studies.

The speakers here are women and men, elders and young adults, those especially known for being able to remember oral traditions, and those who know none of the old stories. None represents a pure encapsulation of "tradition." Like people everywhere, their lives, voices, views, and perspectives are shaped by a mixture of influences that stem from the constantly changing configurations of native cultures and cultures beyond. Those I conversed with are also not "representative" of Central African forest peoples at large, nor even of the populations surrounding the case studies where I worked. But they are not anomalies. Like thousands of others—both past and present—who have made these forests and savannas their home, they are people making a living from the land, and so they carry much knowledge about the land and the forests from which they live. Finally, they are people who agreed to share some of their ecological knowledge with this author.

But knowledge is not all they shared; they also shared their feelings about how the land is becoming degraded and fish and animals are disappearing; their frustrations over the ways in which conservation organizations, in cooperation with the government, are trying to remedy environmental degradations; their struggles to feed, school, and clothe their children under Zaire (Congo)'s decaying economic and political conditions that make any means of living—let alone dreaming and achieving goals—precarious; and their resolve nonetheless to find the means to survive and, amid life's disappointments and frustrations, to find meaning, joy, vitality, and laughter. Despite its difficulty, life for the people I worked with resonates a certain vibrancy revealed, among other ways, in people's art of offering gracious hospitality. I arrived asking to know about their lives, and they received me as a guest— openly. I was shown and taught a great deal, and they accepted me as a partner in conversation rather than someone from whom they must guard their cards.

As I write here in Madison, thousands of miles from the site of my field experiences, recalling such vibrancy sparks within me a certain sense of loss. Often I find myself missing the vibrancy that comes from being intimately a part of a community, the vibrancy experienced in living more connected to the nature upon which we all depend, the vibrancy of living less detached from what Maurice Merleau-Ponty describes as "that world of birth, death, initiation, pain, separation, and struggle, which precedes knowledge, that world 'of which knowledge always *speaks*, and in relation to which every scientific schematization is an abstract and derivative sign-language, as is geography in relation to the country-side in which we have learnt beforehand what a forest, a prairie or a river is'" (Jackson 1995, 172 citing Merleau-Ponty 1962, ix).

How, then, can I write about Central African men, women, and experiences while avoiding utter dissipation of meaning through the words, language, and systematizations of Western abstraction? Can words and language convey the lived-ness of experiences, both my own and those of the people I worked with, from one place to another? Like Robert Coles, my aim is to transmit directly and sensibly a sense of who these people are, something of their ways of understanding the lands and forests from which they draw their livelihoods. Like Michael Jackson, I have found the recounting of narratives, rather than the counting of numbers, a better way of keeping faith with the people who shared with me their world and their experiences. Yet prior to transmitting any story, it is helpful to examine the consequences of using the narrative, its limitations as well as strengths, what the narrative can and cannot do.

A Narrative Approach

A narrative approach may, as Michael Jackson says, serve to tether one's writing closer to lived experience and prevent one from straying too far into the danger of radically reconstructing and systematizing one's own experience and that of others. But no more than any other method of observation and writing can it claim to offer an unaltered picture of reality, a slimmer version recounting only what "actually" happened, the "truth" of the situation in people's own words. Even in a narrative approach,[1] interpretation and analysis always exist. By asking some questions and not others, by the very act of selecting which parts of my taped conversations to include in this book (for I can't include them all), I cast my own mark on other people's experiences, putting things together in ways that reflect my own meanings, purposes, and interests as well as theirs. In his classic study *Migrants, Sharecroppers, Mountaineers*, Robert Coles, one of the people most experienced in accurately documenting others' lives, openly acknowledges both the limitations and the lessons of a narrative approach:

> It is all impressionistic, this kind of work, all tied to one person's mind and body; even the tape recorder can only record the questions I asked, can only convey the trust that can develop between two people over those much advertised and dwelled-upon "cross-cultural" barriers.... Once upon a time (a long time ago, it now seems) I desperately wanted to make sure that I was doing the respectable and approved thing, the most "scientific" thing possible; and now I have learned, chiefly I believe from these people in this book, that it is enough of a challenge to spend some years with them and come out of it all with some observations and considerations that keep coming up, over and over again—until, I swear, they seem to have the ring of truth to them. I do not know how that ring will sound to others, but its sound after a while gets to be distinct and unforgettable to me (1971, 41).

Thus the reader will notice Rick Peterson in the stories that follow. Between narratives I offer my own voice, hoping to open a dialogue with the reader as to the possible meanings and implications of the narratives I am recounting. But the narratives themselves will carry this writing. The advantage of this approach—the reason I have chosen it—is to deliberately avoid making second- and third-level abstractions from the narratives, as if that were the only way for the narratives to yield knowledge. The narratives themselves are people's actual words that I have transcribed and translated from Lingala, Swahili, and French. Generally, all words in quotation marks are taken from tape-recorded dialogue and are not reconstructed from memory. The few cases where I conducted untaped interviews and am recounting narrative from my field notes will be clearly indicated.

Another consideration: In a narrative approach I, as a researcher, am not the only one engaging in interpretation. It is important to underline that the narratives themselves consist not simply of "facts" but include people's own interpretations of experiences past and present. Since much of this narrative material concerns the past—the teachings and laws of the ancestors—it is especially critical to explain here what a narrative approach offers and what it cannot offer. Using such an approach, I do not attempt to show that Central African peoples of the past were primarily despoilers of nature or, in the opposite vein, conservers and good stewards of nature. I do not have the material upon which to base such claims one way or the other. Consequently, with a narrative approach, I am not capturing a historical record of what actually took place in ancestral times. Instead all the narrative approach can offer are contemporary expressions of how people *perceive* the way of life of the past. It is important, therefore, to realize that many of those perceptions present an idealized past removed to a degree from historical reality. Exactly how far removed I cannot say.

Therefore, one might ask: How can such an approach be useful for documenting aspects of Central African land ethics? Its utility lies in its ability to richly convey how local people view their current situation vis-à-vis the natural world in which they live, as well as what they feel ancestral traditions, as they perceive them, can contribute to ease the environmental strains they are struggling against. This is important, because it is these people, with all their actual perceptions, no matter how "factual"—not the ancestors of the past—who will ultimately sustain or degrade the land. Documenting where local people are "coming from" is as important a place to start as any if one is interested in affecting which option prevails—sustainability or degradation. Similarly, modern local people, not ancestors of the past, are the ones current environmental projects in Central Africa must take into account if they are to build environmental sustainability and conserve biodiversity. Hearing and conveying what these people feel and perceive about the work of such projects is therefore another important contribution of the narrative approach.

Thus even though I sometimes compare what people tell me to what Central African historians say about ancestral traditions, achieving "historical correctness" is not my primary concern. Rather it is accurately conveying local perceptions of inherited pasts and the lived present, as it is within and from those *perceived* worlds that actions and ethics regarding the natural world ultimately spring.

I now recount my initial exploration of those worlds with local farmers living near Loko.

Loko: Amid Forest and Savanna

I have to pinch myself to make sure this is really happening. I am again traveling by motorcycle through the forests and savannas of my youth as my wife, Deb, and I head up to Loko from Karawa, the small town where I grew up. The Yamaha's extrawide rear tire keeps the bike straight and steady as we gun through deep stretches of sand eroded and then deposited by the seasonal rains. *Esobe* (savanna, in this area dominated by the grass *Imperata cylindricum*), interspersed with small groves of oil palm, stretches from one arm of gallery forest to the next, creating a mosaic of green hues. Dark green forests wind along the river and stream valleys, giving way to the brilliant green grasslands covering rolling hills. We are in an ecotone, a transition zone between Congo's vast expanse of contiguous rainforest that extends 2,000 miles to the east just short of the Ugandan border, and the northern savannas that border both sides of the Ubangi River and continue on up into the Central African Republic. Intermingling savanna and forest provides habitat for a great diversity of life, both human and nonhuman. The sheer beauty of the land inspires me as I think about the task I have begun—of learning more about that diversity and how its human and nonhuman components interact.

Glancing up from the rutted road, I see that another storm is brewing as dark, anvil-headed cumulonimbus formations tower and rumble on the distant horizon. We manage to stay dry by slipping into the shelter of Mbeti's home just before the storm breaks. A high school friend I have not seen for six years, Mbeti is the director of the primary school in the village we just entered. Over *fuku* (corn and manioc porridge) and *nzombo* (smoked eel), he tells me that he has not been paid in three years, and yet his school continues to function. Parents have come together and

The savanna-forest ecotone near Loko has provided a rich habitat for a great diversity of life.

taken up the responsibility shirked by Mobutu's corrupt and inept government. Now they are the ones who pay their children's teachers, often directly in goods. Three bottles of palm oil for each child each month suffices to keep the teachers teaching and the kids learning. As Zaire (Congo)'s official government infrastructure teeters and sways, local communities must fend for themselves.

Mbeti also tells me that Dedua (another high school friend) recently died from AIDS. Then he names five or six other friends also taken by the disease. I am shocked even though I have known about the extent of AIDS in the region for a long time. Yet these are friends, not statistics. During her week of orientation at Karawa's hospital, Deb has already witnessed the ravages of AIDS, but she has also seen other diseases, some of which have recently resurfaced with renewed vigor. This is partially due, she learns, to the decision on the part of Western donors to withdraw humanitarian aid from Zaire (Congo) as a means of pressuring Mobutu toward democratic reforms. Ironically, Mobutu and his parasitic class have continued to live just fine without the aid while they offer gestures of democratic change. But as medical programs lose funding and supplies, diseases once easily controlled have again begun to wreak havoc on local people. As late as 1991, Zaire (Congo)'s incidence of sleeping sickness was extremely low, the disease effectively controlled through Belgian-funded programs of prevention and treatment. As those programs lost funding, however, Karawa's public health teams began to document soaring rates of sleeping sickness in the area, second only to Congo's Bandundu Region, where rates are the highest in the world. Lacking treatment, many have succumbed to the disease's final stage, which leads to permanent brain damage. In a further ironic twist, the only company (located in France) that makes the arsenic compound used to treat the final stage of the disease recently stopped production due to local environmental protests over the company's leaking arsenic into local watersheds.

We pull out from Mbeti's compound onto the main road after taking pictures and saying good-bye. The rain, hard but short, has failed to convert the characteristic laterite clay into the slick, gluey mud I have slogged through on other trips, and the villages zip by, each sparking another memory. At one point my poncho falls out of the pack and we turn around to look for it. I don't see it anywhere in the road, but then a man comes out from his house bearing it like a gift all neatly folded up, ready to give it back since I had returned, ready to keep it had I not. Soon we are crossing the Loko River and, climbing the last hill, we turn into the road leading to the place that will serve as our base for the next several months.

Loko was originally built as a leprosarium by the Belgians just before Congo gained independence in 1960. The country's early turmoil caused the hospital to lay abandoned for nearly ten years, occupied only briefly by Congolese soldiers during the 1964 rebellion. Although the soldiers decimated the area's fauna for meat, they left the buildings and infrastructure of the hospital for the most part intact. In the mid-1960s, the widow of missionary Dr. Paul Carlson, killed by Simba rebels in 1964, asked President Mobutu if he might cede the hospital and surrounding lands to the Paul Carlson Foundation, a nonprofit organization established in Dr. Carlson's memory. In 1968, the hospital and surrounding concession of 5,600 acres were officially dedicated as the Institut Médical Evangélique Docteur Carlson.

Since then Loko has undergone several name-changes and organizational affiliations, but its mission remains the same: treating the needs of local people in a holistic fashion. From the beginning that has meant combining services in primary and curative health care with work in community development. The latter work began with the goal of improving nutrition through introducing beans into local cropping systems, but it soon expanded to include programs of agroforestry, aquaculture, animal husbandry, and appropriate technology (Almquist 1993, 97–108). Part of the reason we chose Loko for our base was its unique integration of work in both the health and environmental fields. A family doctor, Deb would double the number of physicians serving the 150-bed hospital and surrounding 600-square-kilometer health zone. I would offer what help I could to the environmental programs while carrying out my research on local land ethics that I hoped would also hold some relevance for Loko's work.

"Fieldwork cannot be willed into happening," writes Michael Jackson. "Inevitably it proceeds by fits and starts. Anxieties and doubts beset you, no matter how good your language skills, how thorough your background reading, how extensive your ethnographic experience in other cultures" (1995, 21). Our first days at Loko certainly hold fits and starts. We are graciously provided with a cinder-block house situated in a circle of similar homes housing mostly American and two Congolese staff of the project. We gingerly feel out our places in this community we have joined. At first it is uncomfortable knowing how to balance research with lending a hand in Loko's ongoing work. Trying to help, I spend my early days reading and editing grant proposals the CEUM (the Congolese Protestant church with which Loko is affiliated) is submitting for funding of the community development program it envisions.

The grant proposals explain how the CEUM recently switched from an individual- to a community-based development approach. Previously, development staff sought out individuals with whom they could work or whom they could train at Loko's training center, individuals who had drive, interest, and ability. The CEUM has now set up an extensive hierarchical system of regional and parish development committees that are responsible for the realization of various communally owned projects, such as communal gardens and cattle herds. The goal is to work through the church to establish group-owned projects with benefits that will be shared by all. Later, through talking with some of the villagers seeking to benefit from community development, I will discover some of the difficulties faced by each of these approaches, difficulties rooted in differing conceptions of how the individual relates to the community with regard to land use.

After a few days, my anxieties and doubts begin to mount. They stem partly from the realization that our living arrangements are distancing me from the worlds and people from whom I had come to learn. It is time to get into the *mboka* (village), away from the confines of the compound and the facility of community with other project staff, as enjoyable as that at times can be. The next afternoon, Deb and I walk down to Bogofo, the village closest to Loko, to speak with Pindwa, a local farmer, hunter, and leader in the community. Earlier I had asked him if he might be able to recommend people with whom I might work. I had specified only a minimum of criteria: that they be active land-users with a reputation for their knowledge and hard work out on the land; that at least some of them be elders; and that they come from a variety of ethnic or language groups represented in the village. Bogofo was once a small Mbandja village, but it grew dramatically after Loko's hospital began to offer health care. Having come to the hospital from far away, many people stayed to settle in the village as a means to have, in their words, "health insurance" against future illness. Falling ill in their home villages, they might have to walk days to receive health care. In Bogofo, health care is only 2 kilometers away. Today the village is a mixed community of approximately 1,500 people from numerous language groups.

Deb and I find Pindwa at his house. He warmly welcomes us and sets out chairs for us to join him and another villager with whom he's been visiting. They are quick to mention how much it means to have another doctor at the hospital. Given the paucity of trained physicians, the Congolese doctor with whom Deb works, Dr. Kindi, must fulfill multiple roles that often take him away from seeing patients. In addition to being the hospital's sole doctor, he is chief physician of the Loko Health Zone and the medical liaison for the CEUM. Each position is a full-time job. For his services Dr. Kindi receives a monthly salary roughly equal to US$365.

Dungu, Pindwa's visitor, asks who my father is. I explain that my parents had first come to Congo in 1953 and had started their work in education and health care at Gbado-Gboketsa, just 40 kilometers up the road. Later they moved to Wasolo and then, just after I was born, to Karawa, where they carried on their educational work for twenty-two years before spending their remaining time in Congo teaching at the CEUM's technical high school in Gemena. Pindwa says he remembers my father well. Then I realize that Pindwa is the same person whose hunting skills my best friend from high school used to rave about. The two of them would often go hunting together when my friend came home to Loko on school vacations. Even though Bogofo is a new place for me, having these connections helps break the ice of entering into new relationships. I feel some of my anxiety and doubt beginning to soften.

Then, pulling a piece of paper from his pocket, Pindwa rather tentatively asks if the farmers I had asked him to recommend all need to be Christians. I quickly say they do not, to which he responds, "Good," since among those he has chosen is a Muslim farmer who, I later find out, left Congo for a time to study the Koran in Maiduguri, Nigeria. Pindwa then shows me the names of the eight farmers he's asked to participate in my research based on the criteria I had specified.[2] I copy the names into my notebook as we discuss other logistics and then together decide that the next best step is for all of us to meet. We schedule a group meeting for Saturday afternoon at four o'clock.

The afternoon has slipped quietly away into the half-light of Central Africa's brief dusk. Nearly two hours have passed since we arrived. Before we leave, Pindwa suggests we pray. After hours spent in graduate research seminars and methods classes, I have grown accustomed to beginning research not with prayer but with outlines, logical justifications, and rational defense. Here in the twilight, full of that certain peace and contentment that comes from good talk, I feel the release in hearing myself agree to his suggestion without a thought. It is a moving prayer, a blessing of this research project, a petition that the writing of this book might be a way to aid all those who read it, to bring them *mayele* (intelligence) and *kimia* (peace). As Pindwa accompanies us down the road, I feel again what I have felt before—both awe and gratitude for the chance to be here again in this land, among these people.

Bogofo: An Initial Group Discussion

Saturday's late-afternoon sun casts long shadows across Pindwa's well-swept dirt yard. In the nearby church, drums beating a syncopated rhythm enliven the choir practicing for tomorrow's service. Goats bleat as young boys shoo them from the impromptu stick-goaled pitch laid out for a rousing game of soccer. A yellow-vented bulbul warbles in the *mafuku* (breadfruit) tree; beneath its canopy I have gathered with a small group of farmers to talk about my research and to ask for their participation. I begin by broaching the idea of hiring myself out to work in their gardens in exchange for learning about their knowledge and views of this land of forest, savanna, and field. In reply, Gabi affirms for the group that by gathering together they have already agreed that I accompany and work with each of them in their fields. Libeka murmurs his assent.

But Pindwa raises an objection, "As for me, I can't really agree to take you on as a laborer, but I will take you on as a child who goes to work alongside a father in order to gain the father's intelligence regarding work." With these few words, my coarse and impersonal metaphor of the market is exchanged for that of the family, something much more personal, much more intimate, much more locally appropriate. I gladly accept the role and throw out any thoughts of hired labor.

After setting up a work schedule, we begin to talk about the land, about how to take care of it so that the soil, forests, rivers, fish, and animals are not destroyed. We talk about the ancestors and what teachings they might have given in this regard. "What laws or commands did they lay down for us to follow so that our lives will be good, so that we 'walk well in the world'?" I ask.[3]

Gabi immediately brings my rather abstract questions down to earth, linking them directly to actual practice by comparing different ways of harvesting manioc, a staple crop throughout much of Congo. Whereas some people harvest all their manioc at once, leaving them with nothing, his people, the Kongo, do things differently. "For us, some gardens can last for five years and going out to them, we still find manioc to harvest." Is his point that to "walk well in the world" means to farm efficiently, to live frugally, harvesting only enough for your needs, getting lots of mileage from each plot of land? Perhaps. I do know that Gabi comes from the faraway region of Bas-Congo, much more densely populated than the Ubangi. With little forest left and with land in scarce supply, his people have adopted various intensification techniques such as manuring that enable them to use each field longer than the one or two years common in the fallowing systems employed around Loko.

I return to my questions about the past: "But would you agree or not that the ways of the ancestors still hold valuable teachings that can help solve these problems?" Pindwa thinks they do, but he also points out that for many people such teachings are beginning to lose their hold. "I know that the ancestors had many laws for the purpose of protecting things," he says. "Whether it concerned fields, or rivers and streams, they had many commands. But what I see is that an evolution, or simply all the changes taking place, have begun to throw the things of the ancestors aside. Some of their laws are disappearing. But for other people, especially those like the fathers who grew up hearing such commands, for them the laws of the ancestors have not completely died out. They still guard some of those laws in that they will say, 'In the time of the ancestors, that wasn't the way it was done.' Thus what I know is that some people guard the ancestors' laws while others don't. But they had laws that varied from one group to another. Each group's laws were different."

Turning to Libeka, he asks, "So I don't know what laws the Mbandja had?" A door to dialogue and interethnic exchange is thrown open. Although they are all now living in Bogofo, these farmers represent a diversity of people and languages—the native Mbandja as well as Mbuja, Ngombe, Kongo, Ngbaka, Fulu, and Pagabeti (see Figure 3.1).[4]

Libeka adjusts his giant frame settled somewhat uncomfortably in the wicker-seated mahogany settee he shares with Gabi. His gray hair reveals his status as an elder. Earlier he told me that he used to hunt buffalo with my father back in the 1950s. "The ancestors' ideas for protecting the forest have been around for a long time," he begins. "If you cut the forest [for a garden], after three years or so, leave that forest so that the trees can grow until they attain a certain girth; then you can cut it again.

"And as for fishing, they did not want their children to finish off all the fish or to use poison. Rather they let some fish remain in order to bear many offspring for their grandchildren, for those who would be born after them. Thus the ancestors protected things since they all shared the rivers and streams. Even large rivers they would still divide among themselves—'Here is our area. Their group along with their ancestors have that place.' And so each group was given an area to use. Even if they had birthrights to a certain place, that still didn't give them the right to cross the boundary into their kinsman's fishing area. If they came to another person's boundary they turned around; they could not kill fish there. They went ahead and left it and went on their way. Each group did this. But now, things have gotten mean. Going along river after river, stream after stream, they destroy all the homes of the fish until there are no more fish to be found. You see, some fish seek the protection of holes, staying put there until they go out to feed.

"So then, regarding the matters of the forest—often we don't find very good forest because some people are cutting with abandon, continuing on and on until they reach a certain point and finally let the soil be. Such does not bear good fruit. When

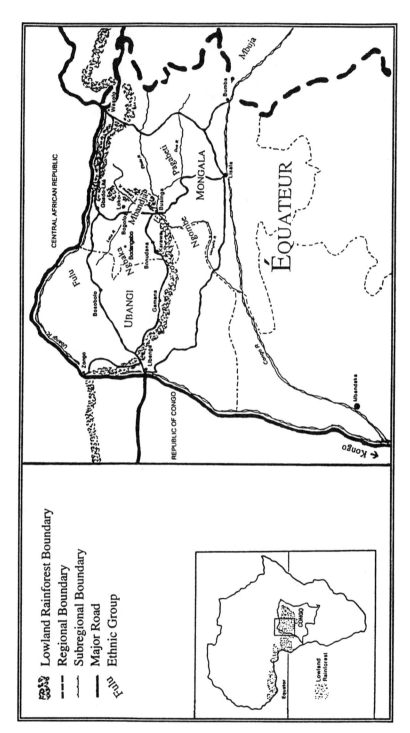

FIGURE 3.1 Northwestern Congo, Équateur Region: Ubangi and Mongala Subregions.

you clear fields from the forest, on the third round leave the field be. Cut another, and then after a number of years return to the one over there and start to clear it again.... Already the soil will have softened and be ready again."

Tambwe, probably the oldest in the group, agrees with Libeka that the ancestors used the forests, the soils, the rivers, and the steams in such a way as to let them replenish themselves. You let "the soil regain its fat. You would never cut all your forest. You cut for a ways and then leave the rest."

He continues with talk about fishing tenure and other restrictions: "By allocating fishing areas according to boundaries, the ancestors made sure that the rivers and streams were not destroyed.... During the dry season, they could fish but not in the holes. You could kill only those fish in that water that filled the basin where you were seining.... They also could not put any things like poisons into the water, for poison will kill all the fish including the little ones, it will kill them all. So then when you seine that area, you will get some fish but you leave those that are tucked away [in the hollows]. Then around June, the rains start again and those fish come out from their holes and begin again to spawn. But our children today are not like this. Now our children have the desire for money. They will go to the waters, to the fishes' holes, they will cut all around the edges, get some poison and if they hear the rustle of fish, they throw in some poison, thus destroying the whole place."

As with fish, so with other animals, Tambwe goes on to explain: "It was to prevent these kinds of things happening today that our ancestors said that elephants and buffalo should be killed only at certain times. Eh? If they killed an elephant, they would leave them for a period in order that they give birth, they would let them be. But look—today our children will not see an elephant. Our children will not see animals like buffalo because these new voices that have come out recently have ruined everything. But if one would pay attention to what the ancestors said about hunting times, then if you killed an animal, you would wait until the rainy season when the animals start to bear their young. But even then you would not kill the mother."

"Were these laws set down before the time of the colonial state? Did the ancestors also make such laws?" I ask.

A chorus of voices assures me that we are talking of matters prior to the colonial imposition of hunting laws. Then Libeka continues, "Yes, yes, the ancestors laid down those laws. Take me for example, a father of children. At the time of fish seining, I should accompany them, to sit and watch. When they begin to seine, little fish like this they couldn't keep. They could only keep the big ones and leave the little ones in order that they remain, grow, and spawn. And that is how children learned to take care of the rivers and streams. There were a lot of fish. Around 1948 or 1950, the rivers and streams were still good. We had lots of fish. But then around 1960, at the time of independence, or 1970—Ehh!" Libeka claps his huge hands as a sign of depletion and then concludes, "The rivers and streams began to be degraded."

Tambwe continues, "The reason being that the laws of the State no longer exist. And some of our ancestors [*bankoko*], those who convey their wisdom and intelligence to us, are no longer here. They've left us. Now each and every person does—[what he wants?]."[5]

* * *

The word *bankoko* can signify the plural form of grandmother and grandfather. Thus, here Tambwe may simply be referring to those of the previous generation who have died and are no longer available for counsel. But more commonly the plural form is used to refer to the ancestors, to all those who have gone before and therefore, by implication, to entire ancestral traditions. According to many of these traditions, the ancestors may have physically died long ago, but spiritually they remain very much alive and involved in the world of the living.[6] Even today, whether acknowledged informally through pouring out a bit of one's drink onto the ground prior to imbibing or more formally through ritual, sacrifice, and prayer, the ancestors still play an important role in the lives of many Central Africans. According to philosopher Kwasi Wiredu, "The influence of the ancestors is as vivid in the African imagination as that of the living elders of the clan and actually more authoritative morally, since the ancestors are, as a rule, supposed to judge the conduct of the living and reward or punish them from a position of moral unimpeachability to which no mortal can lay claim" (1994, 46).

If Tambwe is using the word *bankoko* in this latter sense (and I tend to think he is), then his statement "they've left us" is powerful indeed. Not in a novel but in a matter of lines, he fully captures the pathos of an ethos falling apart. The world in which the ancestors could be counted on for help, guidance, and wisdom in exchange for veneration has broken down. The ancestors are no longer here, Tambwe says. Neither, it appears, does he perceive himself as one day becoming an ancestor who can play a part in the continued transmission of the ancestral laws to those who follow. The chain linking children to adults to elders to ancestors has somewhere become irreparably broken.

Later, when I spend the entire day with him, Tambwe will explain more of what he feels has caused the ancestors to depart and the chain to break; here, his words imply that as a result of their departure people do what they want and the rivers and streams become degraded. Not unlike the idea of covenant profiled in Chapter 2, Tambwe's words indicate that the integrity of the natural world in which human communities are situated is vitally connected to the integrity of their moral and metaphysical worlds.

This theme will come up again in the narratives that follow. So will another theme, clearly signaled in the narratives recounted above: the sense of restraint and limits. According to these farmers there is a limit to the number of times one plants the same field; a limit to the size of the forest one clears for a garden; a limit to how much manioc one harvests at any one time; a limit to the times one goes hunting; a limit to the size of fish one keeps. I hear Libeka mentioning yet another measure of restraint. "And when you left the village to go into the forest, you wouldn't stay there for two to three weeks. You would spend maybe a week and then during the second week you would come out. You would only catch the fish you needed and then come back."

How is one to understand this sense of restraint? Can it be explained simply by the fact that population densities were relatively low, needs were few, and technologies were limited, thus making humans' impact on the land relatively light? Such facts of human demography and technology may have been true for the area and population in general, but can they explain the choices for restraint made by any one group or individual? It is true that having few people around would detract from the relatively large labor supply needed to clear extensive areas of forest, but it doesn't take many people to overhunt or overfish a given habitat. It takes only one person to kill animals before they've given birth or to keep all the fish in the catch, not just the large ones. Might restraint, then, have stemmed directly from moral-ethical imperatives lying within the realm of the symbolic, in addition to stemming indirectly from these more "objective" sources of low population pressure and limited technology?

Of course, it is not a matter of it being *either* one source *or* the other. And it is not helpful to attempt to determine whether the moral chicken preceded the actual egg of ecological practice or vice versa. As anthropologist Parker Shipton writes, "Debates about whether ecological practices and adaptations govern symbolic thought or vice versa will never end, for the links between any such cause and effects, where they exist, are variable, elastic, and recursive" (1994, 367). In fact, he goes on to remind us, the two spheres are inextricable, and for most people and most of human history they always operated together in complex interaction. Thus morality and ethics—often expressed most powerfully through symbol—do play a part in governing human choices and behavior. They cannot just be explained away as peripheral to more concrete material explanations of ecological practice.

A more interesting question than that of cause and effect might be this: If Central African groups and individuals did operate according to what some have called a "moral economy" (Scott 1976, Spear 1997), did that morality ever go beyond the human community so as to be directly applied to the natural world for its own sake? Stated differently, did the various peoples inhabiting Central Africa's forests hold ethics that were explicitly environmental? Later I will come to see how that question also misleads, as it is a manifestation of my inherited Western dualistic bias that separates the human and the natural. For the moment, however, I try to get a better handle on it by posing another question to the group of farmers as we continue our conversation under the *mafuku* tree.

*   *   *

The choir drums have stopped. Somewhere a child starts to cry. "Are there more questions?" Libeka asks me.

"There are," I respond. "Simply to ask, according to your thoughts, what were the reasons why the ancestors placed such laws? What was their purpose?"

"The reason is so that our children and our grandchildren, those who will be born after us, will not lack fish, so that they will also not lack animals," Libeka replies.

"Because you know, if we use them all up, those we will give birth to later will lack these things," adds Tambwe.

As if to indicate that something has gone wrong in the implementation of this ethic of concern for future generations, Libeka poses the rhetorical question, "For today who is being deprived? They will not see elephants; they also won't see buffalo."

Yambo, a Ngombe farmer from the Likimi area to the south, agrees. "And others of our children will no longer know the names of some things."

"Does that mean that it is a bad thing if children don't know the names of all the animals?" I ask.

Yambo responds, "For us it is a bad thing because by not having a chance to see it, they will also never come to learn about it. They will thus not know it."

"And to know it is a good thing?"

Yambo and Libeka quickly agree, "It is good."

When I ask why, they break out into hearty laughter. "Now we've gotten in deep," I quip, evoking more laughs from the others. Then Tambwe tries to explain:

"It's like this—today, you've come here to see us, right? You've come here to Bogofo, eh? If they gave you an animal, say a monkey, eh? And your children have never seen it before. When you bring it over there, they will be afraid, eh? They will ask, 'Father, what is this animal?' Eh? That's what they would ask you isn't it? So, you would say, 'No, I got it in Bogofo. This animal is called a such and such, eh? Since you don't know it, I am bringing it so that you can see it.'

"Among you white people, I think that sometimes in those places where you keep animals like leopards and other animals, sometimes on Sundays you will gather all the children, eh? And you will go stand around their enclosures, eh? Aren't you showing them and teaching them about those animals, in order that they might know each and every animal? That is the reason why the ancestors forbid it [killing fish and animals without restraint]."

* * *

A zoo. Tambwe must be referring to a zoo, a peculiar place found in many Western cities where a bit of wilderness is brought into an urban arena. He is right. On Sunday afternoons, they *are* thronged with families, children, young lovers, elderly couples—all eager to catch a glimpse of wild animals, to learn a bit about their habits from the texts and pictures displayed next to the animal cages. The specialists call this ex situ conservation—preserving species outside of their native habitats. In contrast to the forests, streams, prairies, and savannas where these animals and their forebears once freely roamed, zoo animals, in the worst cases, pace barren concrete settings, safely enclosed behind iron bars. In the more well-endowed zoos, they can be found in elaborate re-creations of their natural habitats, spacious tropical worlds complete with vines, waterfalls, and huge buttressed rainforest trees that look entirely real until, scaling their trunks, one sees the concrete shafts disappearing into a ceiling painted to resemble leafy canopy.

In either case, the artificiality of zoos highlights the separation, the distancing from wild animals and their native habitats so many of us in a modern industrialized world experience, whether consciously or subconsciously. To cope with that separation we bring the animals to us, but no matter what type of home we try to create for them they still remain separate, set apart. We can look but not touch, read the texts but not the tracks. At the zoo, we experience an abstract, alienated wildlife, and part of the reason for its abstraction and alienation is that the very city in which the zoo is located long ago succeeded in destroying nearly all trace of local fauna and the habitat upon which they depended for survival. For all its wonders, artistry, and glorious manifestation of human culture, the city has rarely if ever been designed so as to deliberately safeguard any degree of balance, harmony, or cohabitation with the native landscapes, plants, animals, and peoples that preceded its rise from the land.

But in order to destroy so easily and thoroughly, our city-building mind-set must have already been deeply marked by a long history of alienation from nature. That alienation has deep-seated roots in the West, but it flourished in earnest under a deliberate and radically new way of thinking about the relationship between humans and the rest of creation that arose with the growth of modern science, earth-negating religion, and industrialization (Merchant 1980, Skolimowski 1981). Dualistic separations of all kinds replaced a metaphysics marked by connectedness, unity, and synthesis. What Henryk Skolimowski describes happening within a human being also applies to what happened between humans and nature:

> The separation of facts from values, of man from his knowledge, of physical phenomena from all "other" phenomena, resulted in the atomization of the physical world, as well as of the human world. The process of isolation, abstraction, and estrangement (of one phenomenon from other phenomena), a precondition of the successful practice of modern natural science, was in fact a process of *conceptual alienation*. This conceptual alienation became in turn human alienation: man estranged himself from both his knowledge and his values. The primary cause of contemporary alienation is a mistaken conception of the universe in which everything is separated and divided and in which the human being is likewise atomized and "torn" (1981, 14).

Human separation, division, and alienation from nature are not what I am hearing in the views expressed by the farmers under the *mafuku* tree. According to Yambo, Libeka, and Tambwe, the ancestors had laid down laws to govern humans' proper relation to nature in order that generations to come not lack fish and animals, in order that the children of their children also have a chance to "know the names of things." The ancestral laws ensured that this happened by restraining one's current use of nature so that it would always have a way to replenish itself for future generations. Members of these Central African communities thus saw themselves as an integral part of the nature they related to, the nature they "used" in the same way that other animals used nature. The closest analogue of this—a Central African means of conservation and wildlife education—that Tambwe can articulate within white culture is a zoo.[7] But what a contrast between a zoo and these Central African systems of sustainable use! At the zoo, a father teaches his children about animals by showing, pointing, observing, reading texts from behind a fence. In the cultures of these farmers under the *mafuku* tree, a father teaches his children about animals by going with them into the forest, showing them how and when to trap and hunt, which fish to leave, which to take. A person protected animals not by enclosing them in zoos—bringing them into yet keeping them distinctly separate from the human sphere—but by going out into the natural sphere that one shared with animals and learning ways to balance your needs with theirs, ways to guarantee their survival as well as your own.

Some time following my research experiences in Bogofo, a conversation with two colleagues, Mbella from Cameroon and Yasmine from Afghanistan, further illumined how strange some of our Western perceptions of nature can appear to non-Westerners, even those highly educated in a Western educational system. Our talk was not about white people's zoos but about white people's zoos writ large—national parks. The three of us shared an office in which I had just hung a poster of Yosemite National Park's majestic El Capitan. Both Mbella and Yasmine remarked how beautiful it was. But then Mbella added tales of his Yosemite experience. How strange he had found it to go there with an American friend and just sit and stare in awe of the trees and the quiet, "appreciating" this thing called nature.

Yasmine agreed. "Yes it really is a strange thing isn't it. To go there and stare, not do anything like picnic or—"

"It's a spiritual experience," I interrupted, half facetiously.

Then Yasmine responded with words that really struck home: "That shows the division that's been made between humans and nature—to find satisfaction in going and looking at this 'other.' The two have been separated so one way of treating this other separate entity is to go and stare at it like some object separate and apart from you. Such environmental perceptions come about only after one has cleaved a gulf between humans and nature. If one is part of or one with nature, this type of environmental consciousness doesn't happen."

There is an ironic truth here. As Yi-Fu Tuan has remarked, "An appreciation of nature is most frequently and eloquently articulated by people who no longer live in the midst of nature" (1994a). Conscious, explicit environmental perceptions and concepts such as conservation, natural aesthetics, and wilderness are birthed in cultures most disconnected from nature. The possibility of such concepts exists only to the extent that nature can be set apart, objectified as a distinct other, and dealt with abstractly. In the West, this "othering" of nature has resulted in extensive programs of environmental conservation, yet such programs remain characterized by a dualism that totally removes human beings from the nature picture. Ironically, the ability to abstract ourselves from nature has also enabled the West to destroy nature at an unprecedented rate.

During the early Enlightenment, the isolation, abstraction, and estrangement of nature, which Skolimowski reminds us were preconditions for the successful practice of modern natural science, served not so much to conserve nature as to control it. By dissecting nature with an exacting analytical scientific epistemology, early natural scientists sought to unlock nature's secrets and learn how it worked in order to liberate humankind from the limits it imposed. Later, through an alliance with new technologies, this scientific epistemology was used to exploit nature for the purpose of creating massive wealth. Only after the consequences of this exploitation became readily visible—the alarming signals of deforestation, extinctions, air and water pollution—did a conservation ethic and movement really take hold in the West, leading to the creation of, among other things, such novel institutions as zoos and national parks. Yet the conservation of nature in the West, like its exploitation, remains governed by the same dualism: Society and nature are considered distinct and separate realms.

True, distinctions between the human and the nonhuman, between society and nature, have been made throughout history by all peoples. Yet nowhere as much as in cultures under the influence of modern Western science do we see such a propensity to come at the two with a cleaver, to lay nature out on the laboratory table and so radically dissect out its human component in order to more "objectively" either conserve or exploit it. Natural "wilderness" and natural wreckage, in other words, are not opposite extremes but different sides of the same coin, both logical outcomes of a dualistic metaphysical understanding of the society-nature relationship.

In contrast, seated under the *mafuku* tree, Tambwe, Libeka, and the other farmers clearly perceive that nature very much includes themselves, their ancestors, and their offspring. Their long-standing direct dependence on nature for their livelihood has fostered not a human separation from nature but a strong sense that human society remains within and serves as an integral part of the whole vast natural realm. Furthermore, their connectedness to nature is not something they have sought through deliberate mental, spiritual, and practical exercises, as is common among those trying to reconnect with a nature from which they've been alienated. Rather their descriptions of fishing, hunting, and farming imply an unmediated connection to nature that is lived out in their daily lives.

Earlier I posed the question of whether the various peoples inhabiting Central Africa's forests have ever applied their ethics explicitly to the environment. Some writers have interpreted Central Africans' long-standing direct dependence on the natural world to mean that their ethical relationship to nature is strictly utilitarian. Anthropologist Mary Douglas, writing prior to Congo's independence from Belgium, notes that the Lele of central Congo "frequently dwell on the distinction between humans and animals, emphasizing the superiority of the former and their right to exploit the latter." According to her, colonial laws to protect game "strike the Lele as an impious contravening of God's act, since he originally gave all the animals in the forest to their ancestors to hunt and kill" (Douglas 1954, 9). Many of the farmers I talked with also expressed views that God had given humans animals to hunt and kill. But they also added that this hunting and killing was to be done in certain ways, only at certain times, and only in certain quantities in order that they not despoil the gift they perceived God had granted them.

Nevertheless, such beliefs and practices still strike many as a poor foundation on which to build any kind of environmental ethic. J. Baird Callicott begins the African chapter of his comparative survey of environmental ethics with the common sentiment that "mention of African culture evokes no thoughts of indigenous African environmental ethics" (1994, 156). Later he adds: "One might expect to distill from [African thought] no more than a weak and indirect environmental ethic, similar to ... [an] ... ecologically enlightened utilitarianism, focused on long-range human welfare" (1994, 158). Callicott does succeed in going beyond these initial positions to promote what he calls "African biocommunitarianism" as a vital foundation for an environmental ethic. Although his analysis draws primarily on the example of the San people of southern Africa, Callicott highlights certain ideas that I found to also ring true as I was listening to Bogofo farmers speak about the ancestral laws that guided one's use of nature.

Drawing on the work of Richard Lee, Megan Biesele, and Elizabeth Marshall Thomas, Callicott writes of the San:

> In comparison with their North American counterparts, San hunter-gatherers seem to be members of a less enchanted, more everyday biotic community (1994, 169).

> They behave as though they regarded themselves as plain members and citizens of the biotic community (1994, 170).

> However little the San folkloric record may reflect their respect for fellow members of the biotic community, the ecological and ethological records testify to an accommodation both with the game and with other predators. The prey species remained abundant. They were not overhunted.... Nor did the San attempt to eradicate species such as lions, leopards, and hyenas, which competed with the San for game.... One can only assume that the San, as suggested by their scanty cosmogony, regarded themselves as one with the other fauna and practiced a quiet policy of live and let live with their non-human neighbors (1994, 172).

Under the *mafuku* tree, the Congolese farmers I am talking with express similar ideas and practices within their own heritage: living *with* animals and being *part of* the natural world. For much of Central Africa's rainforest dwellers, the metaphysical perception of being part of nature contributed to practices that, although surely changing nature, were themselves changed in response to nature's changes. This interactive, reciprocal relationship allowed humans and the rest of nature to coexist within a relatively balanced system. For example, one way rainforest peoples coped with elephants' response to their nature-changing practice of farming—marauding gardens—was simply to get out of the way, to move their settlements and gardens out of the elephant corridors (Vansina, personal communication). What a contrast this presents to the image from my own white American heritage of people like "Buffalo Bill" Cody massacring the bison that roamed midwestern prairies, thereby clearing the land for Euro-American settlement.

In Central Africa, society and environment were, and in some cases still are, considered parts of one whole, constantly in interaction, changing and being changed by the other. This interactive relationship may not have produced an explicit environmental ethic or an abstracted environmental consciousness, but that fact should not be interpreted to mean that Central African environmental wisdom does not exist.

Rather the absence of any explicit abstract environmental ethic in Central Africa is more likely due to the fact that until relatively recently there has never been any need for one. For the ancestors of people like Tambwe, Pindwa, and Yambo, environmental degradation was not much of a problem. Even they can recall the not-too-distant years when they could find animals close at hand. As Tambwe would later tell me, "You wouldn't go far to look for animals. You'd be living here and perhaps a guest would come and visit you, and you would go out and get one animal, kill it, and that was it—to fulfill the need to eat."

Furthermore, living as part of nature has made it difficult for Central African peoples to cognitively create a discourse around something like an environmental ethic, as the idea of separating oneself from nature and contemplating one's relation to it in any abstract way is purely a foreign construct. Even my Western-educated colleague Mbella found Yosemite National Park—and by implication the environmental ethics instrumental to its creation—rather strange and puzzling. Perhaps that is because within his heritage, as within those of many other Central African people, the rightness and wrongness of one's actions toward the environment were determined by years of practice living within nature, learning what worked and what didn't work to guarantee the continued existence of both the human community one belonged to and the nonhuman community one depended upon.

Here in America, we are proud of an environmental ethic that has spawned one of the most extensive systems of natural preserves, parks, and reserves anywhere on the planet. Yet it is a rare American who looks deeply enough to see that "creating all these preserves as we do is not so much a sign of success as it is of failure; our preserves simply show that we have not yet learned to live within the natural systems of which we are part. Preserves are confessions that we are unable to control ourselves from destroying the natural fabric that sustains us."[8] Central African forest peoples and the ancestral heritage they hold have much to teach us with respect to learning how to live within the natural systems of which we are part, learning how to control ourselves from destroying the natural fabric that sustains us. Their traditions uphold not the solution of the human-voided natural reserve but rather a human-inhabited natural world in which humans, as part of nature, control their use of the resources it offers them, as doing otherwise is to bring ruin on both themselves and the nature that sustains them.

Later, when I travel on to the Ituri and the Réserve de Faune à Okapis, I will hear from local villagers just how incongruous with their own traditions they find the reserve solution that has been imposed upon them. But these farmers here in the Ubangi still have much to say. Even now under the *mafuku* tree, prior to getting out on the land with each of them, I hear them bemoan the fact that the world—physical, moral, and cognitive—they inherited from their ancestors is breaking down.

* * *

The sputter of a well-traveled motorcycle breaks into the cadence of our conversation. It is likely heading up to Gbado-Lité, the home of President Mobutu. At one time

Gbado-Lité was a tiny village of thatch-roofed rondevelles divided by a potholed dirt road, no different than any other village in the northern Ubangi. After the investment of millions of dollars (a large proportion of which Mobutu skimmed off as the revenue earned from the rapacious exploitation of Zaire (Congo)'s unparalleledly rich natural heritage), the village is now the region's most modern city. In addition to an international airport large enough to land a Boeing 747, Gbado-Lité boasts modern stores, banks, a Coca-Cola factory, a Swiss dairy farm, paved boulevards, restaurants, hotels, conference centers, and not one but two opulent presidential palaces.

One of the latest projects is a brand new university, Université d'Aequatoria. In August, Deb and I visit it after celebrating our first wedding anniversary at one of the city's restaurants. We are surprised to find it completely empty—fully furnished dormitories with no students to sleep on the mattressed bunk beds; brand-new classrooms with blackboards never whitened by chalk; a state-of-the-art refectory, its kitchen containing not a single aroma of local Congolese cuisine. It is not that classes haven't started yet, the lone guard tells us. Rather Mobutu decided having free-thinking university students so close to home might not be such a good idea. There is talk that the shiny new complex will now be used as army barracks; meanwhile the university is supposed to move to Libenge where, fifteen years previous, Mobutu built an elite, European-style high school that never opened. Such examples of Mobutu's wasteful abuse of wealth only scratch the surface of the multitude means by which he has made his country one of the poorest in the world. They are only one facet of the many ways in which the ruthless dictator has whet the destruction, begun by colonialism, of the world whose passage I now hear Pindwa lament as the motorcycle's sputter fades out somewhere down the road toward the city-symbol of Mobutu's corruption:

"My point of view," Pindwa begins, "is simply that in each and every ethnic group, the ancestors had extremely strict rules. At that time, their children really obeyed those rules since they knew that if they didn't obey, their punishment would be severe! Let me give an example concerning the matter of caring for the animals of the forest. I myself was whipped by the elders only because of eating a turtle in the forest. Yet it was to teach me that as a child, I could not eat such a thing; they are things of the elders. So, then in the morning, they beat us, gave us a hard whipping, asking 'Why do you think you children can eat this animal?'

"They were able to agree since they all shared the same way of thinking: If someone was guilty of negligence, they were punished. All of them agreed to carry this out. And because of this, people obeyed such that whether one was caring for forests, caring for rivers and streams, caring for whatever, things would simply go according to the way in which they had agreed upon. If people were negligent, they would bear serious consequences.

"So each and every ethnic group had them [laws], and a few people are still saying, 'In the time of our ancestors, it was this way. It was not the way that we are seeing today.' Thus, perhaps, among us, what you see is an evolution, eh? Learning many different things, they [the younger generation] try to reject you and the ancestors saying, ... 'They are deceiving you. That's the way it used to be, but now ...' Civilization started to come, and it began to destroy these things, these laws."

Earlier in our conversation, Libeka, with a clap of his hands, had lamented the recent depletion of fish in the rivers and streams. Tambwe had offered a partial explanation for this environmental degradation by stating that the ancestors had left them and thus people have begun doing whatever they want.

Like Tambwe before him, Pindwa here is also linking environmental degradation to the breakdown of the ancestral world. His words convey the dissolution of a certain oneness of mind and action (e.g., "they all shared the same way of thinking"; "all of them agreed to carry this out") that in ancestral times had promoted people's obedience to laws conserving the integrity of the forests, rivers, and streams. According to Pindwa, the coming of "civilization" and young people's subsequent rejection of ancestral ways has destroyed the unity and power of communal sanction and discipline. In turn, the community's ability to make sure that the forests and the waters are properly cared for has also been jeopardized.

As mentioned earlier, the narratives that people recount often present an idealized more than a factual picture of the past. For example, it is highly doubtful that in the past, things always went "according to the way in which they had agreed upon." As is the case today, conflict and contradiction were surely present within Central African communities long ago. Again, my point here is not to prove or disprove the "truth" of these farmers' perceptions but to ask what those perceptions reflect about people's present-day understandings of the human-environment relationship. In this sense, both Tambwe and Pindwa's narratives reflect an understanding that the ecological conditions of the natural world are strongly tied to the moral conditions of the human community. In these farmers' minds, ancestral traditions had provided their human communities with a sense of unity, a cause for obedience, a means of discipline, and a source of wisdom and intelligence. When these moral worlds begin to break down, whether due to "civilization" or simply to the fact that the ancestors for whatever reason "have left us," the effects are felt not only intergenerationally within the human community but also out on the land.

Pindwa continues his talk, illumining other ways in which morality and ecology are linked. To him the inability to discipline the young is a crucial element in the social and ecological strain the farmers of Bogofo are currently witnessing:

"Another aspect of it has to do with disciplining," Pindwa explains. "I am especially having trouble with this matter of discipline. During the time of the ancestors, we lived in families like this. Say my child does something wrong; if I beat him, he flees over to you, you beat him; he flees over to him, he beats him; he comes to him, he beats him, until the child flees to the forest. I say, *everybody* will be displeased with him! Well then you see that now the child begins to think there in the forest, 'Eh, not one person likes me. What shall I do?' He will implore his body [*bondela nzoto na ye*] until he is able to refrain from ever doing such a thing again.

"Today, however, civilization is such that if I beat a child, 'Oh, why are you persecuting that child like that? Come.' So, he takes and comforts the child and the child becomes his friend. She will no longer obey like they obeyed during the time of the ancestors. It is something of an evolution that is destroying some of these things that the ancestors guarded."

Gabi, having sat quietly listening to Pindwa, now adds his assent: "It is destroying a lot of the ways of the ancestors."

Tambwe agrees. "The ancestral ways have fallen down because of these things. In the past, if my child and his child were fighting, he would come out and gather and beat the both of them."

Pindwa interrupts with a hearty "Yes." Then Tambwe continues.

"Eh? But today, if he does such a thing, I would ask him why he beat my child. Or I might even bring it before the State. Thus it has ruined us and our children, eh?"

"It reminds me of a certain saying," I respond, "that here in Africa, they say that it takes the whole village to raise a child. It doesn't just take one person. It requires that everyone work together."

"Yes, yes, that is true."

The group is silent. It is nearly dark. The goats have started to settle in under the eaves of the houses. The choir members have long since returned to their homes. Tambwe, as the eldest, draws our conversation to a close: "Well, then, I think that is the end of our words."

"I've brought some tea, if you'll each agree to drink a cup with me," I offer.

"Of course. Good. Thank you."

"Haven't you the guest come to feed us?" someone remarks playfully.

"That's right. It is not a bad thing. You are also serving me."

Dungu, the sinewy Mbandja farmer who tomorrow I will accompany to his garden, responds with his characteristically infectious laugh that, more than once, will delight my soul in the days ahead.

Conclusion

I walk back to the house in the dark, my head filled with all that has transpired just in one afternoon. The conversation under the *mafuku* tree has been a lesson in learning how to ask the right questions, a first step in figuring out locally appropriate ways of speaking about things. I have learned how much I need to unload my abstractions, to refrain in our conversations from words and ideas like "environmental ethic," in order that I might reconnect with "the practical and social underpinnings of abstract forms of understanding" (Jackson 1995, 148).

To these farmers, ideas like "land ethic" in and of themselves likely hold little meaning. It is the practical and social underpinnings that they've been talking about—the proper ways to cut a garden from the forest, to harvest manioc, to fish, and to hunt that have been passed down from their fathers, mothers, grandparents, and ancestors; the important role discipline plays in making sure that these ecological practices are respected; and the social constraints that are placed on individual behavior for the good of the entire community. After this afternoon, it is clear that some of the categories and meanings of "land ethics" with which I had come to Congo are totally inappropriate. In their place, these farmers have suggested other themes that at first I hadn't thought to include in a discussion of the environment: disciplining children, the departure of the ancestors, the destructive desire for money, concern for future generations, and the like—matters one might be more inclined to place into the category of "social ethics."

That is the point, and that is why the question I posed earlier—"Did the various peoples inhabiting Central Africa's forests hold ethics that were explicitly *environmental*?"—is inapposite. It doesn't make sense in a world view such as that described by these farmers, in which ethics, at one and the same time, encompass relations to one's fellow human beings as well as relations to the natural world upon which they and oneself depend. Put another way, right relationship to the natural world to a large extent is governed by right relationship to other members of one's human community.

Moreover, that human community extends into both the past and the future such that one must think of both ancestors and offspring in one's dealings with the natural world. Thus one restrains one's consumption of natural resources in order to guarantee that those who come after will also have enough. At the same time, in using the land, one also remembers the admonitions of the ancestors. In Pindwa's words, one abides by their "laws for the purpose of protecting things,... whether it concerned fields or rivers and streams." Some of the same sentiments expressed by the Ubangian farmers have been echoed in the words of African philosophers. Kwasi Wiredu, for example, captures well the ecological significance of this extension of obligations toward past and future members of the community when he writes:

> Of all the duties owed to the ancestors none is more imperious than that of husbanding the resources of the land so as to leave it in good shape for posterity. In this moral scheme the rights of the unborn play such a cardinal role that any traditional African would be nonplused by the debate in Western philosophy as to the existence of such rights. In the upshot, there is a two-sided concept of stewardship in the management of the environment involving obligations to both ancestors and descendants which motivates environmental carefulness, all things being equal (1994, 46).

* * *

Nearing the summit of the last hill, I see the lights of our house in the distance. The night is still but for the chirping of crickets. With no city lights to dim them, the stars have come out in brilliant display. It has been a long time since I have seen so many stars at once. Before entering into the comfort of home, food, and companionship, I pause alone to take it all in, the depth of both the night sky and the afternoon's conversation. I have already learned much from these few Central African farmers who have befriended me. So far, these are some of the major points regarding human's relation to the land that I am hearing from them:

- Land ethics, if they exist at all, are expressed in actual practices more than in abstract concepts.
- A sense of limits and restraint are key values governing one's use of the land.
- Use of the land is distributed between groups of people, and both individual and group tenures to certain areas are respected.
- A strong ethic of concern for future generations keeps people concerned about the dangers of overexploiting the land.
- This concern goes beyond the physical needs of future generations to include concern for their knowledge of the natural world.
- Human communities are not considered separate from nature but remain within and serve as an integral part of the whole vast natural realm, that is, human beings are part of nature not set apart from nature.
- One safeguards nature not through creation of human-voided "natural" reserves but by learning through practice, in a human-inhabited nature, how to balance human needs with those of the rest of creation.
- Right relation to the natural world to a large extent is governed by right relation to other members (past, present, and future) of one's human community, that is, social ethics and land ethics are inextricably linked.
- The ecological integrity of the land is vitally connected to the relational and metaphysical integrity of the human community, that is, as the ethical and metaphysical worlds of human communities begin to break down, the effects are also seen on the land.
- Causes for moral, metaphysical, social, and cultural breakdown include the coming of "civilization," the fact that the ancestors "have left us," and the desire for money.

In the pages that follow, I will draw on experiences and conversations with individual farmers out on the land to illustrate additional themes, as well as to expound on some of the themes above, adding to them both complexity and contradiction.

Notes

1. By "narrative approach" I simply mean an approach that relies on story—both one's own story of fieldwork as well as the stories of those with whom one lives and works told in their own words—as the primary means to convey both the content and the meaning of one's research or, as Robert Coles puts it, to convey "what there is that seems to matter."

2. This "sampling" method, certainly not random but neither purely one of convenience, holds both advantages and disadvantages. In asking Pindwa to choose participants, I avoid some of my own biases inherent in being new to the community while gaining the advantage of Pindwa's long-term experience, knowledge, and respected status within the community, factors that may well make his criteria for choosing more locally appropriate than my own. However, it is also likely that Pindwa's choices reflect his own biases and preferences, many of which I am not able to know.

3. To "walk well in the world" is a literal translation of the Lingala phrase *Kotambola malamu kati na mokili*, its meaning difficult to translate. The verb *kotambola* means to walk, but it also can mean to run, operate, or work as used to describe a machine, vehicle, or enterprise. Thus my question here infers that laws were given so that our affairs and dealings with the world would run smoothly and not break down.

4. Throughout these chapters, one of the ways I describe people is by ethnic or language group. Although choosing ethnicity as an identifier raises lots of theoretical questions as it is such a fluid and complex notion, I do so here largely because it is a way in which the people I worked with chose to identify themselves. Making mention of people's ethnic identity also reinforces how diverse and complex contemporary African communities are. And, especially in the case of Chapter 5, it indicates the wide-ranging places from which people have immigrated. The current debate on the meaning of ethnicity is very important, but I refrain from entering into that debate here since it would detract more than add to the topic at hand.

5. Tambwe never finished this sentence. Its bracketed portion I have inferred from context. The entire passage quoting Tambwe's words verbatim reads as follows: "*Batiki biso. Sikoyo moto na moto azali kosala— Wana bondoki ezalaki naino mingi te. Sikawa nde Zaire [Congo] ... ezui mayele ebandi kobongisa babondoki. Ntango wana na liboso, bondoki ezali epai na l'Etat moko. Olingi okenda epai na l'Etat, okosomba bondoki bapesa yo mokanda. Ezala baluti, bakopesa yo mokanda nde okoki kosomba baluti wana.*" [They have left us. Now each and every person does— At that time there were not yet many guns. But now Zaire [Congo] ... has acquired the knowledge to make guns. During that earlier time, only the State had guns. You want to go to the State, you would buy a gun, they give you a document. If it was gun powder, they would give you a document and only then could you buy that gun powder.]

6. Vansina discusses the role of ancestral and other spirits in Central African world views (1990, 95–99). A sense of the importance of the ancestors in African world views also infuses the works of numerous African artists; for instance, the novels of Chinua Achebe, the music of Francis Bebey, and the poems of Birago Diop.

7. It is interesting to wonder why Tambwe chose the analogy of a zoo and not other, more experiential ways in which "white people" also educate their children about nature, ways that more closely resemble those of his own people. One can surmise it is because a zoo is the place he has actually observed whites in the area educating their children about animals; perhaps it is the image of Western wildlife education he has received through any number of media. In any case it raises interesting questions about the selective processes by which the West is communicated and represented in Africa and elsewhere.

8. This quote was passed on to me by Dr. Cal DeWitt, who heard it from a member of the Dane County Natural Heritage Foundation at one of the foundation's meetings several years ago.

4

Parts of a Whole: Nature, Society, and Cosmology in the Ubangi

> *The problems of land ethics in the African context should not be viewed as isolated issues of land ownership and land use, rather they should be understood as a search for a better understanding of the relationship between man and God, man and man, man and land which would inform the principles of allocation of land among groups of people, among generations, and among different patterns of land use.*
>
> —**Mutombo Mpanya**

Congolese scholar Mutombo Mpanya is one of the few African intellectuals who directly addresses the idea of African land ethics. Perhaps the paucity of works on the subject is due, as I mentioned in Chapter 3, to the fact that isolating land ethics from the broader ethical realm within which African peoples conduct their lives seems inappropriate and misplaced. As if reflecting his own discomfort with the idea of land ethics, Mpanya is quick to point out that any discussion of the relationship between humans and land must be contextualized within a broader framework of connectedness—both social and metaphysical.

This chapter builds on the preceding one by drawing additional insights from the narratives of individual land-users living in the vicinity of Loko. Although numerous subthemes emerge in these narratives, they do reflect the three intertwined relational themes suggested by Mpanya: the relationship between humans and land (in its broadest sense, the natural realm), humans and each other (the social realm), and humans and God (the metaphysical realm). The word "intertwined" is important, for none of these realms of relationship can be disconnected from the other two. As parts of one whole, each realm encompasses relationships that affect and influence those in the other two realms.

Thus, although I do organize the narratives that follow into three sections loosely corresponding to Mpanya's three relational themes, my primary focus is on how the narratives illustrate the interrelatedness between the themes. As I have hinted already, one of the main lessons I learned from this research is the extent to which nature, society, and metaphysics (or, more specifically, cosmology) are interconnected as parts of one whole within the cultural heritage of the Central African land-users with whom I worked. As in the preceding chapter, I draw primarily on the narratives themselves to bring the nature of this interconnectedness to the surface. This will entail revisiting many of the themes that arose during the group discussion recounted in Chapter 3 (in the bulleted list at the end of the chapter) as they resurface in new ways in the individual narratives that follow. In addition, I discuss some of the implications this interconnectedness holds for current environmental practice in the context of forces and changes that in many ways have succeeded in sundering these interconnections.

I begin with "making a living" from the land as it was described and shown to me by these farmers while we worked together in their gardens. From these descriptions, I reveal how a certain level of conservation existed, and to some degree continues to exist, implicitly within the very means and methods of livelihood. I have spoken about some of these factors in Chapter 3, but in more of an ideological sense. Here I speak more about techniques and actual land use practices, both those of the past as they are perceived by contemporary people, as well as those of the present.[1] I do not want to imply, however, that practice can ever be separated from the world of ideas.

From discussion of livelihood, I turn to a discussion of the relationship between the individual and the community within Central African societies and its implications for the land. Our conversations out in the gardens illumined some aspects of this influential relationship, but they also spurred me to investigate what other observers of African society have to say in its regard.

Finally, I situate discussions of land and society within the broader perspective of cosmology in order to show more explicitly the interconnections between the three realms, as well as the ways in which these interconnections have recently been pulled asunder. Here I draw particularly on the stories of Duateya and Fulani, Ngbaka elders who within their lifetimes have experienced both interconnectedness and its sundering in unique yet exemplary ways.

Making a Living: Implicit Means of Conservation

During my upbringing in the Ubangi, the clear night skies were always filled with stars except when the moon was so bright I could read a book by its light alone. With the closest star-dimming urban glow some 350 kilometers away, the constellations virtually jumped out once we learned to identify them. Spotting faraway stars or cloudy nebulae took more care; I remember learning to discern those ethereal sights not by looking at them directly but by looking slightly to the side.

In somewhat the same way, Central African land ethics and ways of conserving nature come most sharply into view when one looks slightly to the side. Looking directly for an explicit and discrete expression of concern for nature among Central

African cultures may leave one disappointed. But it may also cause one to miss what becomes evident by looking a little to the side within the surrounding context in which aspects of what we call "nature conservation" and "land ethics" are found not in isolation but as parts of a whole.

Chapter 3 provided an initial "look to the side" into the actual land use practices of Ubangian farmers—the proper ways to fish, hunt, cut a garden—and their implications for nature conservation. Here I want to expand on other techniques and features of making a living to show how they implicitly contributed to maintaining the integrity of the natural world upon which people depend for their livelihoods.

Diversification of the Household Economy

For most of the farmers I worked with, tilling the soil is only one of several ways to make a living from the land. In addition to farming, many engage in trapping, hunting, fishing, and gathering various forest foods. Although the extent of specialized production for the market is increasing, historically this diversification has kept people from engaging in any one activity full-time so as to undermine the environmental foundation on which their livelihoods stood. The necessity of engaging in a wide variety of activities has limited the ecological impact of any one of them, as Wasitdo, a young Ngbaka farmer, explains to me:

"To kill a lot of animals at one time was impossible," he states, referring to his Ngbaka ancestors. "With their traps and snares—as the trap would catch, that is how they would kill. And at other times, they might go out with their bow, shoot once or twice, kill two monkeys ... like that."

"But they would not go beyond that?" I ask.

"Not very often."

"Why?"

"Because, often they didn't have just one occupation. If they got a little to eat, that sufficed and they would return in order to go and cut their gardens and do other work."

Each of these means of livelihood was itself diversified and remains so today, consisting of a combination of techniques correlated to particular seasons and to particular species of plants and animals. The number of different species harvested from the forest is truly amazing. During my earlier research in the Ituri, members of one household named forty-one different items they obtained from the forest (Peterson 1991, 78). To take gathering alone, the Aka living in the forests of the Central African Republic are representative of many other forest peoples in the variety of species they routinely harvest: nine types of fruit, eight types of snails, thirteen types of termites, twenty-two types of caterpillars, two types of honey, and thirty-two types of mushrooms (Vansina 1990, 89).

A Mbandja man gathers caterpillars in the forest near Loko. Many different forest peoples routinely gather a variety of caterpillar species as a protein supplement to their diet.

One result of this two-tiered diversification—in the variety of economic activities engaged in by any one household, as well as in the variety of species utilized by any one activity—is a wide distribution of the weight of human impact on the natural fabric of life. If all that weight were concentrated in only a few activities, as is the case in economies of specialization, or if people sought out only a few particular species, as happened in North America's early fur trade, inevitably the fabric would tear. Industrialized agriculture has indeed ripped natural ecosystems to shreds in much of the world. Commercial fishing of a few key species has left a huge hole in the marine biology of certain seas. In contrast, by distributing their weight across numerous points of contact, Central African peoples increased the chances that the fabric of life would be able to bear their weight without suffering too severe a rent.

The farmers I worked with are very aware and disapproving of the damage that comes from concentrating that weight, for example, through the current use of imported technologies (guns, cable snares, spotlights for nighttime hunting, agricultural poisons) for the commercial hunting of wild meat. With dismay, Pindwa tells me of the changes in the quality and quantity of hunting technologies being used today by his people the forest-dwelling Mbuja, who live north and east of the Congo River port of Bumba:

"Things are different today especially in our village regarding the killing of animals. The thing that has come to finish off the animals—and yet they don't end, I am amazed the way God takes care that some more people survive—it is because of two things: one thing is using cable instead of twine. [Among] our people, one person can set up to 200 snares in the forest. In addition to the 200, he also sets nylon snares to trap porcupines, perhaps 150. So in total he has more than 300 traps, and most certainly he will not lack. Why? Because it is no longer just to eat, it is also for selling. In order to sell it is necessary that his entire life now be within this business of hunting animals. That is one thing that has finished off all the animals.

"The second thing is the matter of headlamps or flashlights. Because in one night you might find two *mboloko* [*Cephalophus monticola*—blue duiker, a small forest antelope] sleeping.... And sometimes, having found these two sleeping *mboloko*, you kill both of them entirely. So it is these things that have finished off the animals. And it is only because of God's grace and the vast size of the forest that the animals do not run out, that they aren't all finished."

Integration and Periodicity of Resource Use

Engaging in a diversity of household economic activities may simply spell increased levels of busyness and exploitation if those activities are left unintegrated. The genius in these Central African ways of making a living from the forest is their total integration into a single system of food procurement. Farming was, and in some cases still is, correlated with trapping, hunting, and gathering in a manner bearing enough flexibility to dramatically decrease the risks (much greater in unintegrated or undiversified systems) of going hungry. Vansina describes well how this holistic system worked:

> Activities to procure food formed an integrated system because the available time and labor for each component of the system was limited. These factors had to be carefully allotted according to season, according to technological and production levels, and according to the division of labor. Thus farming aimed at producing only some 40 percent of the food supply. Because certain crops attracted particular species of game, specialized traps were set up around the fields to protect them and to provide a steady supply of meat. During the one- or two-month-long fruiting season, gathering provided most of the vegetable food. The proper short seasons for the harvesting of caterpillars, honey, mushrooms, and termites were taken into account. In the dry season, villagers went to live for a month or so in outlying camps for collective fishing, hunting, and more gathering (1990, 83).

The complex and intricate system of correlating shifting cultivation with trapping developed by Central African forest dwellers over hundreds of years of experiment and experience illustrates the ecological benefits of the holistic system particularly well. Not only did such trapping systems serve to control animal damage to people's crops; they also allowed farmers to obtain sources of protein close to home, thereby indirectly leaving large tracts of forest farther afield unexploited. Vansina (1992) and Koch (1968) both document how complex these systems were. Throughout the gardening cycle, farmers would change the type of traps they used according to the type of crop being planted and harvested. For example, the planting of forest yams was correlated with setting large traps around the perimeter of the gardens to catch forest pigs, the yams' primary ravager. Smaller and larger animals might very well escape these specialized traps, just as pigs would escape the smaller traps set for monkeys and baboons during other periods of the gardening cycle. Trapping's specialization, periodicity, and complex correlation with farming contributed to the sustainability of these human-inhabited forest ecosystems.

Such systems also decreased the risks of overutilization of the resource base because they engendered a periodicity to the manner in which any one resource was used. During the period given to cutting gardens, little hunting took place, and the forest "was given a rest."

Inside her brother's rondevelle, sheltered from the rain that cut our garden work short, Mama Malabo, a Mbandja farmer from Bogofo, describes this hunting periodicity to me in the following words: "But they would avoid killing in a disorderly fashion. Say he killed an animal. He then returned and stayed for awhile. Thus the other animals, like the ones that were pregnant, could give birth. Then he could go and get another."

During my conversation with Nani, a Pagabeti farmer and hunter, I receive some clarification as to the reason why a hunter might have "stayed for awhile": "They would go into the forest according to its time," he tells me. "They would enter, I would say, in January, to cut gardens; February and March to burn them; and April to plant. In May they were still planting. Then in June or following in July, they were able to go into the forest [to hunt].... [They would go in] individually, each with their own camp. In that way they would go in and kill a few animals, not far away from the village to which they would come out, leaving the forest free. The animals then also began to grow. They had these two seasons so that the animals would not disappear.... The way of giving rest was to return to the village to do some other type of work. This is what caused them to rest in the village. They didn't give the animals rest [directly]."

A few days later, Pindwa also brings up periodicity during our conversation: "Regarding hunting there in our homeland … to hunt animals there are periods. Still today they are keeping them—the ones of the ancestors! It means that when it comes time for gardens, they make it known that those who are hunters, those who are people with cable snares take them all out so that they do not remain in the forest.... Such a thing is like a law begun by the ancestors, that a man who has a wife, a child, not lack a garden. Thus it is absolutely necessary to come out [of the forest]. Thus that conserves the animals."

Curious as to what Pindwa attributes such action, I ask, "Do you think they made that rule in order to conserve the animals or because of the shame if one lacked a garden?"

"Well, for me I see that it really wasn't in order to take care of the animals, but because, like in our village, they saw the vital importance of farming. Even the ancestors ... they wanted no one to lack a garden; because if you lacked a garden, with what would you eat that animal?"

"But did people abide by it? Even today?"

"Yes, still today. You will see that they even have songs, our ancestors have songs for it."

"Will you sing one?"

With a laugh Pindwa continues, "Those songs are about a woman who was caught up in looking for snails, and thus she didn't plant anything in her garden. Then she began to cry out her hunger when it came time to look for food. She wanted to ask someone, and he said, 'No, you were only looking for snails, so eat your snails alone without anything else.'"

Switching from Lingala to Mbuja, Pindwa begins to sing: "'She set her heart only on snails, but her garden has nothing.' It means that when it comes time for gardens, quit looking in the forest. Return to cut a garden, for you will eat the things of the forest with the things of the garden."

Some weeks later when I speak with Tata Fulani, a Ngbaka farmer and elder, I learn that the opposite was also true: When it came time to trap in a certain forest, people were warned not to go in there to gather or farm. Tata Fulani explains: "In others of our forests, the ancestors would hunt using three main techniques: One is setting snares, another is digging pit traps, and a third is using arrows and spears. They would then put a ban on those forests so that people would not enter them because they had set traps to kill elephants there. Therefore, people would not gather mushrooms there, nor would they look there for caterpillars. Because they could quickly be endangered and best keep out."

"They wouldn't farm there as well?" I ask.

"No, they wouldn't cut their fields there either. It remained there in order to protect the animals so that they would bear offspring for them to hunt.... So that was their way of protecting the forest so as to obtain game."

Rotational Harvesting of Game and Fish

In addition to rotating their use of resources over time (periodicity), Central African peoples also rotated their use of the forest across space, from one area to another. In

Chapter 3, Libeka, Tambwe, and others described very well the system of shifting cultivation they inherited from their forebears as a means of maintaining food production on poor rainforest soils. Such rotational agriculture also allowed the forest to regenerate and, where sufficient fallow periods were observed, kept the forest from reverting to savanna. In fact, researchers have found that rainforest rotational agriculture has created a patchwork of primary and secondary forest habitats that can be even more productive than primary forest in both plant and animal species (Bailey and Peacock 1988, 92, 113).

The details of shifting cultivation have been written about extensively and are well known (Carneiro 1960, Miracle 1967, Jurion and Henry 1969). Less well known are the rotational systems that governed hunting, fishing, and even village settlement itself. Central Africa's various groups of hunter-gatherers have perfected this rotational style of hunting, shifting forest camps from one area to the next when hunting yields begin to decrease, as one of them explains to me:

"The system of moving around—for instance, in the way we are here today in the village. We will take off and go into the forest and make our camp over there. There we might stay for perhaps two weeks. Then the hunting starts to not be so good there, and it's necessary that we go on to look for another area where there is peace. Meaning, after we've been there for two or three days disturbing things, the animals will then try to flee from there. Then it is necessary for us to also leave there and look for a place of peace, a place where the animals might be walking around not knowing whether a battle will befall them there. So we leave where we were and … return to the place we were before. Another month, we go to another area, and so on and so forth. Within one year, we go in and hunt on this side of the river. After we see that side is no longer very good, we go over to the other side of the river. We then see that side of the river is no longer very good, and we return to the place we had been and so on."

In the same way that fallow periods allow soils to regenerate, abandoning hunting grounds permits fauna to replenish an area prior to the band's return another season. Of course, low population densities make such rotation possible, but they do not cause it. Hunting peoples choose to move for a variety of reasons, not least of which is to optimize returns on their labor, an explanation of mobility favored by Western anthropologists. But many of those I worked with also indicated that they observe rotational harvesting so as to preserve animals for the future. In any case, the result again is that human impacts are more evenly distributed across the land, decreasing the chances of total degradation.

Sitting together on the banks of the Lokame River after transplanting young *makemba* (plantain) shoots, Yambo offers me some of the *safu* (*Pachylobis edulis*, wild plums) he has gathered and boiled. He begins to describe how the Ngombe followed the same rotational harvesting in their fishing. They would plant sections of rivers with certain grasses or reeds (*sonlu*, scientific identification n.a.) to attract fish. Each year the clan would decide which section of river they would fish, cut the *sonlu* they had planted there, and harvest the fish, leaving other areas untouched. Once the decision was made, individuals could not cross over the boundary to fish in next year's territory. After an area had been fished, they would replant it with *sonlu*, then abandon it to let the fish reproduce. When I ask Yambo why they followed such a method, he answers: "In order that the fish also receive a rest.... It is good that they receive a rest; perhaps for an entire year. Then they have also forgotten. They think that not one thing will happen to them. And then you fish once, and you get fish."

At a larger scale of organization, whole villages would shift from one locale to another when, after years of rotational use, resources of soil, game, and fish began to be too depleted or too distant. Also it was common for villages that became too populous to split. People would move to new, less congested areas, resulting in a more even distribution of environmental impacts across the landscape (Vansina 1990, 79, 257).

Sanctions Against Waste

Using the forest rotationally takes lots of skill, planning, and thought. To engage in such rotations *pamba pamba* (without purpose, in a disorderly fashion) is considered a waste and heavily discouraged. Especially frowned upon is cutting but not planting, killing an animal but not using it.

According to Yambo, cutting gardens *pamba pamba*, "to cut here and leave it [without planting] and to go and cut over there, and leave it ... only crazily like that" is bad practice. "That is ruining the forest, no?... For if one does that, one goes and goes until now the forest is far away and there is no way for a person to carry their loads all the way to the village. This will give them problems. It is better if they cut only close by and proceed slowly, slowly."

"Thus, when it comes to cutting, let's proceed in a line," Yambo continues. "Let's follow a line according to real rules. Like you see me—I am heading in that direction. After cutting this area, for me to leave here and go cut again over there, return and cut again like this—that is what causes trouble. Better to proceed with some order following a line ... like this—as I've now cut here, when it's finished, I will begin to return this way, so that that area over there can still remain.... [This area I've cleared] will become secondary forest, and the plantains that remain in it will be there. And now, I will begin to return this way bit by bit by bit until one sees that over there, a lot of forest still remains.... I will not return over there quickly.... I will cut only here.... When this is done, I will start to clear this secondary forest.... I will not go to another primary forest."[2]

"Why?" I ask.

"Because I need to conserve that primary forest over there yet."

"For whom?"

"Because my child is here. My children exist. At some time they may cut a garden there. But if you do not follow such ways, the children will also lack, and they will not have anything to eat."

Pindwa echoes some of the same concerns over wasting the forest among the Mbuja. "Since our people live in the heart of the big forest, the idea that the forest could end didn't exist. But there were rules for cutting gardens: If you cut a garden but did not plant it, it was considered very bad luck. They did not approve of things like that. You would really be laughed at in the village. You would be seen as a person who has no meaning among the people because, what wisdom is there in this going to ruin forest for nothing like this, to leave it useless and to keep from using it? ... If people cut but did not plant anything, it is like they had ruined that forest for nothing, because they were in a position to receive something, and they didn't receive a thing.... So, they said that if you do not want to cut a garden, leave the forest to sit there like it is."

"For whom?"

"For the future, meaning, when you first thought that you wanted to cut a garden, the hope was that you would reap the produce from it. But when you cut and left it useless, it meant that you ruined it for nothing.... For example, according to what our fathers told us, before the white man had yet come, they cut in order to help themselves to eat. The idea of selling didn't exist. Now, each thing they did was for the good of their survival, their livelihood. So if you, a member of the family, cut a garden in order to get food by which to survive, and you did not plant one thing, this meant that you ruined them, ruined their opportunity to receive that food. You only expended your force uselessly, for nothing. Thus they will be angry with you."

Optimizing, Not Maximizing, Production

Among Central African forest peoples, sanctions against wasting force, forest, and game were part of a larger strategy of resource use characterized not by maximizing but by optimizing production. According to Vansina, Central Africans were constantly improving their tools for harvesting nature's provisions. "The drive to improve," he writes, "was rooted in a desire to achieve higher returns, but not at any cost. The available evidence shows what almost amounts to an obsession to achieve optimal, not maximal returns" (1990, 92).

Unlike the conventional directive within market economies to maximize production and profits oftentimes regardless of the costs, optimizing production required balancing a variety of costs and benefits, not all of which were material in nature. As I have already indicated and will clarify below, maximizing production in Central African societies held significant social and cosmological costs. Wealth gained through such production made one stand out within the community, increasing one's vulnerability to accusations of witchcraft; witness the example of Tata Fulani in Chapter 1.

At other times maximization was rejected because it resulted in a less efficient use of labor. Drawing on Koch's research among the Djue of southeastern Cameroon, Vansina offers the following example of this principle of optimal returns:

> The Djue ... had the choice between building highly effective trapping systems with palisades crossing whole valleys from ridge to ridge or forgoing the barricades and placing only flimsy traps instead. The former solution yielded more meat but took more time and collective labor to set up than the alternative. A single expert could do that in a day or two. Most Djue used the first technique, yet the virtuoso trappers used the second. The loners obtained a higher return of meat per person working than their fellows, and they spent less time on the project. The expert solution was the optimal one. In terms of returns, the flimsy work made more sense than the fine trap and the sturdy palisade (1990, 93).

Whether to economize labor or to avoid certain social and cosmological costs, the drive to optimize rather than maximize production meant that more of nature was left unexploited and implicitly conserved.

Benign Technologies

In contrast to the technologies that have been introduced into the forest through contact with other parts of the world (guns, cable snares, nylon fishing nets, spotlights, manufactured poisons), the harvesting technologies Central Africans have invented out of locally available materials are relatively benign in their impact on the environment. Bows and arrows and spears are not as accurate or as deadly as guns; nets and snares using twine made from the *kusa* vine (*Manniophyton fulvum*) and other botanical sources can rot and thus are not as durable or as strong as those made with cable or nylon; botanically derived poisons used to stun fish are not lethal and wear off relatively quickly; and torches made from the sap of the *ndibili* tree (*Canarium schweinfurthii*) provide enough light for walking but not for hunting at night.

Furthermore, owing to the search for optimal returns, flimsier, less damaging technologies are sometimes chosen over those that kill more thoroughly, as shown in the example above. Technologies, such as the trapping systems described earlier, are also built to be selective—specialized to catch certain species while allowing others to go free. The same is true of fish traps, as Tata Fulani describes: "Sometimes they make a large fish trap to catch larger fish. Thus that sort of trap automatically selects for fish since the small fish that enter it can pass right on through. But the large fish will be caught inside."

The day I work with Libeka, he makes it clear how, unlike the guns used today, locally produced hunting technologies limited the number of game a hunter could kill in a year. I trail the Mbandja elder as we wind through his garden harvesting maize from the windblown stalks planted not in rows but here and there in order to facilitate intercropping with a myriad of other species. Such intercropping in and of itself represents a relatively benign technology. Rather than planting monospecific fields that violate the patterns of biodiversity observed in the natural forest, generations of forest farmers have developed complex and diverse cropping systems that in many ways are modeled after what they see working in nature. Monospecific fields could be wiped out entirely by locusts or some other bug in a single season. Here, if something catches the maize, you've always got the squash and peanuts. Peanuts also put back into the soil some of the nitrogen that other crops remove. As in the forest, plants, even domesticated ones, exist here in some degree of symbiosis.

Passing up several choice ears of maize, the Mbandja elder playfully calls back, "I left you two women there—did you see them? Choose which one you will marry and leave the other. Which one did you choose?" We move up into the shade of his *malako* (garden shelter) and snack on the sweet apple bananas Libeka cuts from the nearby tree. Looking up I see what looks like a sling made out of *kusa* twine and leather stuffed between the *malako*'s smoke-darkened stick rafters and *esobe* (savanna grass) thatch. It reminds me of the stories I heard as a little kid, of the Hebrew David slaying the giant Goliath with his sling and "one little stone." Whirling it over his head, Libeka sends a rock hurtling across the garden to show me how the Mbandja use the sling not for slaying giants but to scare away the birds that love to nibble on the maize.

Settling into a long conversation there in the quiet of the distant forest, I ask him if the ancestors ever put a discrete limit on the number of animals a hunter could kill. "Or could you kill as many as you wished, perhaps twenty, fifty, or seventy?"

"No, we didn't have the means to kill a lot of them. Only a spear. If you threw it … sometimes you would hunt and hunt and in the course of an entire year you would only kill one of them.… The Ngbaka, they had their arrows and bows. That's what they would use to do their hunting. They would kill some game with them but not a lot. For a person to kill five [animals] in the course of a year was pretty difficult. Because we didn't have the things made for killing a lot [of game].

"And that is why these rifles, these guns that have come now—they have finished off the animals since they can kill beyond any limits. It kills simply to kill.… If it were in the time of the ancestors they would not condone it. Their words, what was often spoken between them was that if there are no more animals, how will our children and our grandchildren ever be able to know animals? Such things were clear.… The thing that has finished off the animals is the gun. It is the gun; nothing else but the gun."

Production for Consumption More Than for Markets

Many of the farmers I worked with emphasized that their ancestors farmed, hunted, trapped, and fished primarily to meet the immediate needs of their households. Buying and selling were uncommon. According to Wasitdo, the diversification of each villager's occupations meant that "you have some meat in your house, and I also have some meat in mine.... There was no need for someone to come and buy from his friend."

In their minds, this absence of production for commercial markets is one of the key reasons why until recently forest, fish, and animals have remained abundant. Conversely, they link the entrance of commercial markets for forest and garden products to numerous other changes that have degraded the land: much larger fields, longer traplines, full-time hunting specialists, no limits to the number of game killed, and the use of agricultural poisons to kill fish. As Pindwa states above, the one thing that has depleted all the animals is the market-dictated necessity for people's entire lives to become committed to the "business of hunting animals." And Tata Fulani distinctly links more extensive forest clearing to the coming of the whites and their initiation of markets for garden produce.

However, the historical research reveals that trading has been occurring in these forests for centuries—according to Vansina, since "well before the Bantu speakers arrived" (1990, 93). Even prior to the impacts of the Atlantic slave trade and colonialism, people exchanged items between themselves and between regions. Such exchange stemmed partly from the fact that necessary resources such as salt and the rosewood powder (*Pterocarpus soyauxii*) used for ritual and beautification were not ubiquitous. But exchange also took place as different forms of gift-giving between communities and individuals. One early Bantu word for exchange has the connotation of "division of spoils" or "to distribute," a popular activity of village and house leaders who used it as a means to gain additional followers (1990, 295).

In contrast, Bantu words for more distinctly commercial activities such as markets, currencies, and trading caravans arose only later. Vansina writes that based on the linguistic evidence "markets—whether on the beach, in a village, or between villages—were poorly developed institutions until well into the present millennium" (1990, 297). The Lingala word for "to sell," *koteka*, may have arisen in conjunction with the early trade in copper along the middle and lower Congo River. It is likely that its use expanded significantly during colonial times. What is interesting is that it may be derived from an older word meaning "to set a trap," a meaning bearing quite different connotations than do earlier words associated with trading, such as "to give away" or "to distribute" (1990, 147, 295).

In any case, although trade, selling, and buying certainly existed in Central Africa prior to the Atlantic trade and the colonial era that followed, they were of a significantly different style and scale from the new trading systems that arose in conjunction with those external forces. The Atlantic trade not only increased the volume and types of goods traded; it completely rearranged the political and economic organization of much of the region. Previously far-flung and unconnected areas became integrated through a network of trade routes along which passed a multitude of food and forest products, in addition to the trade in human beings to supply the plantations of the New World. Regions started to specialize in the production and trade of certain goods and services in contrast to the more diversified household and village-based economies that preceded regional integration (Vansina 1990, 197–237; Harms 1981; Harms 1987, 157–175).

Such trade based on specialization had serious impacts, both ecological and social. Among the Nunu of the middle Congo, it "exploited the differences between environments instead of conforming to the imperatives of a particular one. Perhaps its most significant feature was that it allowed individuals to amass wealth at a previously unheard-of pace and scale" (Harms 1987, 174). Furthermore, the "challenge from the Atlantic" led to the growth of new types of economic and political institutions characterized by trading "for trade's sake, something absolutely foreign to the old tradition." Leadership in such organizations "rested on wealth and violence alone, in contrast with ancient leading roles" (Vansina 1990, 236).

But the production for market that Pindwa, Fulani, and others farmers say dramatically changed the subsistence strategies of their ancestors is most likely that associated with colonialism and what followed. Colonial practices such as cash-cropping, designed to force people to adopt the foreign currencies needed to pay colonial taxes, opened the door for more modern capitalist markets to enter the region. It is these markets, which continue to thrive today, that are having the most significant impact on the region's forests and fauna. Markets, coupled with new invasive technologies such as guns and poisons, can have absolutely devastating impacts.

Bill Lundeen, a missionary working at Loko, tells me about a man hunting for market near the town of Bumba. The man happened to be a soldier. Armed with a machine gun, he killed more than sixty red colobus monkeys in a single day. When I bring up this example in my conversations with farmers, they invariably respond that it is not good; it is not the way the ancestors taught. One farmer in particular (whose story I will tell in Chapter 5) responds, "Even in one year he couldn't eat all those!" It is striking that in an age when market hunting is on the rise, this farmer still expresses disgust over the waste of killing far beyond what one can eat rather than admiration or envy over how much money the soldier will earn from selling his booty.

Controls on Population Growth

Local sentiments such as these are often left out of the hubbub of environmental conferences being held today in many corners of the world. More often such conferences are marked by heated debates between natural and social scientists over the so-called P-problem. Many natural scientists stress a burgeoning human population as a key cause of global ecological crises and tend to lump all types of people into their emphasis on sheer numbers. Social scientists, in contrast, often downplay the numbers in favor of differentiating people according to class, emphasizing how the numbers in the poorer classes are forced into population growth due to a maldistribution of the world's resources and the poverty that ensues. Left out of the debate is how poor people themselves view population growth. Contrary to images of naive and ignorant peasants reproducing blindly, many of the farmers I worked with are very aware of how an overabundance of people is causing severe ecological stress in their homelands.

When I ask Mama Malabo for her assessment of the current environmental conditions in Bogofo, she offers the following: "Only that I see that the world has now been ruined by us. They are no longer killing animals like they used to."

"What do you think is the main reason why it's been ruined?"

"It is only that we people are now abundant, on top of the animals, killing them. Young men now, they have their guns. They go into the forest and kill animals in their own way. So [there is no] way for those things to grow.... The ancestors used to go quickly into the forest and kill an animal. It's no longer like that. Now there are lots of people."

Three days later, Mama Kagu, a Ngbaka farmer, expresses a similar assessment. She shows me to a chair in the small mud-and-wattle house she shares with her children. We have just returned from planting peanuts. I wielded the *wala* (Ngbaka long-handled planting stick with a small shovellike blade), peppering the soil with small pocklike holes, while Mama Kagu followed, throwing three or four peanuts into each, covering them over with a sweep of her foot. Peanuts now cover half the garden, awaiting only the rain to begin pushing up between the manioc plants. The other half will wait for another day.

Mama Kagu was born in the Ngbaka region but grew up with her uncle's family among the Mbuja near the city of Bumba. When her mother died, she came to Loko and has stayed on, supporting herself and her children through farming, an occupation she learned from her uncle. She begins to describe to me various farming and hunting methods of her Ngbaka ancestors, their use of three- and four-year fallows, and the pit traps they would dig to catch the forest pigs that raided their gardens. I ask her about the relative abundance of game now as compared to ancestral times. Today "there aren't very many animals," and the reason, she is quick to point out, is "because people are full in number and are greater than them.... At the time when they were abundant, people did not kill many of them like they do now. Now, there are so many people, way more than the number of animals.... [During the ancestors' time] they didn't kill a lot like they do now. Because people weren't abundant like they are now."

It is clear that both these women view population growth as a primary cause of local environmental deterioration. Also interesting is the fact that both Mama Malabo and Mama Kagu, in contrast to local norms, are single heads of household, having left their husbands and chosen not to remarry. Such marital status, increasingly common in many African villages, may have implications for the population problems they mention. One possible impact is that single female heads of household may choose to have fewer children, for a variety of reasons. First, they may be more careful to avoid pregnancy, as being pregnant without a husband can carry some negative social connotations. Second, without husbands they may be hesitant to have additional kids, as the burden of raising and schooling children falls primarily on them alone.[3] Such has been Mama Kagu's case, she alone bearing the burden of feeding and schooling her children. Yet Kagu makes it clear that she is not eager to remarry.

Soon after beginning our conversation, Mama Kagu's son pulls up a chair and joins us. As a single mother she has worked hard to put him through six years of schooling. When I ask her if it is difficult not having a husband, she replies, "No, it's not difficult. I tell you, you work alone, come back, and can just rest. But when you have a husband, sometimes you work hard and he doesn't work at all. You may work for a long time like we did, and when you come home he says, 'Oh! Bring me some water for bathing! Give me something to eat! Do this. There's money? Give me some of it!' While he doesn't earn a cent.

"But if you are single like this, you do your work until … say today you work, tomorrow you rest. Or, if you work long like today, you return and if you have children, they fix you some food, you eat it and rest."

Similar to Kagu, Mama Malabo also indicates that she really has no desire to return to marriage, the conventional arrangement for bearing and raising children. Ironically, she attributes many of the problems of marriage to the same ancestral tradition that earlier she had extolled for its environmental friendliness. Her views underscore the fact that the past—the ancestral tradition—is not all rosy but, like life today, holds both positives and negatives.

Through the doorway of her brother's rondevelle, Mama Malabo and I see jagged bolts of lightning race through the bank of dark blue storm clouds on the distant horizon. The reverberant thunderclap that follows momentarily stops our conversation about how garden labor is divided between men and women. As the rumble carries beyond the hills into stillness, I hear Malabo and her sister-in-law, who has joined us, saying that it is women who do the majority of the garden labor since men's only task is to fell the big trees. But such an unequal division of labor is not limited to the fields, as Mama Malabo points out.

"What man will go to the garden, return, go to the water source, cut firewood, return and cook manioc greens—whose husband would do that?" The women fill the room with laughter. Earlier, when our conversation first turned to issues of gender, they had told Mama Malabo's brother to leave the room. Now Malabo continues: "A man does not do such things. For him to do those things means he doesn't have a wife—he is a bachelor."

"But do you think such a way is just?" I ask.

The women laugh again before responding, "It is nevertheless just, is it not?"

"But I see that the weight is not equal."

"Eh. The weight falls on the women, no?"

"And is that just?"

"There isn't justice."

A group of women farmers plants peanuts in the Ituri

"But was it also so among the ancestors?"

"Yes, it all started with the ancestors. It started with them and it remains until this day. The ancestors did just the same."

"And you are not able to protest, to refuse?"

"To refuse would not work. Since it has been that way from the beginning, refusing it would not work. We are just following the way the ancestors began with."

"If you tried to resist?"

"Your husband would beat you. [More laughter.] He keeps beating you like that until … you cry and cry,… until you go and draw water. He washes. He alone tells you to go and cook something. You go and cook, give to him and he eats. All that with the pain from his beating you."

"So do you think it's better for a woman to have or not to have a man?"

"In the way that I am alone here, haven't I made it just fine? What can cause me pain or give me troubles?" Mama Malabo and her sister-in-law break into another round of laughter. "I come and sit—" Without finishing her sentence, Malabo nods in the direction of her sister-in-law. "She is really the one with troubles. If she comes from the garden, they say, 'Not until your husband has eaten.' But as I have no husband, I go to the garden, cut my garden, come home, go and bathe. I return, cook a little of my food, eat, and am done. There is no husband to bother me."

* * *

The experiences of women like Malabo and Kagu certainly add complexity to the picture of African women as players in the demographic changes taking place on the continent. Nevertheless, population densities around Loko *are* significantly higher today than they were when Mama Malabo's Mbandja ancestors moved to Bogofo from across the Lokame River sometime during the previous century. In fact, the low population densities of the past played a huge role in enabling people's ways of making a living in the forest to be sustained without causing severe ecological damage. But low population densities did not just happen. In the rainforest as elsewhere people had to cope with the problem of becoming too abundant. And so they developed ways to control their population growth. "Conscious demographic policies were followed in equatorial Africa, and population was controlled," writes Vansina. In some societies, after giving birth a woman was not allowed to resume intercourse until her child had mastered walking at (plus or minus) two years of age (Vansina 1990, 257, 372). Breast-feeding, a natural form of birth control used by women in many different societies, also provided space between one child and the next.

Further, as mentioned above, when an area became too congested some people chose to leave and find new settlements in areas of lower density, diffusing the pressure large settlements placed on the land. The causes of such relocations were not limited to changes in the natural world, that is, people did not simply wait to move until the soils, game, and fish began to be impoverished. Settlement divisions could also be spurred by social events such as conflicts between senior and junior members of the community (1990, 257). Stemming from multiple causes, relocations and population divisions may likely have been frequent and may have often preempted severe ecological degradation.

More traumatic, warfare between different groups also had the effect of limiting population growth within Central African societies. Lila, an informant to historian Robert Harms in his work among the Nunu, puts it bluntly: "In the past people didn't die as much as they do now. The people noticed that they had become too numerous. It was necessary for some to go and die in wars" (quoted in Harms 1987, 147). Although popular images of African "tribal warfare" depict violence and devastation, the actual numbers of people killed in such wars is not clear. Oral traditions of warfare may overestimate the number of deaths to stress one group's superiority over another. Thus Harms notes that whereas Nunu oral traditions recount wars between settlements involving deaths of up to 160 men on a single side, one of his informants stated that a typical battle would end after "about two people had been killed on each side" (Harms 1987, 146).

* * *

Gabi, a Kongo farmer who arrived in Bogofo relatively recently, concurs with this view on warfare. One of Gabi's friends has given him land to farm, but it is not close by—an hour-and-a-half walk from Bogofo. The day we work in his garden we travel by motorcycle, heading north for a kilometer or so before we turn east on the road everyone has been telling me about. It is *balabala ya President*—Mobutu's road—stretching out 7 kilometers from the main drag into the middle of nowhere. Opening up the throttle, we spin the tires on the limonite gravel. The speedometer quickly reaches 60km/hr, but I see there is no reason to slow down—there are no potholes, no washboards, no sand traps. This is the best road I have ever seen in all my years in the Ubangi. Gabi tells me of all the work that went into building it, the big graders and Caterpillar tractors they saw tearing up the savanna, burning up gallons of precious fuel by the minute; the culverts and the cement that allowed for good drainage; the truckload upon truckload of limonite that was hauled in to provide a durable surface. In contrast, most of the public roads I've been traveling on have not been repaired in years. They are more like 4x4 competition courses than roads. The same is true throughout most of the country.

I am curious to see there are no houses, no villages, no gardens, nothing alongside as we sail past. This road was not built to get produce out to market or to ease the transportation bottlenecks of local villagers. The story I am told is that Mobutu wanted to have a fishing spot, a place to get away from it all—"it" being his palaces, farms, and plantations in nearby Gbado-Lité. So he had his own "development" company build him a road—and a fine road it is—out to the Lokame River, so he could go fishing.

After 7 kilometers, the road begins to peter out as we near the river. We stop and get off and continue on foot down to the riverbank. Expecting to see some sort of fishing lodge, restaurant, or guesthouse, I find only a riverbank, a little clearing under some trees, a few logs strewn here and there apparently to provide a place to sit. Gabi tells me no one has come here for some time. After the road was finished, the president came maybe a couple of times and has not used it since.

We turn around, park the bike at the edge of the forest, and proceed down a narrow path to Gabi's garden. After walking the field with me, showing me his various crops, Gabi picks some maize, builds a little fire, and roasts it, turning it on the coals from time to time with his bare hands. We munch on it while we talk. He begins to tell me about ancestral times and ways.

"The things of the ancestors are big. I really agree with much of their ways.... Their ways, the ones they gave, some of them are very good.... Look at us now, we are ruined, us people; our abode has not become good. During their times, there were more good things.... During the old times of the ancestors, you could walk and when you met a brother, he would give you some food and you would go on. His house, he would leave it like that, wide open. He would only place a small stick across the entrance and then leave. But today, you lock it with a metal lock and while you are away, they come and break in. And then, there is not really any understanding or trust. Because nowadays, a person comes here and I will think, 'Maybe she's a thief. She does not know my ways, my village. Maybe he's a thief. Maybe he will kill me.' So, there is not good understanding and trust. But in the time of the ancestors it wasn't so. In the time of the ancestors, if he came, I would welcome him. We would eat well, he would stay, sleep, get up, and be on his way."

"But weren't their wars during the time of your ancestors?"
"There were wars. They had their wars. They would fight between themselves. These people would fight against those and those against these—there were wars. But they would come to an understanding quickly. If they fought [a bit] they would then come to a peace. They would fight and then end it declaring one side the winner ... and it was over."
"And they wouldn't fight any more?"
"No, they wouldn't fight again."
"They would live in harmony?"
"They got along."

* * *

Vansina documents that two types of warfare existed in the region. The more frequent "restricted" war was governed by a stringent set of rules that limited the extent of fighting. Elders called a halt to these conflicts usually after one or two men had been killed. The purpose of "destructive" war, a second type of warfare, was to truly destroy one's enemy. It involved burning villages, plundering land and wealth, and often taking captives. However, Vansina notes that "in recent centuries they were infrequent, and in early times they may have been even rarer, because competition for resources was less marked" (1990, 80).

Thus one may surmise that warfare reduced local population growth as much or more through the threat of killing than through actual killing itself. The very real threat of warfare meant that one needed to be ready at all times to rapidly mobilize and flee. Having lots of young children to carry would definitely slow you down. Thus although he likely exaggerates the spacing of children, Wasitdo informs me that among his Ngbaka ancestors, "a woman would give birth to a child, and after the child had reached ten years of age, she would have another. They did not give birth to a lot of children like we do. It was necessary that the first child know how to walk and know how to flee from the wars before they gave birth to another."

Colonial campaigns to end such conflicts, misnamed "pacification campaigns," employed warfare, often of a much deadlier variety. After the many years that it took for colonial powers to subdue native peoples, the "peace" these campaigns gained, through such tactics as resettling people in roadside villages and appointing erroneous leaders as "chiefs," not only opened the door to new types of conflicts; it also removed a relatively benign means by which Central Africans had limited their population growth for generations.

In addition, Belgian colonial policy regarding Congolese child-rearing practices generally favored an early weaning of children while discouraging various indigenous birth-spacing customs. The reason for such policies was, in part, to increase the labor population of the colony after a perceived demographic decline in the 1920s. Nancy Rose Hunt provides some dramatic documentation of the ways in which the colonial regime "entered into some of the most intimate aspects of African women's lives: the birthing process, breast feeding, weaning, dietary choices, and sexual activity" in order "to increase the birthrate, promote infant and maternal clinics, and socialize African women as biological reproducers and mothers." Although the extent to which Congolese women cooperated with such programs varied, "by the end of the colonial period ... the population crisis had diminished, and in fact reversed itself" (1997, 306–308). Any discussion of Central Africa's current population problems must not ignore such historical factors.

* * *

These eight features of making a living from the land represent only a few of numerous other aspects characterizing Central Africans' means of livelihood. By "looking to the side" into even a few of these ordinary patterns of living practiced by Central African forest dwellers, we discover some striking examples of the ways in which these peoples implicitly conserved their environments. I now continue with another "look to the side," this time into the social context within which people interacted with nature. Particularly, I examine Central African understandings of the relationship between the individual and the community, especially in regard to land use, in order to reveal other insights on culture, values, and the environment in Central Africa.

Living with Neighbors: The Individual-Community Relationship

We humans are for the most part social beings. An important aspect of our sociality has always involved defining and enacting how the individual and the community relate to each other. Different societies and cultures have crafted different ways of working out this relationship according to their different values, beliefs, knowledges, and material conditions. Common to all is the fact that the manner in which a society defines and enacts this relationship carries far-reaching impacts that spill over from the society onto the land. Through this research project, I have come to realize more clearly than ever before how closely intertwined are our relationships with others and our relationships to the land. The ways in which we understand and enact our relationship as individuals to our community hold very real implications for the human-environment relationship and for the environment itself.

In the West as elsewhere, the relationship between the individual and the community has carried a certain degree of tension. At times these two aspects of our humanity have been held in balance, but in more recent history Western societies have gravitated toward one or the other of two opposite and extreme definitions. Capitalist and liberal democratic societies have placed an extreme emphasis on the individual to the detriment of community. Partially in response to this, Marxist societies have striven to create a society in which the community holds sway over individualistic concerns. For many years, countries grounded in these two opposite approaches moved farther and farther apart, each seeing the other as evil, each remaining convinced of the moral rightness of their own half of the dualism.

Today capitalism's individualistic emphasis is flourishing, and its community-destroying tendencies are wreaking havoc on remnants of community in the West and elsewhere as economic globalization increasingly comes to define our postmodern world. Conversely, Marxists operating under the same dualistic metaphysic as capitalists have pushed their communal emphasis to such an extreme under communism that they have destroyed their societies. Today the former Eastern bloc is as open as anywhere else to the same brand of free-market capitalism that its leaders had previously viewed as the enemy.

I do not mean to imply that by gaining the victory over its nemesis capitalism has thereby proved itself superior, a common sentiment among business and national leaders alike. Rather the point I wish to emphasize is that our Western story of working out this important and complex balance between the needs and aspirations of the individual and those of the community has been marked by dualistic "either/or" thinking that places an extreme focus on one side at the expense of the other. Its enactment in the world has wrought and will continue to work disaster. Indeed, the global enactment of this extreme emphasis on the individual now threatens to seriously damage the wider community of the biosphere on which we all depend for our livelihood.

I believe we can learn something of how to achieve a balance by looking at the ways in which other peoples have defined and enacted the individual-community relationship. More specifically, I think Central African peoples, some of whose stories are contained in this book, have much to teach us about how to meet the needs and aspirations of both the individual and the community without undermining either. In this section, I first draw on the voices of African and Africanist scholars to address some theoretical aspects of the individual-community relationship as it has been defined by various African peoples. I then offer some illustrations of how these definitions have been enacted on the ground, drawing on my conversations with the Ubangian farmers near Loko. It is interesting to note that although both the theoretical and practical accounts that follow bear witness to a relative balance between the individual and the community in and among many African societies, the latter accounts—much more than the former—also reveal that the individual-community relationship in Africa is certainly not free of conflict and discord.

Central African Individual-Community Relations in Theory

In his recent history of the Central African forest region, Jan Vansina reveals how the human communities that inhabited the rainforests of equatorial Africa were geniuses at maintaining a balance between their needs for both individual autonomy and communal security. Historically, myriad groups were involved in a repeated dynamic of decentralizing in order to maintain their individual autonomy and sense of group identity, simultaneously working to promote good relations with outsiders in order to reap the benefits of communal security and cooperation in the face of large-scale threats. Although some groups did tend toward centralization and experienced rapid growth, many more creatively intertwined both decentralization and cooperation. Vansina draws from this a lesson:

> One must admire the ingenious solutions found to allow for both almost total local autonomy and cooperation where needed. That is the distinctive contribution and the special lesson to be learned from equatorial Africa in the world's panoply of political institutions. The inexorable march from local community to chiefdom, to principality, to kingdom or state, proposed by unilineal social evolutionists was not the only possible option (1990, 253).

The same balancing of individual and communal aspirations played out at the intercommunity level is also enacted within the community between individuals and the group to which they belong. Any attempt to understand the relationship between the individual and the community in an African context must first gain a deeper sense of the nature of African communalism, the base that supports all African traditions, according to Jimoh Omo-Fadaka. The African community is not composed of a group of individuals "clinging together to eke out an existence" (Omo-Fadaka 1990, 178); nor is it, as Malidoma Somé describes community in the West, "a conglomeration of individuals who are so self-centered and isolated that there is a kind of suspicion of the other, simply because there isn't enough knowledge of the other to remove that suspicion" (van Gelder 1993, 33). Rather in Africa the community is imbued with a certain bondedness. Harvey Sindima captures the fullness of the African understanding of community in the following passage:

> The African idea of community refers to bondedness, the act of sharing and living in one common symbol—life—which enables people to live in communion and communication with each other and nature. Living in communication allows other's stories or life experience to become one's own. Sharing of life experiences affirms people and prepares them for understanding each other. To understand is to be open to the life experiences of others, to be influenced by the world of others, and this is fundamental in living together as a community. Bondedness to others and the cosmos makes one aware that there are selves other than ourself yet all are united by one creation, one life. This calls for a sense of justice or "just-ness, a correct insertion and reciprocity, in respect toward what is 'ourselves otherwise'" (1995, 153).

Bondedness entails respect, which in turn entails taking responsibility for one's fellow human being, not as an atomized individual but as a member of the common fabric of life. Since life's fabric is of one piece, connections within the fabric have to be maintained. If there is social or personal disharmony or illness, something has become disconnected and needs restoration. Therefore, for the good of the whole, the responsibility to restore this broken connection falls on everyone. In such a manner African communalism provides a strong source for individual morality. As Nigerian philosopher Innocent Onyewuenyi writes,

> In his judgment of his conduct, the African takes into consideration th'e fact that he is not alone; that he is a cog in a wheel of interacting forces.... In the life of the community

each person has his place and each has his right to well-being and happiness. Therefore, what to do and what to avoid in order to preserve, increase, and strengthen vital force in himself and others of his clan constitute morality (1991, 43–44).

Indeed, African communalism contributes a great deal to social ethics, but at the same time its influence does not necessarily mean that the individual is smothered or ignored as some Western writers have been wont to believe. Again as in the situation with autonomy and cooperation, individualism and communalism do not exist in dualistic and oppositional relation but in mutual interaction whereby each reinforces the other. As Kwame Gyekye explains:

> In African social thought human beings are regarded not as individuals but as groups of created beings inevitably and naturally interrelated and interdependent. This does not necessarily lead to the submerging of the initiative or personality of the individual, for after all the well-being and success of the group depend on the unique qualities of its individual members—but individuals whose consciousness of their responsibility to the group is ever present because they identify themselves with the group (1987, 210).

In regard to the global environmental crisis all of us face, the skill of balancing seemingly opposite but equally necessary components, as Central African understandings of the individual-community relationship illustrate, will be crucial for steering our ship with its human and nonhuman cargo through some tumultuous waters ahead. The lesson and hope Central African experience offers us is that the option of cognitively creating and physically enacting a "both/and"[4] balance rather than continuing to race headlong toward an "either/or" extreme has existed and has been fulfilled in the past. It may be difficult to re-create, but it is not impossible.

In regard to actual environmental projects working on the ground in Central Africa, I found this relative balance between the individual and the community to be a possible solution to some very real community development quandaries facing Loko and the CEUM. Many of the development difficulties Loko and other nongovernmental organizations (NGOs) working in Central Africa experience seem to be rooted in fundamentally different metaphysical understandings—on the part of Westerners and African villagers—regarding the individual's relation to the community and the relation of both to the environment upon which they depend. I will delay until Chapter 7, the concluding chapter on applications for practice, to explain how that is so. Here let me briefly offer some examples of how individual-community relations are enacted on the ground.

Practical Examples of Central African Individual-Community Relations

Pre-Westernized systems of land tenure in Central African forests illustrate well certain aspects of African "both/and" ways of thinking with regard to individuals, communities, and land. Unlike our Western emphasis on individual ownership and on seeing land as a commodity, under Central African tenure systems the goals, aspirations, and property of the individual and those of the community exist hand in hand within a total system in which the two ideals are held in some degree of balance. Land is neither private property nor communally owned and worked in a socialist sense. Rather land, in most cases, is held in communal trust; it belongs to the group, to all members of the community, extending usually at least to the level of the clan. In general, such a communal arrangement means that no one who wants and needs land goes without. As Parker Shipton states in a review of African land tenure systems, "Access to land should go to those who need and can use it, and no one should starve for special want of it, at least not within a group whose members consider themselves the same people, which usually has meant a kin group or ethnic group" (1994, 350).

However, within such forms of common property ownership, each individual at the same time has their own piece of land that truly "belongs" to them. It is land to which they and they alone have usufruct and for which they and they alone (including family and extended family) are responsible. One's rights to that land are to be respected by other members of the community, and one is prohibited from crossing the boundary into another person's land and fishing area without their permission.

* * *

A group of women sits around a large basket filled with bright green manioc leaves, harvested that morning and soon to become part of the afternoon meal. With great speed and dexterity they separate the leaves from the red stems. Dungu, a Mbandja elder, explains that it takes a lot of skill to clip the leaves just right—if too much of the stem gets in, the *mpondu* (cooked manioc greens) is too bitter, and if too little of the leaf makes it in the pot, all the harvest work literally boils down to nothing. We sit outside his house in *goigoi* (lazy) chairs, the African equivalent of middle-class America's recliner. Bright yellow weaver birds chatter overhead, busily building their hanging nests out of the fronds they've stripped from the oil palm whose branches, although quickly being denuded, still offer us a bit of shade. We are talking about land tenure. I ask Dungu how he obtained the land on which he now farms.

"It is because they gave birth to me here and I grew up here. It is my village here. All the space ... is for us the natives of the village. So people can choose their own place to cut a garden. Thus I chose and cleared it."

"Did you have to inquire of someone else?"

"No, I needn't ask someone else because this is my village, and I was born here. Thus, while walking, the places I see that are good for clearing, I can clear them.... I cut only a little at first. My thoughts then were to extend it and I did. Since it is my village, the soil left by the ancestors, I can work according to what fits me. That is why I cleared like I have."

"Okay let's take the example of people who are not native to this village. What will they do?"

"If they want to get some land, they could ask the *kapita* [neighborhood headman] or the chief [of the village]. Okay, they then receive a plot of land on which to build a house. They build a house and if they ask, they are then given a place to farm, which they begin to clear. And thus they are able to become like a person of the village."

"Will they themselves choose the garden plot?"

"No, if they ask, the leader of the village will then go and show them, 'You, you can have this garden.'"

"Will they have to pay?"

"Well, recently, some people are selling garden spaces; they sell them outright."

"Which people?"

"The native leaders of the village. For instance, the garden that I have cleared, eh? If people come to me, if they are family members, I will give them a place to farm for free. If they are another person, I will show it to them and they will buy it."

"Will they pay you only once?"

"Yes, one time and it is finished. They will not pay periodically."

"I see then that it is something that doesn't really involve the State?"

"The State? If you begin to fight over these matters, or if you argue, then the problem can be taken before the State. If it comes before the State, in judging the case, they will ask who gave you that land."

"But for you the Mbandja natives of Bogofo, for you it really doesn't concern someone else; it's only up to you, the place you choose you can cut?"

"Yes, the place you choose is the place where you do your work."

"But you can't take another person's land?"

"No, no. Like now, the land I cleared there, when I am no longer here, my children will use that same land. They will just work there. If another person tries to infringe, they will say, 'No, the boundary between our father and your people is right here, and you shouldn't cross it like that.' They can refuse such actions."

"And do people respect it?"

"Yes, [if people enter another's garden] they can refuse to let them. [They can say], 'This is not the true boundary between us.'"

"Okay, now what about the garden you've opened up over across the Lokame River—how did you get that one?"

"That one, it's like I told you how the ancestors dwelled over there at one time. That forest and those waters are those of our families. Thus if we want to do something over there, we don't have to ask. Since I am a village native, and that place is the place of our family, I will do my own thing. Thus I cut a garden over there like I wanted to, but I did not ask anyone."

* * *

The "both/and" manner with which Central African societies treat both communal and individual drives regarding land, as illustrated by Dungu's account, allows the two to play themselves out in tandem. Under such methods, the community does not forego the benefits of individual responsibility, effort, care, and motivation that come through individual "ownership" (very different from our Western sense of private property). At the same time, the community keeps individualism from getting out of hand by preserving a communal sense and communal systems whereby the land belongs to everyone. With individual usufruct comes communal responsibilities and various social leveling mechanisms that keep individuals mindful of their obligations to others. If an individual or another group is found violating these responsibilities, the community heavily discourages their behavior and can take action to set things right, as Libeka explains.

"But other people at times came into our forests in order to ruin them," he recounts. "These things happened a long time ago, not during the Colonial State, but a long time ago during the time of the ancestors, when the State had not yet entered the picture. They made an effort on behalf of their forests, that people not go in and ruin them. Because sometimes people would come from other places and go into those forests and ruin them by … putting in snares and killing lots of animals. And so they would often go after them and make them return to their home since they didn't want all the animals to be finished off. The same way with fish."

"They would force people to leave, they would tell them to get out?"

"'You, get out.' But if they had come in a good manner, then they wouldn't kick them out. But those who came with bad intentions, to only ruin the forests, to kill animals … people really protested against this."

* * *

Dungu and Libeka are both men. What about Central African women's rights to land? Certainly, in many Central African societies women do not enjoy the same degree of access to land as men. Often their claims to land are obtained through their husbands (in which case they work the land to supply the needs of their families) or through a male relative. However, even single female heads of household most often have some sort of access to land to meet their needs. It is very rare to find them landless. Mama Kagu, for instance, purchased one of her secondary forest fields near Bogofo from a Ngbaka woman whose husband had left her after he had cleared the primary forest. The other field, also in secondary forest, she purchased from a Ngbaka man who works on Loko's cattle-raising project. In both cases, the price was minuscule. Mama Malabo obtained her land in Bogofo from her brother, who cleared the primary forest from the fields she now farms. If she chooses to return to Lisala, the town where she grew up, she will also be able to get land for free through her connections to her aunt, who cut the primary forest.

In recounting the story of her land tenure, Malabo also emphasizes both the individual and the communal aspects of landholding. "Whose land is this, this land around Bogofo?" I ask her.

"Is it not ours, the people of Bogofo?"

"Which ethnic group?"

"Is it not the Mbandja?"

"So if a person comes and she is not native to this place and she wants to get some land to farm, how will she get it?"

"Well, if she and the leaders of the village agree, they will give her a small field to work."

"Even primary forest?"

"Yes, even with primary forest."

"Even secondary forest?"

"The same. If she is a good person and the two agree, they will give her a bit of secondary forest and she works it."

"And will she pay?"

"Not usually, since they themselves gave it to her, she will not have to pay."

"In your forefathers' time, did they consider it all right for someone to sell forest?"

"Not really. They refused to sell it."

"Why?"

"They refused because they wanted it to remain theirs alone; that they alone be the ones to work it."

"Did they sell secondary forest?"

"No, neither did they sell secondary forest."

"Even savanna?"

"Even savanna they didn't sell. Because they didn't want to sell gardens [in order] that they remain like that, they alone being the ones who work it."

"So this matter of selling has come in rather recently?"

"Just now."

"It came only now with this democracy that we now have," Malabo's sister-in-law adds. Her statement startles me.

"Really? It started with democracy? But what really is this democracy?" I ask, anxious to explore this spontaneous opportunity to learn more about how local people understand the concept that has now come to be heralded, supposedly without any ambivalence, as a universal good.

"Democracy is like this: Regarding the earlier ways of the ancestors, these were ways that were good. If they said let's do this, people would behave in a good manner and do it. But now, everybody is fending for themselves in order to do things in their own way, sometimes only in order to survive, to do something in order that they alone can survive. In such a manner, they are following only their own way, their own way alone."

"Such things are democracy?"

"Eh." Both women nod together before Malabo's sister-in-law continues, "Now things have really become crazy here. The person who obeys her friend can no longer be found." These women make it clear that for some people in the world democracy has come to be associated more with rampant individualism than with community well-being and collective freedom.[5]

* * *

In addition to instances of land tenure, examples of this "both/and" relationship between the individual and the community can be found in Central African systems of labor. Common among many societies are various communal institutions created to allow for a group sense of cooperation, helping each other out and making sure the whole village survives. One particularly important organization of such type is known in Lingala and Swahili as *likilemba*—communal labor groups whose labor rotates from one individual's project (in this case, usually garden clearing) to another's, somewhat on par with the Amish or early pioneer ideas of "bees" or "barn-raisings." With most *likilemba* no payment is involved; only the obligation to feed the group of workers.

"The *mohai* [the Mbuja word for *likilemba*] is very meaningful to us," Pindwa explains with regard to his people back home, "especially for matters of development. If people have livestock, say they raise pigs, or goats, or ducks—fish ponds haven't really come to us yet—among us, they really encourage, facilitate people working together. That is, if you have a goat or a pig, today you call people together, even as many as fifty, they will eat and do everything you need done, especially when they work as a *mohai*."

It is unlikely that a *likilemba* would ever include the whole village; rather they commonly remain within extended families or among neighbors. One finds variations of the *likilemba* in urban centers—mutual aid groups (*mutualités* in French) whose members all contribute to a common pot and then rotate the use of the pot between individual members as specific needs arise.

One concrete example of a *likilemba* in the region near Loko is the group Ntembe na Mbeli, a name difficult to translate but which means something akin to "Awe of the Knife." Ntembe na Mbeli is a group of about twenty-five young men, ages twenty to forty, who have banded together in a cooperative to help cut each other's gardens. They also cut a communal garden belonging to the entire group. For a fee, they will also cut gardens for other people, but their main focus is to help each other. The structure of the cooperative is quite formal, with rules and regularly held meetings. Ntembe na Mbeli offers a good, living example of how individual and communal objectives can be realized in tandem.

Similar to the case of land tenure, we see that the *likilemba* or *mutualité* are a means to provide space for both the individual and the communal to exist and be lived out simultaneously. What belongs to the individual is preserved—each person has their own garden or project—but at the same time, a sense of community, mutual help, and cooperation is fostered by people coming together to cut each other's gardens or contribute to a common pot. By allowing for individual projects or individual property to exist, one enhances the possibility of achieving the motivation and responsibility necessary for endeavors to succeed. But by surrounding such endeavors with mutual aid and cooperative labor, the community places checks on the individualistic drive and preserves and fosters community and a sense of helping each other.

With regard to ecological benefits, the fact that Central African communal tenure systems allow room for land to "belong" to individuals also enhances a sense of responsibility to take care of one's land and to manage it properly. Furthermore, since one "owns" a particular area of land within the communal domain, one is entitled to pass that land on to one's children, adding extra incentive to leave the land in good shape for one's future offspring. If all land was "owned" and worked communally, it might be easy for people to depend on others to do the caretaking; or there might be less motivation for any one individual to invest in proper management of the land for posterity (not to mention the confusion it would leave for matters of inheritance).

In addition, the boundaries placed around individual pieces of land as well as around each group's communal area of forest may also provide some order to land use and distribute human impact more evenly, both on individual and community levels. As quoted in Chapter 3, Tambwe makes the point that "by allocating fishing areas according to boundaries, the ancestors made sure that the rivers and streams were not destroyed." The norm of respecting such boundaries places limits on how far an individual can fish the river, hunt, or extend a garden in the forest. In the past such limits also applied during communal hunts when a group reached the boundary between their forest and that of another clan or ethnic group. Libeka recounts how different clans generally respected the boundaries of others' hunting territories. He offers the example of hunting elephant. Once a group of hunters reached the end of their territory, they would not keep hunting into another group's territory. Reaching the boundary, they would turn back. Such observance of boundaries may have offered elephants and other animals a form of refuge unless, of course, members of another clan happened to be hunting elephant just across the boundary at the same time.

By providing a check on individual "ownership" of land, the communal nature of Central African land tenure in turn discourages individuals from abusing land that is considered the trust of the entire community. For instance, there is a great disdain within the community if an individual starts land-grabbing, laying claim to areas of forest by clearing here and there but then leaving the area unworked. Not only do such practices disturb the balance of land distribution within the community, they are also ecologically wasteful, clearing forest for nothing.

Wasitdo, Mali, their children, and I talk together in the shade of a tree after planting a portion of their garden in peanuts. When issues of land tenure come up, they angrily describe this relatively recent phenomenon of people grabbing land. Mali traces it to the changes in land tenure that were initiated by President Mobutu: "Now, people have come to cut gardens according to their own wishes, because since Mobutu came to power, he made a policy that the land is for everyone. It is necessary that people do things according to their own way. A person should not forbid a friend from having a garden, nor access to rivers and streams, nor say that here is our land or here is not your land, because in the end we are all one. Thus because of that people have begun to cut gardens as they wish, in their own way."

Her husband continues, "Sometimes ... people will get a huge area and they will not want others to enter it. But meanwhile other people lack gardens. They have gotten the whole place and sometimes they don't cut it, while their friends see pain trying to find a garden, and they don't want to help them."

"So do you think that ... the way in which people cut wherever they wish is ruining a lot of forest?" I ask.

"[It is] ruining a lot of forest," says Wasitdo. "Because they themselves will cut here, and then perhaps after two years they leave here and go to cut some more in order that they can secure areas of forest. They will make a small sign and then leave it. They make—"

Mali interrupts him, exclaiming, "And they enter way, way back into the forest and cut there and then say, 'This place is mine!'"

"No one else can enter," Wasitdo continues. "They go on and enclose some more so that when other people come there later, they have already taken all the land. And if the latecomers ask them [for some of the land] they will not agree [to give them some]. Or perhaps they will want people to buy the land outright, even plain unworked forest in which they have not yet put anything. Sometimes people get that scheme...."

"They will buy it from whom?"

"From them, no? Yeah, they will do that in order to start a sort of commerce, to start selling forests."

"And in the time of the ancestors, such things did not happen?"

"Uh-uh. Only if you've grown something in the garden can you sell it. Otherwise you give it to your friend or relation for free. Only if there is manioc or plantains in the garden will you be able to sell it."

"But selling primary forest is not done?"

"Uh-uh."

Parker Shipton reminds us that throughout much of Africa land-grabbing is referred to as "eating someone—the person as well as the possession" (1994, 360). Although land-grabbing, the resultant unequal distribution of land, and the host of negative environmental impacts that ensue (Thiesenhusen 1991) are surely taking place in Central Africa, compared to other parts of the world they are generally much more limited (Peterson 1991, 22; Drennon 1990, 4; Bruce 1988). Such a situation can be attributed to numerous factors, but it is well to include among them the strong communal sanctions against individual abuses of tenure that can be found within the heritages of many African peoples.

* * *

I began this chapter by considering several implicit means of conservation "built-in," if you will, to Central African ways of making a living from the natural world. I have just finished discussing various social and environmental repercussions stemming from the "both/and" manner in which Central Africans define the relationship between the individual and the community. I now wish to situate these discussions of land and society within the broader realm of cosmology—how various Central African peoples view or make sense of the world in all its manifestations, physical, social, and spiritual. In this final section, I hope to show more explicitly how intimately connected are the ways in which people view their world, act toward their neighbors, and impact their natural environment.

Living with Meaning in an Interconnected World

By the time we arrive in Gbuda, the sun long ago burned off the cool morning mists we had seen hovering above the Loko's slowly rolling waters. Leaving the road, we skirt several compounds before joining a footpath that will lead us to Duateya's house. The path winds through five-foot-tall *esobe* grass that in places has been cut and laid out for drying, to be used as thatch. In the distance, a border of dark green gallery forest traces the course of one of the many small tributary streams of the Loko. A goshawk whistles from an unseen perch, a familiar sound that immediately conjures up many memories of walking these Ubangian savannas in earlier days.

We find the elder sitting outside in the shade of a papaya tree, deftly weaving strands of raffia together into baskets, which now in his old age he sells to earn a little cash. Looking up he greets his grandson Nzongba and me with a toothless grin. There is a spark in his eyes, a certain spirit of playfulness, and I can tell I have come to the right place to hear some of the old stories. A few days earlier I had asked Nzongba about Ngbaka oral traditions, and he told me he didn't know any *gbagulu Ngbaka* (important Ngbaka sayings). "When we sit in our *malako* [fishing shelters], we only speak the meaningless words of us new children of today. Those of the ancestors, we don't know." But he also told me about his *nkoko* (grandfather or grandfather's brother)—Duateya—who could recite the long stories "from beginning to end." Now, sitting together in the shade of the papaya tree, we begin to talk.

Gbuda: Duateya, Leopards, and Seto

Duateya begins by telling me about leopards and how they are not commonly eaten by the Ngbaka of BoKarawa, his native district. "Leopard—she is a person. She is not just a person, she is a 'photo' of a person."[6] Perhaps he is referring to the fact that among the Ngbaka, as among virtually all the peoples of the Central African rainforests, the leopard is the quintessential emblem of leadership (Vansina 1990, 74). Duateya goes on to describe the ritual dancing and ceremony that are required if a leopard happens to fall into the hunter's trap. The spoils are paraded throughout the village to be given up to the village headman; or they are sent on to the leaders of another village of superior rank. Strips of its fur become symbols of power to be worn only by leaders, but the meat of the animal is distributed among the people of lower status.[7]

But then Duateya begins to tell me about how others living next to them (i.e., outsiders) had turned themselves into leopards and had killed many of his people, including his mother's first husband. Such shape-shifting (belief in which is not limited to the Ngbaka; see Jackson 1989, 102–118) represents another way in which a leopard might be referred to as a "photo" of a person.

Nature Imbued. In either case, Duateya's narrative on leopards reflects a common facet in the cosmologies of many Central African peoples: the belief that certain natural phenomena—in this case specific animals but, in other cases, artesian springs or waterfalls (see story in Chapter 1), areas of forest, and species of trees— are imbued with symbolic meaning, be it political, spiritual, personal, or otherwise. Often an indirect result of such imbuement is some degree of conservation of the natural phenomenon. The totemic animal of a clan may incarnate the spirit of an ancestor and therefore must not be killed; a waterfall or area of forest may be the habitation of various nature spirits and therefore must not be visited; certain animals may be emblems of leadership and power and therefore must not be hunted and consumed in any ordinary fashion.

Granted, the extent of nature conserved through such practices may differ significantly from that conserved in the reserves and national parks that hallmark Western conservation. But the beliefs behind such practices intimate a more general cosmological consciousness among many Central Africans, the ecological significance of which may indeed play an important role. As Kwasi Wiredu expresses it, "Any object, living or nonliving, may be within the immediate province of a superhuman force or power, and one has to avoid reckless and, in some cases, unsupplicated appropriation and use of it" (1994, 46). Thus, according to such consciousness, nature is certainly not considered to be dead matter, or even simply a living material system, but in many ways it is spiritually and symbolically charged. Therefore, one must take caution in relating to nature to avoid suffering both social and cosmological consequences. Unlike Western Enlightenment thought, which removed much of this symbolic meaning from the natural world and went so far as to reduce it in some instances to dead matter, in Central African thought and practice the natural world is alive both physically and spiritually, and therefore one must relate to it with discretion.

Duateya's comment that leopards are persons reflects other beliefs held by various African peoples. According to Jackson, for the Kuranko of Sierra Leone "personhood is distributed into the natural world and not fixed within the margins of the village" (1989, 107). Thus other aspects of creation may be not only symbolically or spiritually charged; they may also be considered as kin and lie within the very important realm of personhood. Such personal entitlement also affords nonhuman nature a degree of respect in many African societies.

Implications for Practice. Beliefs such as these, grounded within local ontologies, may be able to provide a surer foundation or serve as more effective catalysts for current environmental projects than, as mentioned in Chapter 1, theories and practices bearing little local resonance. By stating the reasons for conservation and sustainable use of a resource in locally understood idioms, people may be better able to comprehend and cooperate with a project's purposes. For instance, an interdiction against killing certain animals based on their local ontological significance may be more widely adhered to than one based solely on the authority of the State.

This juncture of ecology and cosmology broadens the field from which those doing environmental work in Africa have traditionally drawn their lessons and insights. For example, Ghanaian Emmanuel Asante argues that environmental issues are not limited to the natural sciences but are also the concern of theology. He calls on African theologians to enhance ideas within their own cosmological heritage in order to address some of the current ecological problems facing Africa. Specifically, Asante suggests building on the widely held African idea that personhood and kinship extend into the natural realm as a means of fostering environmental responsibility:

> Kinship, in an African context, implies interdependence which in turn implies mutual responsibility among the members for the well-being of each individual. To be kin to nature, therefore, is to accept responsibilities towards the well-being of nature, just as nature exists for the well-being of humans.... The African's understanding of his or her interactions with nature between subjects—not between subjects and objects—is a waiting potency crying for theological enhancement in the face of the problem of Africa's ecological mismanagement (1985, 292).

Though such theoretical work is important, some of the best examples of fostering environmental sustainability through incorporation of culturally embedded beliefs come from actual religious practice, especially that found within the fastest growing religious movement on the African continent today: "indigenous" churches—currently some 5,000-plus. Conservation and sustainable development projects could learn much by observing ongoing practices in these churches, which combine elements of Christianity with indigenous African cosmology. For example, the "Africanized" theology of such churches has reinterpreted the Christian idea of the Holy Spirit to encompass "the Earthkeeping Spirit," offenses against which are considered serious sins in need of confession. Thus, for the growing number of believers within such churches, there is strong incentive against engaging in activities that increase soil erosion, foul water supplies, and destroy trees without replacing them. Conversely, activities to enhance environmental wholeness, such as planting trees, have been enveloped within rituals that draw upon indigenous African ideas and inspire continued care and commitment to environmental concerns.

A liturgy developed by the Association of African Earthkeeping Churches of Zimbabwe for use in tree-planting Eucharists is exemplary of the ways in which ecology, indigenous African religious beliefs, and African Christianity are being woven together so as to encourage an earthkeeping that resonates in the hearts and minds of many people. Excerpts of the liturgy include the following passages:

> *Look at the stagnant water where all the trees were felled.*
> *Without trees the water-holes mourn, without trees the gullies form.*
> *For the tree-roots to hold the soil ... are gone!*

These friends of ours, give us shade.
They draw the rain clouds, breathe the moisture of rain.
I, the tree ... I am your friend.
I know you want wood for fire: to cook your food, to warm yourself against
   cold.
Use my branches....
What I do not need you can have.
I, the human being, your closest friend, have committed a serious offense
as a ngozi, *the vengeful spirit.*
I destroyed you, our friends.
So, the seedlings brought here today are the mitumba *[bodies] of restoration,*
a sacrifice to appease the vengeful spirit.
We plant these seedlings today as an admission of guilt,
laying the ngozi *to rest, strengthening our bonds with you,*
our tree friends of the heart.
Indeed there were forests, abundance of rain.
But in our ignorance and greed we left the land naked.
Like a person in shame, our country is shy in its nakedness.
Our planting of trees today is a sign of harmony between us and creation....
This is the water of purification and fertility.
We sprinkle it on this new acre of trees.
It is a prayer to God, a symbol of rain, so that the trees will grow,
so that the land will heal as the ngozi *we have caused withdraws....*
You, tree, my brother, my sister, today I plant you in this soil.
I shall give water for your growth.
Have good roots to keep the soil from eroding.
Have many leaves and branches so that we can breathe fresh air, sit in your
   shade and find firewood.

—cited in Deneel 1994, 257–259

The strength of such creative actions comes in part from the fact that they are grounded within local understandings of nature, humans, and the spiritual world, as theologian Harvey Cox describes:

> On a continent plagued by the loss of woodlands and arable land, a religiously based ecological ethic is appearing. More specifically this ethic is based on a spirituality that mixes ancient African religious sensibilities with modern environmental awareness.... The prohibition against the wanton destruction of trees, for example, is not the result of importing an idea from modern Western forestry. It is the extension of an age-old sanction against cutting the trees in a sacred grove, one that was inhabited by spirits. Now, however, the whole earth and all the trees are understood to be sacred (1995, 245).

Lest these pictures of drawing ecological ethics and practice from local African cosmology seem too simple and neat, listening to more of what Duateya and others have to say makes it clear that people on the ground may not always exhibit the ideas, beliefs, and practices ascribed to them by intellectuals writing from a distance. There are many untapped resources for ecological well-being within the cultural traditions of Central Africa, but the implications, practices, and uses to be drawn from them are often muted by other factors—contradictory beliefs about God and the purposes of creation; the destruction wrought by colonial legacy; and current realities such as population growth and the exigencies of poverty that force one to exploit nature's gifts beyond what tradition at one time allowed. Duateya discusses some of these factors in response to my questions regarding purposes and origins.

Origins, Purposes, and Change. During our conversations issues of cosmology were often broached with my asking by whom and for whom people thought the world was created, or why they thought God had made animals. When I ask Duateya about such matters, it invariably leads to talk about *Seto*, a rather ambiguous figure in Ngbaka oral tradition who sometimes can resemble the trickster and, at other times, be referred to as one of the Ngbaka gods who can be seen in the stars of the constellation Orion. Duateya begins his talk about *Seto* with a story:

"One man who's name was Bindolo built a house to sleep in, but instead of cutting the elephant grass around it, he only cleared the area in front of the door. The elephant grass grew thick around the house, nearly to the point of breaking it. Yet the only area Bindolo would keep clean was the special place where he kept his shield and arrows.

"When Bindolo would go out walking, he would leave a young child behind. One day, a fire started to burn towards the house and the child called out, 'Bindolo eh, Bindolo eh! The house is burning, the house is burning!' When Bindolo heard the child's cry, he hurried home, took his shield and arrows and laid them on top of the roof. The fire immediately died down and didn't burn the house.

"When *Seto* heard about this, he came to see Bindolo and asked if he might give him his shield to use to protect his own house from fire in the same way Bindolo had done. Bindolo replied, '*Seto*, if you take my shield, after you've finished building your house and placed the shield inside, when fire comes you must call my name singing, "Bindolo eh, the house is burning! Bindolo eh, the house is burning! … " If you don't sing in my name and instead use your own, then your house will burn.'

"*Seto* agreed and returned, built his house, placed the shield inside, and when fire came close he called out in song using Bindolo's name. The fire immediately died out. But then *Seto* said to himself, 'I'm not going to give my friend any more fame. Instead I'll sing out using my own name!' So he made another house and put the shield inside. One day a fire started to come toward the house and soon the flames were licking at the roof. *Seto* began to sing, '*Seto* eh, the house is burning! *Seto* eh, the house is burning!… ' And his house burned completely to the ground."

The children who have gathered nearby to hear the elder's tales giggle in delight. We laugh along with them at *Seto*'s folly-laden pride.

"Who is *Seto*?" I ask, curious to hear Duateya's perspective on this enigmatic figure.

"Ah! *Seto*, our ancestors of long ago would praise him like this *Nzambe* here [*Nzambe* commonly used to refer to the Christian God]. Eh! *Seto*! A little thing, they would name the name of *Seto*. Eh-heh [yes], a matter like going hunting in the forest, they would call out in *Seto*'s name. Even little things—in *Seto*'s name ... because our ancestors in those earlier times didn't know *Nzambe*, but only and always *Seto*."

"Where does he live?"

"Ah! Even me, who they gave birth to long ago, I haven't seen him nor his place—what can I say? I have only heard about him and his exploits. Ever since we were little kids, always and only the name of *Seto*!"

"Some people say he is up there and you can see him at—"

"No, that matter of being up above,... that they pulled him up and placed him there above, and thus, as those people who were before us would say to you, 'Tonight *Seto* will cover both you and us.'" Duateya breaks into a hearty laugh. "That place up above where they've pulled him to—eh! But—"

"You don't agree? You can't believe something like that, that *Seto* is up above?"

"He's up there. What do they call them?... They've named three [stars]. Eh! They say that is him. But his entire body is spread out only as they've imagined. Thus those three [stars], they say that is his machete sheath. And the little ones in the middle, those are his genitals. [Hilarious laughter.] His one buttocks sticks out over there. One of his arms is up there, one of his legs is stretched out over there...."

"That is what the ancestors said?"

"Eh! Didn't they say that to us?"

I tell them the somewhat similar story of Orion within my own Western tradition. Duateya's response indicates how the Ngbaka have linked the cosmology of *Seto* to agro-ecological practice.

"He always comes up in the dry season. Yes, in the dry season he comes up like the moon. He travels on and then when he gets over here, they start to plant corn."

"Do you know some more stories about *Seto*?"

"About *Seto*? No, I haven't heard anything else. His affairs are gone just like that. Now that *Nzambe*, the news of *Nzambe* has come out, the children of today have completely abandoned those kinds of things. We are struggling with the matter of returning to the place of *Nzambe* only." I wonder whether among Christian converts such as Duateya, struggling to prepare oneself for "the place of *Nzambe*," the future heavenly world of the Christian faith, has completely taken precedence over the cosmo-ecological struggle to live well in the present world here on earth.

Later, during another visit, Duateya tells me more about *Seto*, the one the Ngbaka ancestors knew as the one who had created the world, the one who along with the spirits of the ancestors helped them to live well by providing for them from the bounty of the forest.

"*Seto*, *Seto*, they said that *Seto* simply made us. They would say in their language, '*Seto* will give me these things.' Or if they wanted to go hunting in the forest, they would first go and set up a piece of wood, sit at its feet and say, '*Seto, Seto*, to the spirits of our ancestors, you who have died, won't you give me an animal?' These were their prayers. They would always only address *Seto* in the spirit of the ancestors, those who had died before. They said that their spirits were there in the forest. They know everything: Let them put an animal in the eye of my trap; let them set an animal before me so that I can shoot it with my gun; let them set an animal before me that I might shoot it with my arrow. And that's how it was. They did not call on *Nzambe*. It went on like that for a long, long time, until afterwards, the matter of *Nzambe* began to come out."

I am curious to know Duateya's thoughts on recent changes inspired by the news of *Nzambe*. What happens when "children of today" abandon the cosmology he has described in which the ancestors play such a key role? Duateya speaks of himself learning by sitting at the feet of his fathers asking lots of questions in much the same way as, he tells me, "you have come to ask me questions."

"And do any of your children or grandchildren follow your example—sit with you and ask you about lots of things?" I ask the elder.

"Oh! Those things are finished."

"Thus will that wisdom die out if young children quit asking their fathers questions?"

"It very well might die out."

"And how do you see that? If such wisdom dies, is that good or bad?"

Duateya dips his hand into a bucket of water and splashes it onto the reed he is working in order to keep it pliable. The basket has begun to show its form.

"It is not good, it is bad. I tell you, it is bad for a reason—the things of the ancestors, we cannot afford to abandon them. But regarding the things of today, they have started to fight with this money, this money they keep putting out from time to time. They have left all those good things and only think about worthless matters. They didn't pay attention very well to their mothers along with us their fathers. Instead they just fight like that and do other worthless things."

I wonder if this "not paying attention" to mothers and fathers might be linked in any way to the recent changes in the natural world that Nzongba has been telling me about, changes such as the fact that the rivers no longer teem with fish? Before exploring that question, I try to get a sense from Duateya of what type of attention the ancestors themselves may have paid to the natural world.

"In the time of the ancestors, did they make laws or give advice that would help people live in harmony with the natural world, be it a matter of farming or hunting or fishing?" I ask.

After some thought Duateya replies, "Yes, it's like I am telling you. During the time I was growing up, perhaps my father or my uncle, or if I traveled to another village to visit a friend, his father, would start to talk: 'Know how to cut a garden, to hunt in the forest, to do good work, to hate no one, to be a friend to people, to visit them, to eat at their house, to come home and likewise receive this brother, exchanging visits with him, treating him the same way when he comes. In the same way, my father along with his brothers and others, when they circumcised us at night in the forest, they would talk and I would hear. I took it all and hid it in my stomach and thus it has remained with me forever, like this. But then there were others who were circumcised with me, whose fathers birthed them at the same time as me. They refused their father's words and did not do what I have done. They will not obey. They will not care for their fathers and their mothers. It is in this way that I still have this wisdom."

"Your child Nzongba has said that there are few fish in the rivers today. Was it the same during your time?"

"Eh! No, in our time there were many fish. And also animals. Here in BoKarawa, there were lots of animals—elephants.... But today, no elephants remain after us. The same with all the buffalo. They have ... crossed to the other side of that river there ... into the big forest on the other side.... But here in our land, there are no longer any big animals; only smaller ones like sitatunga [*Tragelaphus spekii*, a type of antelope] and *mboloko*, here within these small forests."

"Why do you think they are all gone?"

"Because of us, no? We kill a lot of them."

"And how do you see that?"

"Do you think they ran because they think death is a good thing? Death comes in order that ... I, as my name is Duateya Duateya, this thing death surely eats. So I leave my things and flee fast in order that death not chase me and kill me someplace over there."

"And if the ancestors were still here today, if they saw that all the animals and all the fish had fled, what would they think?"

"They would not be happy. And we their children who are also here, we are also not very happy."

"Why aren't you happy about it?"

"Because those are our things to eat, are they not? They have their meat, and now if they all flee from us, what will we do? We eat only these worthless manioc leaves that we have. If you get a little, like my kids here who have gone and picked a little, they serve with this little bit of stuff we bought in the market. We eat only once and that's it."

"So among the Ngbaka, why did they think *Nzambe* or *Seto* created the animals?"

"He made them for us. He made them in order that all of us, you and us, should eat them. Okay, then they took some others, they put them in the village in order that we raise them, so that at the time of the New Year, we kill them to eat. You see, not all of us are hunters. Some sit around empty-handed, they are not hunters. Others simply walk around in the village, and when they see that their brother who hunts a lot has returned, they go to him, to our ancestor: 'Where's mine, where's mine?' Give to him. But the people of today, they don't give a person anything."

"Really?"

"Oh!"

"But in the past, people shared things well?"

"They shared things really well. Like I told you, I have killed a lot of pigs and also sitatunga with my traps. Okay, if I kill an animal, and my brother comes, I will give to him. I catch something in the forest and when I return, I will cut perhaps the leg and go and give it to him in order that his wife might cook it for him to eat. Because I am his brother. But today, ooh! We don't do it like that!"

"And if you give to him, would he then be obliged to give you something in return?"

"Once you have given it to him, it is finished; but to give to him in order to get something back in return, that is bad. He might buy his own [food] from someone else. Once they've finished cooking it, he calls me and I will go and eat with him and it is finished. Then, when we begin eating, he will cry, 'Ah! You my brother, you went hunting in the forest and have returned well and I have been rescued. Now you haven't hunted, and so I bought this food from someone.' Isn't that what he would say? He is mourning for my body. That's how all of us Ngbaka lived together. But now, those who are here today, uh-uh, they don't know how to do these things. Those that have grown up now—eh!"

"Each and everyone is looking out for themselves?"

"Ehhh!"

"But that's not what the ancestors taught?"

"Uh-uh. Our forefathers did very good things. Among the true Ngbaka, once everything was cooked, I would call the head of this house, this person, that person.... They all came. They put a huge bowl in our midst in which we all joined our hands in eating. We would eat until it was finished.... Then after we had washed our hands, we would want them to stay and visit a bit and ... would say, 'All of you come again.' So then once having left and gone walking, perhaps hunting in the forest, or wherever, you would already be full.

"But if you catch [an animal] and eat it alone while your brother is sitting there like that—you have truly killed him. That's what our Ngbaka ancestors used to tell us. They said not to do that. They called it greed. They taught us not to be stingy toward our brother."

Nature, Society, and Cosmology Interconnected. Duateya's narrative poignantly illustrates the vibrant interrelatedness that existed between the natural, social, and cosmological realms within the world views of Central African people such as the Ngbaka. The very words and ideas the elder uses to speak about God, others, and nature flow between the three realms unimpeded by strict categorical boundaries. When I ask about the ancestors' laws governing use of the natural world, he speaks of ways of farming and hunting, but in the same sentence he goes on to talk about how one should treat one's brother. When I try to learn more about *Seto*'s place in Ngbaka cosmology, Duateya recounts the important roles *Seto* played in guiding the agricultural calendar and in providing for people from the bounty of the forest. When I inquire why animals were created, Duateya is quick to mention their purpose of providing sustenance for humans, but he also talks about the social necessity of sharing what one receives from the natural realm with others. Duateya's words reflect a world view in which nature, society, and cosmology are woven together in a web of interrelationship.

Thus changes in any one realm dramatically influence conditions in the other two. The effects of a shift in cosmology, for example, rippled throughout the society and spilled over into the natural world. As the Christian *Nzambe* entered into Ngbaka cosmology, replacing the ways of *Seto* and the ancestors, more attention was given to preparing for that extraworldly realm where *Nzambe* lived than to venerating the ancestors, an act that among other things helped procure one's sustenance within the natural realm of this world. Along with the passing of the affairs of *Seto* "just like that," so went many of the ancestors' teachings concerning how one was to relate to other members of one's community.

More than once, Duateya mentions how they were taught to share, underlining sharing's central place in the life of ancestral communities. Much of what one had to share came directly from the natural world. "If you catch [an animal] and eat it alone while your brother is sitting there like that—you have truly killed him," says Duateya. It is not a jump to conclude that one result of this social ethic of sharing what one procured from nature was a decrease in the level of exploitation required to maintain the sustenance of the community. By giving a leg of the sitatunga one has killed to one's brother, one frees him from having to go out and kill another animal to get his own leg of meat. In the same manner as today, sharing meant that a little bit goes a long way.

Thus the repercussions of Duateya's observation that the "people of today don't give a person anything" are seen not only within the village but outside it as well in the surrounding savannas and forests, as people more and more have to fend for themselves because they can no longer count on receiving part of nature's harvest from others. In the end, "there are no longer any big animals," and the reason, Duateya is quick to point out, is "because of us.... We kill a lot of them." Killing animals in and of itself is not the problem according to Duateya. The real problem is not sharing what you have killed with others, a factor contributing to the overhunting that has caused the animals to flee.

Anthropocentrism or African Humanism? In spite of the fact that he is not happy with this scarcity of game, Duateya remains unequivocal in his belief that God made animals in order that humans might eat them. Others of the farmers I worked with echo similar understandings, reflecting, it is true, aspects of that belief so commonly attributed to African peoples: the primacy of humans over the rest of the animal kingdom.

Libeka responds to my question of God's purpose in creating the world with the following: "God put fish, animals, and forests here for us in order that they be there to help us. They are ours. Because, God made the world, and thought of humans."

Tambwe, a Fulu elder, tells me that their supreme God, *Satongé*, created the world "for them really, eh? It wasn't made for the animals, because animals were made so that we could eat them, eh? God made chickens and us and them to sleep in the same house so that we might eat them. But animals like dogs, we will not eat, because dogs we use to hunt the large savanna rats which they lead us to that we may eat them; but dogs themselves we do not eat.... Because God made the world for us humans that we gather these things to eat."

Yambo, a Ngombe elder, speaks of *Akongo*, the name given to God by his ancestors, creating the world "for people, no? So that people could live from it, no?... The world is for people, no? No, [it is not for animals]; is it not for people? Yes, [only for people]."

When I ask Gabi why the animals are here on earth, he replies: "Ah, the animals here on the earth . . . they have a good work here on the earth.... They care for us. We will eat them; they too go on living. Each person will find a way; even though they are poor, they will receive something to eat.... For whom did God make animals? Did God not make animals for us to eat?"

And contrary to the idea of nature holding symbolic meaning, the view of Wasitdo is that the ancestors "really gave the most thought to their friends. The animals didn't really bother them with meaning, nor did the forest really have meaning. The thing they saw as holding meaning was really humans.... You see, if you killed an animal it wasn't a problem; if you cut some trees, it really wasn't a problem. But a person, if you bothered a person a difficult affair followed."

Such sentiments might lead one to conclude, like J.S. Mbiti, that in Africa "man puts himself at the centre of the universe.... It is as if the whole world exists for man's sake. Therefore, the African peoples look for the usefulness (or otherwise) of the universe to man. This means both what the world can do for man, and how man can use the world for his own good. This attitude toward the world is deeply ingrained in African peoples" (1975, 37–39).

Yet such sweeping generalizations obscure more than clarify the issue. Above all they tend to confuse the deeply ingrained humanism of Africans' world views—their concern with human life, its increase, vitality, and wholeness—with a later Western-styled anthropocentrism, in which humans as the center of the universe believe they can do whatever they please with the rest of creation. Such generalizations also obscure the fact that African beliefs, such as the world and its various creatures being created for the sake of human beings, are couched within a communal rather than an individualistic utilitarian ethos. J. Baird Callicott is correct in his assessment that

> the anthropocentrism characteristic of Western utilitarianism drags in its train a lot of conceptual baggage that seems out of place in an African context. African peoples are traditionally tribal, and the typically African sense of self is bound up with family, clan, village, tribe, and more recently, nation. To such folk, the individualistic moral ontology of utilitarianism and its associated concepts of enlightened self-interest, and the aggregate welfare of social atoms, each pursuing his or her own idiosyncratic "preference satisfaction," would seem foreign and incomprehensible (1994, 158).

Within the context of a communalism marked by strong social ethics such as the requirement to share with others what nature provides, the belief that the purpose of nature is to serve humans was not translated in such a way as to condone practices that exploit nature beyond limits. The necessity of sharing provided one restraint on nature's utilization, but so did other beliefs whose purpose was to preserve a sense of equality and cooperation within the community. Too much exploitation and accumulation of nature's wealth could give cause for envy and make one susceptible to accusations of witchcraft that bore serious consequences one normally would want to avoid (Vansina 1990, 96). Among the Pagabeti people of northwestern Congo, certain techniques believed to increase the number of kills for the individual hunter are still discouraged today, as it is believed their use also leads to a "corresponding increase in deaths among the human hunter's kin" (Almquist et al. 1993, 65). Such "social controls" on humans' use of nature significantly dampen the ecological harm that may result from a cosmology in which nature's purpose is to serve human beings.

Equally significant in dampening the ecological effects of such an ascribed anthropocentrism is the very nature of the deeply ingrained humanism for which the term "anthropocentrism" has often been mistakenly substituted in Africa. The very fact that "to conserve people's lives was [and to some degree remains] a primary goal in these societies" (Vansina 1990, 98) encourages ecological practices that guarantee the conditions for life's continuation, both in the present and in the future, while discouraging those that maximize current human life at the expense of the rest of creation and future generations.

Thus Yambo, for example, immediately follows his statement on the purpose of creation, quoted above, with the following: "Yes [*Akongo*] did place rules [concerning this earth].... Well, according to the way in which [the ancestors] made the decision that they not ruin the waters—such rules came from ... the way in which God said the fish are for us to eat. We go to pains to catch them. But if you [ruin the rivers and streams] and lack fish, you have then in other words killed your own bodies, right? That is the rule of God, because God gave us things. Since God has given us these things, let us use them in a beautiful way. If we use them in a bad manner, then it becomes our own pain and troubles. In the way that God has given to us with tenderness, let us also do the same so that [the world] can work in a beautiful manner."

And after Wasitdo tells me that animals and forests held no real meaning for the ancestors, he offers the following response to my question of whether we have a responsibility to care for the earth: "Yes, because to care for the earth, whether the forests or the animals, one should care for it. It is now in our hands ... to care for it."

"And how do you think we can take care of it?" I ask.

"We should clear [the forest] in the right manner, in the manner that fits. You can clear in order to help you with food, but do it in the right way so that it helps us and our children. Because if we finish everything off, our children will return later to cut gardens, to return like that, and some of the trees, or some animals they will not see. Because if we finish off all the forest, the animals will flee and be finished and children will not see them and some of the trees they will no longer see."

"And if they don't see some of the trees, is that a bad thing?"

"It is bad, for how will they know? The things of their world and they don't know them ... before they have all vanished; and so they won't be able to give them to their children later. Is that not sad?"

"And so do you see it as a loss?"
"Yes, because they need to know the things of their home."
"If children don't know the things of their home, it is bad?"
"It is bad.... They should know the animals, see them with their own eyes, see the trees like *mbangi* [*Chlorophora excelsa*], *kpa* [genus *Afzelia*], *gbodo* [*Triplochiton scleroxylon*] and so on in order to say, 'Ah, look this is it.' But for them to say, 'Look before we had a tree here whose name was—'? But if they haven't seen it, how will they know?"

"They Have Forgotten Everything". According to Duateya, the young people of today have indeed abandoned much within the cosmology and social ethics of the ancestors. "They have started to fight with this money, this money they keep putting out from time to time. They have left all those good things and only think about worthless matters," he laments. Indeed, history has dealt the integrity of the cosmological and social worlds of Central African peoples some serious blows, to the point of leading some seasoned observers to speak outright of the "death of [the] tradition" taking place sometime around the 1920s (Vansina 1990).[8] Yet it is insightful to be reminded by the work of historians such as Vansina just how resilient aspects of the tradition were in the face of momentous changes brought to the region by the Atlantic trade and, later, by the entire colonial enterprise. The resiliency stemmed not from obdurate resistance to change but from just the opposite—the ability to flexibly and creatively adapt aspects of the tradition to changing circumstances while still maintaining enough of the tradition's integrity to hold cognitive worlds together on both individual and societal levels. Writing about the impact of the Atlantic trade, Vansina concludes,

> In the physical world the impact ... might be predominant, yet in the cognitive world the old tradition firmly held the reins.
> Seen from this angle, the integrity with which the tradition was maintained is most impressive.... Despite the new ways of living brought in the wake of Atlantic trading, no foreign ideals or basic concepts were accepted and not even much of a dent was made in the aspirations of individuals. For all that wealth was sought by traders, wealth for its own sake did not acquire followers. The spirit of capitalism made no genuine inroads (1990, 236).

However, the tradition was not able to resist the making of those inroads during the years of colonial subjugation that followed. Accompanied by forty years of violent warfare to subdue Central African lands and peoples, the deliberate destruction of institutions such as initiation ceremonies vital to the transmission of the tradition from one generation to the next, the sowing of seeds of doubt in the minds of young people as to the validity of their ancestral world view, and the recruitment of converts to a new Christian world view (of a form that was often market-friendly), the spirit of capitalism succeeded in opening up Central Africa's natural realm to unprecedented levels of extraction. Forests were cleared for timber. The earth was mined with huge new machines in order to claim its hidden mineral wealth. Agriculture took on a whole new face as the shifting cultivation so well adapted to forest soils and vegetation was subjected to a variety of attempts at modernization, and growing cash crops for market added new stresses to both the environment and society (Jurion and Henry 1969, Jewsiewicki 1983). Duateya's story reflects some of the impacts of these changes within the society. The story of Tata Fulani reflects others.

Bodangabo:
Tata Fulani's Story of
Changes in the Homeland

"All of it began to break down when colonization came bringing with it its civilization," Tata Fulani tells me as we sit together in the shade of his *baraza* (outside shelter used for meetings or to receive visitors) to escape the sun's noontime heat. Deb and I had left Loko early that morning, traveling by motorcycle up the back roads to Bodangabo, a large village situated at the top of one of the highest hills in the area. Stopping at the summit to catch a glimpse of the view, I can see nothing but savannas of *Imperata cylindricum* stretching as far as the horizon. In the old days they used to refer to this village as "Bodangabo *ti-i-i nguluma*—Bodangabo beneath the *nguluma* [*Gilbertiodendron dewevrei*] trees." *Nguluma* are a large, monodominant primary forest species common throughout much of Central Africa's rainforests. Fulani grew up here, not in the open expanse of grasslands to which he has returned to settle today but in the shade of the big forest that once covered the entire hill. It has all been cleared for agriculture, which has stretched beyond the subsistence level as "the intelligent people, the Europeans began bit by bit to change people's ways of thinking, and due to that everything began to be transformed," Tata Fulani explains.

"Can you give me an example of how they changed things?" I ask.

"Yes, of course. Let's talk about it in reference to fields. You see, when they came to the work of farming,… people farmed and they [the Europeans] began to buy. People began to buy the produce from the field. Because in the old days, our ancestors did not sell many things. Instead they had a sort of barter system that did not require them to have very large fields.… But when the whites came, … they began to sell things. Thus if you planted corn, someone would buy it like that. A person, these ones, the strong men began to get the idea, 'Ah, if that's the case, I will clear and plant a large field, enough to give me a little bit to eat, and the rest to sell in such a way as to give me some [money].' Thus that was the way in which agriculture was transformed, by oppressing the soil a great deal and finishing off the forest. However, in the old days of our people, they never burned the savanna meaninglessly."

"Thus, at the time of your ancestors, there was already savanna here?"

"There were a few places with savanna, but the rest of it was covered by a big forest … of *nguluma* trees. But then when troubles came … I saw that forest while I was growing up."

"It was still here?"

"It covered all of these hills. It has only disappeared before my own eyes."
"Really?"
"Eh-heh [yes]. What that means is that they [the ancestors] went a long ways in taking care of things."

* * *

According to Fulani, the deforestation of his homeland is a relatively recent phenomenon standing in stark contrast to the limited ecological impacts of his ancestors' ways of making a living. Much of the deforestation he links to the colonial introduction of agricultural markets and the creation of a need for cash within a money, rather than a barter, economy. This in turn created new opportunities for aggrandizement—both that of the fields people cleared from the forest and that of the wealth of individuals within the society.

True, aggrandizement in the form of garnering wealth and prestige was not something Europeans had brought to Central African societies. For many generations young boys had been taught from early on the virtues of becoming great men. Part of that was to garner many "objects of wealth," as an invocation from a puberty ceremony of the Djue people of Cameroon illustrates. Placing an ivory bracelet on the arm of his grandson, a grandfather would say:

> This elephant which I put on your arm, become a man of crowds,
>   a hero in war, a man with women
>   rich in children, and in many objects of wealth
>   prosper within the family, and be famous throughout the villages (Vansina 1990, 73 citing Koch 1968, 46).

Yet, as I have mentioned, Central African societies also had means to check this aggrandizement, and wealth rarely if ever was sought for its own sake. Even after the Central African basin had been opened to the impacts of the Atlantic trade, "goods still retained their value as items for use rather than for exchange. Whenever possible, wealth in goods was still converted into followers. The principles underlying markets for land and labor were still rejected" (Vansina 1990, 251). Furthermore, "being rich in many objects of wealth" was not the only definition of "greatness" taught to young people in the course of their initiation into adulthood, as the following narratives on *Gaza*, the name given to the puberty rites of many Ubangian ethnic groups, illustrates.

Gaza: *Initiation's Role and Demise.* When most of the farmers I worked with were children, their initiations had lasted anywhere from seven to twelve months. Many could vividly recount details of what they had learned from the elders who accompanied and lived with them in the forest initiation camps. Much of that learning centered not so much on becoming rich but, as Tata Fulani tells me, on "being a person who does beautiful things," a person unafraid of hard work, who respects and serves those within the community who are older.

"When they would take us into the forest during *Gaza*, they would give us an abundance of knowledge and wisdom," Fulani explains. "Look, they would give us counsel such as this: 'We want to make it clear that you a boy have become a man. Grab your machete alongside your father. Be a person who hunts animals. Seek out a woman of significance in order to have a wife of your own. And also, take special care [literally, "be of strong eyes"] to defend and take care of yourself.... Prepare things well so that you will become a person of beautiful deeds. Do not be a person of war.... Order your life well. Be of good character so that if your child or grandchild travels to another village and they want to cause her troubles, she can say, "I am so-and-so's child," and someone will recall, "No, that person was a good man. You cannot harm or kill his child before my eyes!" So be careful of your life that it not bring ruin to your descendants.' They would impart to us a great deal of wisdom and intelligence."

"Powerful advice," I respond.

"Yes it is powerful. 'Take care of the elders, respect them. If you see an elder carrying the burden of work, help him. If an elderly woman is carrying something, carry it for her all the way to wherever it is she wants to go.' They gave very fine counsel."

Several weeks earlier, after having spent the morning working together in his garden, Tambwe and I had returned to Bogofo to talk under his *baraza*. He recounted similar teachings among the Fulu north of Bosobolo, where his initiation in the forest lasted seven months. "In many different ways they taught us to know work, so that, coming out, you would be a child who knows work.... If you want to have a long life on the earth, obey your fathers and obey your mothers. If you see an elder passing by who is carrying a heavy load, you must help her by carrying her load, accompanying her all the way home.... If you see an elder in the midst of some work and you aren't doing anything, help him complete it."

Initiations such as *Gaza* were key not only for giving the young moral instruction; they were also the means by which the heart of the ancestral tradition, including its social controls on aggrandizement, was passed down from father to son, from mother to daughter. As colonialism took hold throughout the Central African forest region, these circuits of transmission were deliberately shorted, being declared illegal by the colonial state or severely indicted in the teachings of certain Christian missions.

> Missionaries directly inveighed against the old tradition in all its manifestations....Local leaders sought to fight back. Where collective initiations existed, they used these as weapons to defend the tradition. Thus a struggle between a mission near Kisangani and the local *lilwá* initiators lasted for decades before the missionaries succeeded in breaking this alliance between the old and the young. By the 1930s many initiations of this sort had come to be outlawed in Zaire [Congo] under one pretext or another, often because they supposedly harmed production quotas set for the rural population. Some initiations survived, but the time devoted to instruction was radically reduced. The campaign against initiations was particularly significant because it directly attacked the formal transmission of the old tradition's essence (Vansina 1990, 245).

Some initiations survived longer in the Ubangi, but "beginning around 1957, they began to suppress them, making it imperative that circumcision happen only in the hospitals," Tambwe informs me. "To continue learning in the forest was bad because it was something of the ancestors. Now, when the [independent] State emerged, it continued to forbid these things. It especially forbade girls' initiations."

Girls' educational programs started by single female missionaries, such as the one Tambwe goes on to describe, proved efficacious in aiding the State to abrogate initiations. "Mademoiselle Agnes came out with many programs[?].[9] She would travel in the villages teaching some of the girls, gathering them together. They would sing, '*Gaza eh-eh, eh-eh mawa mingi eh* [*Gaza*, a lot of sorrow]; *Gaza eh-eh, eh-eh soni mingi eh* [*Gaza*, a lot of shame].' They continued singing that song, and then they would hold flowers in their hands—beautiful flowers in their right hand and wilted flowers in their left. And then they would sing, '*Makambo ya kala masili koleka* [the old things have passed away]', and they would throw down [the wilted flowers]; '*Makambo ya sika tosimbi seko* [the new things we take hold of forever],' and the beautiful flowers they would continue to hold. Then they would say, 'We no longer want girls to take part in *Gaza*. Circumcision is something for boys.'"

Back in Bodangabo, Tata Fulani explains how the introduction of formal schooling also cut into the time available for initiations. "But today, where are you going to find children in order to teach them? When we send them away to elementary school, they will take them to the hospital to circumcise them. Or even while they are still small, still in the hands of their mother, they may have already circumcised them. So there is no other place or time to gather them together. Thus it is school—the rules of school—that are engendering their own [instruction]—their lessons and so forth. But these do not include the teachings I've been talking about.... They [the ancestors] gave us lots of instruction, but today we are giving our children only a little bit. As far as school—children will live with you for two months, meaning that for ten months of the year they are living someplace else."

"In order to go to school?"

"Because of school. They come home to you for two weeks, and leave again. They go there for three months. They return to you for two months, they go there for ten."

"And in your time, when they cut you *Gaza*, how long did you stay in the forest?"

"It was for nine months."

Consequences for the Community and the Land. As for the various consequences of initiation's diminishment, Tambwe relates how "today, you will see that many of our children are not getting advice. They no longer listen to us well.... Now we've come to the point where even small children, their whole desire rests on having wealth, on getting money, and they have forgotten the laws of the ancestors."

During another day spent working with Wasitdo and his wife, Mali, planting peanuts in their garden, I hear the same thing. According to Mali, "Now, everyone wants to be rich, to have many things. Everybody is struggling to become rich, even though God has not chosen them to be wealthy. They will struggle long and hard like that for nothing." With a quiet laugh she concludes, "They will never become wealthy."

"And in previous times it wasn't so?" I inquire.

"No, it wasn't like that."

Wasitdo joins in, "They would only think of getting a little, enough to live on. Because there wasn't a great deal of wealth."

"But this desire for wealth, where did it come from? How did it come to us?" I wonder aloud.

Wasitdo's reply is telling. "Is it not because of civilization? It is necessary that we have Yamahas, or bikes, or that we buy some things in order to live well, to eat well, sleep well, and also know that we have money."

* * *

Initiation was a time to instill the "laws of the ancestors" within the young. In the wake of its diminishment, children learn from other sources. One key source has been the Western culture Wasitdo refers to as "civilization," the source of a plethora of material goods marketed as needs to all of us, including Africa's young people—Yamahas, bikes, and the like.

Mali's statement that "God has not chosen them to be wealthy" implies an understanding that wealth is limited and divinely partitioned, just the opposite ethos from that of capitalism—constantly seeking new markets for its goods under the guise that wealth is unlimitable and democratic. Just how much a lie this is is shown by the fact that levels of disparity in wealth tend to be the highest in those countries such as the United States, where capitalism is most advanced. Under a capitalistic new world order, this disparity now characterizes a global arrangement in which the gap between the haves and the have-nots is increasing dramatically. In Congo, the structures and dynamics of capitalism have contributed to the opulence of a Mobutu reaping benefits from the neocolonial exploitation of the country's natural resources existing side by side the impoverishment of a populace whose per capita income is the equivalent of roughly US$150.

At whatever scale, from the local village to the planet, such an ethos of the unlimitableness of wealth and natural resources is also having drastic impacts on the natural world. Tambwe speaks to the impacts this desire for wealth is having at the local level in Bogofo:

"Sometimes, people will go and see a small fishing hole, and they will clear the area, cutting around it, ruining the whole thing, spoiling it entirely. But our ancestors would go only with a hook. The fish are in the fishing hole, and they would catch one; the rest still remain there. But they would not cut or ruin the place."

"So what that means is that the children of today do not obey those limits?" I ask.

"They don't heed them. Today children don't heed them."

"And what do you think is the reason why they don't obey them?"

"It comes from money, their desire for money now. Eh? To illustrate this desire for money: Their friend dresses sharply or is selling things, and they say, 'We will also go into the forest and do the same.' But if God doesn't want it to happen, they will not reap the same benefits as their friend. And so they ruin the forest. Today you see that if people go to look for animals, they will walk really, really far. But before it wasn't so. Before, sometimes you would go only like to that little forest over there [Tambwe points], and you would kill an elephant.... You wouldn't go far to look for animals.... Because our ancestors didn't really ... think too much about gathering the riches of the world. But now, having received our independence, ah! Children's whole desire is for money. And they no longer want to listen to their fathers."

The day I work with Nani and his wife, Mputu, Pagabeti farmers who recently moved to Loko because it proved a better place to market their produce and wild meat, a huge storm keeps us stranded in the garden *malako* (shelter) until nearly dusk. There is plenty of time to talk over the fried plantains and *mpondu* (manioc leaves) that Mputu prepares while we wait for the rain to stop. Theirs is not a story of people trying to "get rich" but of simply trying to get by, to find enough money to send their children to school, enough food to supply their needs. In addition to farming, they hunt various game with Nani's locally made shotgun. The exorbitant price of shells coupled with the lack of other economic opportunities have forced them into a pattern of spending long hours (*six heures à six heures*—"six o'clock to six o'clock") in the forest to procure wild meat for market. Their case, Nani tells me, is replicated by many others in Bogofo and the surrounding villages. It is a case illustrating how most of the burden of years of economic and political mismanagement of the country is being carried on the shoulders of the rural poor, and from there, transferred directly onto the land. Sharing the load, both land and people suffer at monumental long-term costs.

In contrast to the way in which the Pagabeti used to "rest the forest," Nani informs me that "in this time we have come to, in this hour, there is no longer a way to give the forest a rest, and the animals are starting to be finished off."

"Six o'clock to six o'clock!" Mputu exclaims.

Nani continues, "Six o'clock to six o'clock but ... the animals are growing scarce and it is getting dangerous, something meriting attention."

"And how do you view this?" I ask them.

"Well, we simply see that animals are decreasing," Nani replies. "A way to protect or conserve animals in the forest doesn't exist. There is no way to forbid it. So, instead we must look for a way to raise livestock in the village like chickens, cattle, and so forth."

"But how do you see this disappearance of forest animals?"

Nani's reply is striking in its frank admission of culpability. "Okay, normally, let me say that we see this as something bad. Nonetheless, we the very ones who see it as bad, spend from six o'clock to six o'clock killing animals regardless. Because, today ... I know that monkeys used to be close to the village. I could go in and come out with one right now. But [now] the monkeys are finished. They've fled and they've been exterminated with bullets. But me here, just like I am saying, tomorrow I will rest; but the day after, I will enter the forest only to go kill them again. Even though I see it as bad, I continue to kill."

"What for?"

"In order to get a little means to eat or to sell.... The thing that is really finishing off the animals is just this. Because today I am here, but off in the forest, people are packed in." Nani slaps his fist in emphasis. "Others are coming out since it has rained today, but others will in turn go into the forest. There ... is no other way. Resting does not exist."

Mputu agrees. "There is no rest. We are not talking about months but, instead, day after day people are going into the forest."

"In the course of a year, will they pick one or two months when they don't go in?"

Mputu and Nani chorus—"No!" Then Nani continues, "Every single single single single single single single day there is someone going in. Even as we speak some may be coming out in the rain, but others are going in. That's the way it is."

"And many of them are hunting animals with shotguns?"

"Animals with shotguns, and fish in the rivers and streams, everything."

"And now how far do you have to go to find animals?"

"Sometimes it takes me nine hours. If I walk hard, nine hours, but if not so hard, I will arrive late in the afternoon.... The things we began with in our childhood all the way up until this day are cutting gardens and going into the forest."

"That is all of it," emphasizes Mputu. "In order to enroll the kids in school. To enable them to do well in school, there is nothing else to do."

With his various enterprises, Tata Fulani, unlike Nani and Mputu, is not struggling to get by. He represents a case in which the new opportunities for aggrandizement opened through contact with the West have lifted him above others in the community. Compared to many of his neighbors, he is a wealthy man. Compared to my family's income, his is relatively meager.

In any case, Fulani's efforts to get ahead have not come without costs, both social and environmental. Aggrandizement bears consequences on many fronts in a world where cosmology, society, and nature are deeply interconnected. Some of the social costs have been clearly revealed in the tale of the microhydro and the Mami wata recounted in Chapter 1. To be accused of killing your father and to be seen far and wide as "someone who kills people" is no small burden to bear.

As for the burdens Fulani's level of accumulation places on the environment, of these he is fully aware. When I ask him "So according to your thoughts, does the destruction of the ancestors' means of conserving nature stem primarily from this matter of civilization?" he replies thus: "Yes, that is where it comes from. But I want to say that owing to how I have worked hard I have arrived at a position of being a destroyer of the environment. Because I have a sawmill and I cut lumber. This gives people a reason to cut the large trees, felling them in order to ask me to come and saw them up.[10] I have become a destroyer of [the environment]."

As in Nani's account above, most striking to me is Fulani's full acknowledgment of complicity in causing environmental degradation. What a contrast to the scores of enterprises operating outside personal and cultural constraints that often successfully hide their severe ecological devastation through corrupt means and even condone that devastation by buying off politicians to push through laws that sanction their actions. Much has been written about African people heedlessly destroying their rainforests, but Fulani is far from heedless both of the consequences of, and his responsibility for, his actions. Earlier he told me of a conversation he once had with a European priest. The priest had observed him using no sugar in his tea[11] and, being surprised, remarked, "You must not be of the same race as other Zairians [Congolese] who when given a cup of sugar finish off the entire thing." Fulani replied, "But *mupe* [priest], you've gotten it all wrong. With your cakes, cookies, and ice cream, it's you white people who really use a lot of sugar. We really use very little."

The exchange captures well how prejudice and blindness to one's own actions can easily misconstrue the state of things. Third World peasants are often viewed as the immediate and most pressing cause of much of the world's ecological devastation. They are the ones accused of "using lots of sugar." In reality, the most serious and extensive ecological harm is caused by a total habituation to an overly sweet consumerist lifestyle that blinds people to its ecological effects and their own complicity in environmental destruction. There is much to learn from the frank admission of responsibility by people like Fulani, Nani, and Mputu.

Possible Solutions to Social and Environmental Breakdown. Awareness of his complicity, along with other factors, has inspired Tata Fulani to make a conscious effort on several fronts to try and heal the social and environmental breakdown taking place in the region. In Bodangabo, he has encouraged people to grow trees. "I speak out a lot to the children of Bodangabo, encouraging each child of Bodangabo to plant even five *nguluma* trees.... Or if they don't want to do that, they could grow two *nguluma* and three fruit trees. Then things will improve." He is very supportive of Loko's agroforestry efforts and feels that more people need to follow the example of the woman down the road in the village of Bodjokola Moko. She has planted nearly a hectare of *Leucena* trees to provide her with income while supplying her clients with building poles in lieu of having to cut them from the forest.

Fulani has also openly challenged State environmental officials on their mismanagement of funds that could be used for such things as reforestation. He recounts a recent experience he had with one government official: "One time I went to the State Offices to pay my taxes and the State official there was telling me, 'Pay this, do that.' ... I said, 'I agree, but look, what are *you* going to do? You are asking me to pay and I agree to pay. But what are you going to do to follow up on my work that has already ruined things? You are getting money from my hand, but what are you going to do with it? Will you replace the forest? You want me to pay you while you also want me to cut a lot more. Will you then demand that I replace [what I have cut], or what will you do so that we can replace the forest after people like me have ruined it?'

"He replied, 'Oh, well, we really can't do anything more.' I said, 'No, you can do a lot more! Long ago the Belgians brought in the system of forced labor.... But now what you need to do is lay down some strict laws. A person like me will follow them. You should demand that if I want to cut lumber I must plant as many trees as I cut. Demand also that those who fell the trees and call me to saw them must plant as many as they cut. And if you find they have not planted anything, you will give them another fine. You haven't done that? Get out of this office, let me sit here, you get outside! Because they have sent you to do these things and you have not done them. If you simply ask me to pay for the damage I've done and leave it at that, I will not agree!'

"Lowering his voice, he had nothing more to say and we left it like that. But you see, it is a way for government officials to get money. Because now with this crisis, they don't think. They only desire that which will yield them some quick cash. But as for repairing things in their wake, uh-uh."

Keeping tabs on State officials is one thing. Depending on them for possible solutions to social and environmental problems is another. In fact, many of the people I worked with realize all too clearly that the State is a chief accomplice and instigator of ecological damage, as the following account illustrates.

In the shade on the edge of his garden, Pindwa and I begin to pull the leaves and stems off the peanuts we have just finished uprooting. He begins to tell me about the various fishing techniques of the Mbuja. One technique is to place various botanically derived stunning drugs inside sunken logs in the stream bed. Feeding in the logs, the fish "come out drunk" and are harvested by hand or machete. The effects of the drug quickly wear off in those fish not caught, and all traces of the drug disappear a matter of yards downstream from where the drugs are placed.

In contrast, the use of Aldrin, an industrial pesticide introduced into the area by European coffee planters, can have serious consequences for all forms of life, including humans. I had heard numerous reports of people in the area around Loko and farther north toward Gbado-Lité placing Aldrin into the rivers to kill fish. It has also killed people, something Pindwa knows about from first-hand experience. According to him, it is State officials who are supplying the Aldrin.

"Such things as throwing poisons into the rivers and streams to kill fish, causing many other problems,... such things are far from the ideas of our people there in our homeland," Pindwa comments. "Here, we are truly amazed about it—that people would put Aldrin into the water!" With a slap of his palms, he exclaims, "Why, didn't I almost die here? Yes, I almost died here ... from drinking water. I came here to this garden and worked for a long time until I went down to wash. When I was done bathing, I drank some water and when I came up, I ran into the mother of this child, the one to whom I just gave some peanuts. She told me, 'Be careful of drinking this water because they put'—*eh Nzambe*—'they put Aldrin in upstream but it has not yet reached here.' Ah!

"So I was really afraid. Here I was drinking water and they had already put Aldrin in it upstream, but it had not yet reached here. Yet during that time, two people died ... here in Maniko [a village 20 kilometers north of Loko] ... from drinking that water. Neither do they warn their friends since it is a bad thing, a thing of thieves. They go and put it in, pouring out a whole jug into the water, while people don't know and don't take precautions."

"But doesn't the State apprehend them?" I ask him incredulously.

"Does the State have any strength left? Who is that State? You should know that now the State is also an accomplice. Who could supply that Aldrin? Is it not the State? The State will say, 'Go, and if things get tough, I will arrest you but you won't stay in jail long.' They are the ones setting it up. The State is corrupted now. Because, you see ... like these poachers, some of whom even kill people and are put in jail—where do they get their guns? These big guns, where do they get them? From none other than the State.... These big-time thieves, now the State has cut a deal with them—that State being the majors, lieutenants, captains, the officers in charge of the soldiers. They give these thieves guns.... So, it is no longer a State. I was pleased with a soldier who admitted to me, 'It is us who have wrecked this land; take heed of it.'"

Given this state of affairs, many farmers believe that actions for change will have to come from them and be worked out at the local level of the village. Such is the belief of Tambwe, who replies when I ask his thoughts on solving the social and environmental problems he described (i.e., young people's desire for money that causes them to overfish and overhunt):

"Okay, if someone asked for my advice concerning the fact that the fish are no longer plentiful in the rivers, and not many animals remain, it is necessary that we tell our children, eh? That we teach them and say to them, 'You see now we live in a troubled world.' But if this advice is not given in the circle of elders, if no elders are present, it will not hold. For among our ancestors, we would wait until evening so that boys not walk about, girls not walk about. All of them would sit together and one of the elders, the grandparents, would begin to give them advice:

"'My children, look at these things we are seeing—they are breaking down. Now let's have some rules about the forest. If you come across someone who begins to cut all the logs [in the streams] with his machete [i.e., destroying fish habitat], bring him out to us that we may instruct him and give him the advice that we must take care of the forest so that those we give birth to tomorrow will also have some benefits. And stop killing those animals that are yet small. Leave them, don't kill them. If you come across a young animal leave it be; even though it means money, leave it be. If you don't take such care, those we give birth to will lack a place. Let's say you keep some animals; ... perhaps you keep chickens. If you sell all those chickens and a guest comes to visit you what will you do? If in playing with [your] friends, you happen to kill someone's chicken, is it not necessary for you to go to her household and replace it? We have to take care of the animals of the forest just like we take care of our livestock.'

"You could give this type of advice to the children of your household," the elder explains. "In each *groupement* [political administrative grouping of several villages], all of you, all of the *kapita* [neighborhood headmen] would gather in such small groups. It could work. If not, things will not get better."

"If you sit like that, do you think the children will obey?"

"They will obey. If some of them refuse, then those who have listened to this advice will come and tell the elders when they find the hard-headed ones doing something [wrong].... We will give them advice that they not do it."

"Who gives the punishment and advice?"

"It is us the elders."

"Do you think it would work for the State to punish and give laws, or is it better that this be done by the elders?"

"It is better that the elders do it. Because the State—eh? If, for instance, we are still eating well but the State comes in, it messes things up within the family. If we have a problem, we need to bring in elders who really know things. They will ask you and you tell your story; they ask me and I tell mine. And then they say, 'Look, don't treat your brother like that. That which you have done to your brother, stop it and don't do it again.' Then they will bless us, and I will no longer do such things. But I will not take you in front of the State.

"If I take you to the State, it means that I have split the family. Because the State will want some money to come out of your pocket. Or if it doesn't come out of your pocket, out of your father's since he is also my father. Then that money will come from his hand and he will go pay it. That money was supposed to go to helping us and the family here in the home, but instead it went to the State."

Like Tambwe, Tata Fulani also feels that much of this social and environmental healing must begin with today's young people and the instruction they receive. Thus he has responded to the collapse of the *Gaza* initiation tradition by calling for its revitalization, ironically enough, within the CEUM church, fruit of the very same missionary efforts that to some degree have discouraged *Gaza* in the region. As a member of the church council, he began to bring up the issue a long time ago but has had little success.

"It is a great idea," he says, conveying a certain sense of pride, "but up until now, I have not succeeded. And soon I will no longer have any strength left to speak about it. So it will end. Because I have talked about it a lot. I have spoken to the heads of the church,... but not one person has agreed to it.... But I have given it a lot of thought. And I really want to bring such things back in order to put our children on the right path.... Many times I have asked about making it part of our camp program."

"Why do you think those in charge continue to refuse?" I ask.

"I am not able to tell you why. They hear what I am saying and they agree to it in principle, but as far as really taking action?"

* * *

Back in the Ituri several weeks later, I find out that Fulani and Tambwe are not alone in their desire to reinvigorate traditional models of moral instruction to address the problems confronting Congo's youth. While waiting at Mambasa's catholic mission for a ride to Epulu, I come across a copy of the magazine *Renaître* put out by the Jesuits of the Province of Central Africa. An interesting article reports that in Kinshasa, the Catholic youth organization Bilenge ya Mwinda (Young People of Light) is using the traditional initiation ritual as a model of instruction to help today's urban youth cope with the moral and ethical challenges they face. According to Abbé Katende Edouard, Bilenge ya Mwinda is not simply another movement to indoctrinate the young but "a school for life, an initiation into life based upon the methods of the black African tradition.... Bilenge ya Mwinda initiations retain the practice of seclusion, various tests for the purpose of emulating certain persons of distinction and to encourage the emergence of one's own personality, while not forgetting to instill a climate of mystery surrounding the initiates through the use of certain traditional symbols" (Misambo 1994, 13).

In July 1994, the organization sponsored a "scientific colloquium" with various speakers from the university and the church. One speaker, Abbé Mpati Noël, acknowledged that Bilenge ya Mwinda holds a Christian ethic but stressed that it is one of enculturation to African traditions. "The young pass through the greatest moments of transformation within their existence by following the practice and symbols of initiation into life found within African traditions. It is a question of permitting them to interiorize the criteria for acting well and for giving meaning to life. Besides the theological virtues of the gifts of God, the '16 mysticisms' [part of Bilenge ya Mwinda's initiation] indeed contain the moral and human virtues (friendship, identification, etc.)" (Misambo 1994, 13).

Another speaker suggested that the movement needs to collaborate with *les initiateurs décaneaux* (the deans of initiation of the past), those who specialized in instructing the young during their period of seclusion. He stressed the need for Bilenge ya Mwinda to take concrete initiatives, to be not just a corpus of teachings but a way of life.

It is encouraging to see how the church, after so many years of being involved in the deliberate destruction of African traditions, under African leadership is changing its approach. In addition to Africa's "indigenous" churches, the Catholic Church has taken great strides in the "Africanization" of Christianity. By incorporating aspects of traditional culture into its programs of teaching and ministry, it has opened new avenues to transformation and healing that resonate more clearly with local people. Perhaps such programs offer opportunities for instilling within today's young people some of the ancestral values that—according to the elders whose stories I have recounted—helped to sustain the land as well as human communities.

Conclusion

"Things fall apart," says the great Nigerian novelist Chinua Achebe, referring to colonial Igbo society. In much of Central Africa things have fallen apart, as many of these farmers' narratives reveal. Yet it is remarkable how many of the ideals of the past are preserved in the farmers' minds and hearts, how much they do remember of the ways taught by the ancestors. The existence of that memory provides some hope, enough to legitimate efforts to recover and build upon the remnants of tradition that persist in people's consciousness in order to repair and transform that which has fallen apart. Worn and torn as they are, these remnants remain the building material of choice; as Harvey Sindima reminds us, "For a true transformation to take place, the paradigm for change must emerge from within the people themselves because a paradigm imposed from outside can never include people's hopes and fears" (1995, 197).

Such means of drawing on the past in order to transform the present and prepare a better future is exactly what people are doing in many places across the region. *Sankofa*[12] is happening within the work of thinkers like Sindima, who recovers two key African symbols—*moyo* (life) and *umunthu* (personhood)—to build a model of transformation that will make sense to African peoples (1995, 198–214). Drawing on the work of another contemporary African thinker working for transformation, Fabian Eboussi Boulaga, Sindima describes the work of *sankofa* in these terms:

> The African quest is for fullness of life. To speak of fullness is to imply that there is a level of life that is realized when people take control of their own future as they appropriate their past. In living in the present, they share and enjoy the material and spiritual benefits available and take responsibility in the shaping of their future. To participate in creating one's destiny or that of one's people is being the subject of history. Subjecthood is fulfilled humanity, or personhood at its fullness. Boulaga puts it in this way: "Personhood is perfectly itself when it is the presence both of what is no more and of what is not yet, when it fulfills the former and proclaims the latter all at once. Personhood arrives at self-realization in accordance with the ancient figures. Better, it actualizes and renews these. It is a process of activation of the past and invocation of the future" (Sindima 1995, 197 citing Boulaga 1984, 19).

Sankofa is also happening within African "indigenous" churches; within groups like Bilenge ya Mwinda; and, in the Ubangi, within various indigenously formed nonprofit organizations such as the Communauté pour le Développement Intégral de Bodangabo (CODIBO), recently created by Tata Fulani and other natives of Bodangabo. According to CODIBO's charter, the group came together conscious of "certain negative effects modernism has had on our ancestral values"; aware that "the Ngbaka people, their ancestors and descendants, remain distinguished by unity, dignity, respectability, solidarity, and mutual aid"; "guided by a common will to reinforce and preserve these virtues, the pledge of our identity in whatever circumstances"; and "convinced that this firm determination can only be realized within a well-defined framework" (CODIBO 1995, 1). One of the group's social and cultural objectives is "to push the work of *Ngambe* [a Ngbaka cultural organization promoting Ngbaka traditions] in such a way as to make the ways of the ancestors clearly known in order for people to follow those that are needed" (CODIBO n.d.)

Perhaps initiatives such as these, built on culturally embedded ideals, institutions, and understandings, hold promise for instilling within today's young people some of the traditional African values that Ubangian farmers find are missing. Reinvigorating the traditional periods of initiation indeed may be one crucial step with which to begin, for it was through initiations that the young were taught the ideals that *materially* (the use of nature within limits so as to guarantee the continuation of its providence to future generations), *socially* (the skill of balancing individual and communal needs so as to promote both personal initiative and the welfare of the community), and *cosmologically* (the belief in a spiritually imbued world wherein humans fulfill their purpose of increasing life's fullness by respecting and honoring the ancestors through whom life from the Creator proceeds) governed a holistic and interconnected way of life.

But many might argue that such traditional ideals hold no weight, that they have no place in a contemporary Central Africa marked by the vicious realities of poverty, civil strife, ethnic cleansing, massive violations of human rights, and increasing environmental deterioration. Few have replied to that argument better than philosopher Mary Midgeley in her essay "Idealism in Practice: Comprehending the Incomprehensible." I conclude this long chapter with her eloquent prose in order to remind us that ideals, such as those expressed by Ubangian farmers, are vital for the struggle of practicing transformation.

> There is a fierce tension between ideals and reality, especially, of course, for people who are trying to do something to bridge the gap. Again, this is nothing new. Two thousand years ago Cicero, who was desperately attempting reform but depended on very corrupt allies, cried out that "the trouble is that we have to work, not in Plato's *Republic* but in Romulus's pigsty."
>
> In this mood, it is natural to write off all expressions of ideals, including books like the *Republic*, as simply irrelevant to reality. But this very natural response misses the point. We have to have this longer perspective—the perspective that includes ideals—if we are to make sense of what is immediately before us. These distant visions are an essential part of the scene. They form its necessary background. If we try to work with a world-view which shows us nothing but the present horror and complexity, we lose our bearing altogether and forget where we are going. By their very nature, ideals have to be remote from practice. A society ruled by ideals which were no better than its current practice would be an absurd condition. This remoteness does not mean at all that the ideals have no influence.
>
> And of course, deliberately limited views that are supposed to be realistic are not actually realistic either. They are as selective as any other view of the facts. Every account enshrines some particular emphases and expresses some dream. All are shaped by our value-judgments, our particular fears and ideals. That is why human imagination needs to be stimulated, not from one point on this spectrum, but from many. If we are to keep our power of responding to what goes on, we need to be struck constantly from different angles by different aspects of the truth. We need to consult many conceptual maps—many attitudes, many ideologies—and be ready to use one to correct and supplement another (1995, 43).

Among those many points on the spectrum Midgeley speaks of, one is represented within the narratives of the Ubangian farmers whose voices shape this chapter. There is much to learn if we allow ourselves to be stimulated by their ideals.

One among many of the things to be gained from the narratives of people like Duateya and Tata Fulani, Tambwe and Mama Malabo, and Pindwa and Mama Kagu is the knowledge that the world is an interconnected place—materially, socially, and cosmologically. Nature, society, and cosmology are of one piece. Thus the lesson to be learned is that conditions in any one realm can damage or heal conditions in the other. Applied constructively to the work of transformation—the work of *sankofa*—such knowledge teaches us that to heal conditions in any one realm will require treating conditions in the others.

With regard to actual environmental projects on the ground in the region, this means that work to foster environmental sustainability and conservation in Central Africa will not fully succeed unless it incorporates actions to address the rifts that have been opened up within the social and cosmological worlds within which people use and understand nature.

With regard to our entire human species and its relationship to the earth, home to us and to myriad other forms of life, Central African understandings of the interconnectedness of things, especially the close connection between social ethics (i.e., justice) and environmental integrity, confirm what I only occasionally but powerfully grip or, to be more accurate, am gripped by. The last time I was gripped was during a lecture by African American scholar Cornell West. After listening to Cornell, I went back to my office and penned the following words in regard to what I had chosen as the focus of my doctoral studies, environmental ethics:

> I am struck again by the priority and immediacy of working to heal the human diseases that are killing us as well as the planet. Environmental ethics, policies, or efforts that begin and end only with "the natural world" per se, or that seek to "conserve or preserve nature," or that seek to "heal the planet" or "save the species" unfortunately miss the mark. It is within the human species, within the deep wounds to its soul and psyche, to its abilities to make and be community, to empathize and care, to forgive and be forgiven, to love and be loved, to satisfy and be satisfied within limits, wherein the roots of the dis-ease lie. Before we heal the planet we must heal ourselves. The planet will take care of itself in its own ways. Our disease certainly sickens it but ultimately we hold only the power to destroy ourselves, and in the process, some of our kindred species. But we do not have the power to destroy all of life. If we want to be environmental caretakers, we would do well to first take care of ourselves, to try to heal the sickness that is ravaging our souls, communities, and societies. Working toward such healing is perhaps the most efficacious way to help the planet in its own healing from its human-inflicted wounds.

* * *

Five days after our trip to Bodangabo, Deb and I hop on the Yamaha and say goodbye to friends. Rain from the storm that thundered in the middle of the night is still falling lightly. The roads will be slick but passable. We are headed back to Karawa and then on to Gemena, from where Deb will board a DC-3 to Nairobi. The summer's experiences have filled her head and heart with plenty of fresh material for the course on "Primary Care in Developing Countries" she must return to teach. I will stay on, leaving the Ubangi and returning to the Ituri, to some of the sites where I had conducted my master's research six years earlier.

Our time in the Ubangi has been short but full. I leave with mixed emotions: sad to have to yet again say good-bye to the land in which I was born; uncertain about how I will ever be able to give back to those who have given me so much; excited about the new research adventures awaiting me in the big forest stretching to the east; dubious as to whether Pindwa's prayer that my book might help all who read it will be fulfilled—a whole gamut of thoughts and feelings goes through my head as we slalom around gargantuan mud puddles and ease across slippery log bridges. Pindwa's comment during our final meeting together as a group still rings in my ears: "To learn is one thing; to serve is another." In our culture, research and writing are primarily understood as records of one's learning. To Pindwa, such learning needs to be linked to the greater good of helping others.

In Chapter 7 I will discuss how the lessons learned from these farmers might serve to improve the ways in which Loko and the CEUM are working to meet the local development needs of these people in an environmentally sound manner. But such service is really very minuscule, and who knows what changes recommendations can ever effect. I leave Loko feeling deeply in debt to those who have shared so much of their lives with me.

The sun starts to peek from behind low clouds slinking among valleys as the Yamaha motors us farther and farther away from Loko and all the people who have blessed our lives. Thinking about what I might have to offer them in return, I am heartened by what Tata Fulani shared with me during our time together in Bodangabo. His many travels include a visit to the United States, and he regaled us with stories about his impressions of this land from which many of his *mindele* (white) friends come. Besides telling us that he found Americans to be slaves to two things—talking on the phone and eating—he also spoke of how previously he had always wanted those who came to serve in Congo for short periods of time, or those like myself who had grown up there, to return some day and continue to help his people. But after visiting America he realized that "if you all come back here and work, who will speak for us over there? We need people to speak for us, not just for one year but for the long term."

If this work, this learning experience, can be linked in any way to helping others, it is likely to be here, in trying to "speak for" these farmers, trying to make their worlds and understandings and needs more real to those "over there," in spaces, hearts, and minds that can so easily become far removed from the realities of places like Central Africa and from hearing the voices of local people. I turn now to sharing more of those voices from another place—the Ituri Forest of northeastern Congo—where local people have much to say regarding the reserve model of rainforest conservation that has dramatically affected their lives. It is a model they find quite incongruous with their own understandings of humans' relationship to nature.

Notes

1. My explanation at the beginning of Chapter 3 clarifying the difference between peoples' perceptions of the past and what actually may have taken place in the past applies equally to this discussion of land use practices and techniques. I am operating under the assumption that current understandings of ancestral practices point to the ethic, the ideal—which is the subject I am after—and not necessarily to what the ancestors actually did.

2. Yambo may be referring here to cutting fields along a survey line (*jalon* in French, *molongo* in Lingala, *mandima* in Swahili), a practice greatly encouraged by Belgian colonial agricultural agents (see Jurion and Henry 1969). During precolonial times, it is not clear to what extent fields were cleared in the linear manner Yambo describes. Nevertheless, his and other narratives do point to ancestral sanctions against clearing but not using an area of forest, as well as to the orderly rather than disorderly manner in which farmers rotated the use of their fields so as to let the soil regain its fertility.

3. Thanks to Debra Rothenberg and Fran Vavrus for offering feedback and insights on these issues.

4. See Chapter 2, note 6.

5. Bill Lundeen, an American missionary at Loko, tells me of another case that captures very well how definitions of democracy are influenced by contextual realities. One day while walking through the forest he stopped to rest near a stream. Pulling out his thermos of coffee he offered a cup to the Congolese man who he'd happened to meet along the way. After apologizing for the thermos being nearly empty and not having much to offer, Bill realized that the coffee was also sugarless. The man's response to Bill's apology over the sugarless coffee was a simple two words—"That's democracy!" The man associated democracy with the economic decline and reduction in availability of goods that had coincided with the legalization of multiple parties in Zaire (Congo) around 1991. The economic downturn was caused in part by the withdrawal of Western aid to Zaire (Congo) in protest over Mobutu's slow capitulation to democratic reforms.

6. It is interesting that Duateya uses the personal pronoun *ye* instead of the impersonal pronoun *yango* (it) to refer to leopards. His use of the word "photo" for symbol or emblem is also interesting and provides an example of how items of modern technology are incorporated into the metaphoric language people use to describe ancient beliefs.

7. Vansina discusses the symbolic and political significance of leopards from a historical perspective in detail; see Vansina 1990, 74, 78, 104, 109–111, 116–120, 149, 193, and 276–277.

8. I remind the reader of Vansina's argument for speaking of a singular tradition for the equatorial forest region of Central Africa. See Vansina 1990, 5–6; and the section "Environmental Ethics in Central Africa" in Chapter 2 of this work.

9. Unclear from tape; word is either "programs" or "problems."

10. Fulani obtained his small mobile sawmill through a development project run jointly by the Evangelical Covenant Church's mission in Congo and the CEUM. In clearing their gardens, most people leave the largest trees standing, as the added sunlight the garden would obtain is not worth the labor it takes to fell these buttressed giants. A portable sawmill turning such trees into lumber does make the labor of felling worthwhile and increases the extent of deforestation.

11. Fulani was diagnosed with diabetes some time ago and has adjusted his diet accordingly.

12. *Sankofa* is an Akan word meaning "to draw on the past to prepare people for the future."

5

Reservations About Nature Reserves: Local Voices on Conservation in the Ituri

[Daniel] Janzen exhorts his colleagues to advance their territorial claims on the tropical world more forcefully, warning that the very existence of these areas is at stake: "If biologists want a tropics in which to biologize, they are going to have to buy it with care, energy, effort, strategy, tactics, time, and cash." This frankly imperialist manifesto ... seriously compounds the neglect by the American [environmental] movement of far more pressing environmental problems within the Third World. But perhaps more importantly, and in a more insidious fashion... the wholesale transfer of a movement culturally rooted in American conservation history can only result in the social uprooting of human populations in other parts of the globe.
—**Ramachandra Guha**

With the appearance of a textured Berber carpet, forest—big forest—stretches in every direction beneath the wing of the Cessna Caravan as we soar east from the small mission of Wasolo. After leaving Karawa, we had stopped there briefly to drop off a missionary doctor and accountant. While the pilot refueled the plane, I had run up the hill to see the house in which I had been born. It had been twenty-six years since I had last been there, thirty-six since that day when Dr. Arden Almquist had carried me outside and held me up in front of the people who had gathered to await news of the birth. Then, in what this good doctor describes as a "spontaneous act of joy," he told the workers to "join in the celebration by taking the rest of the day off" (Almquist 1993, 26). As we climb to our cruising altitude, I feel a certain pleasure in thinking about the sense of celebration and community surrounding my birth, grateful that it happened to take place here and not in the cold, cordoned, sterile delivery room of some 1950s-era Michigan hospital.

By the time the plane levels off, it is clear that we are no longer on the edge of the equatorial forest but flying over its very heart. No roads, settlements, or clearings break the contiguous landscape of virgin rainforest for miles and miles. Eventually a brown road comes into view, looking no more or less conspicuous than a thread on a living room carpet. Between the places where overarching trees swallow it up, the road is dotted with tiny houses, flanked on either side with the meager garden clearings of far-flung villages. From 8,500 feet in the air, the clearings, in comparison to the vast greenness around them, appear only as minor abrasions, tiny spots where the carpet has worn thin. The panoramic view from the plane makes it easy to understand the perception that I will hear from many of the Ituri's villagers in the days ahead: "Forest does not end. Forest is forever." The road winds northward, leads to a river, then abruptly ends, leaving no trace on the opposite bank.

At one point we fly over spectacular waterfalls. A quick study of the pilot's charts indicates the falls to be either on the Nava or the Nempoko River. Their confluence is just north of where, now joined, they flow into the larger Ituri. We are nearing the edge of the Ituri Forest. Another twenty minutes and the terrain starts to rumple, rising up to the steeply sloped hills that characterize this forest region lying on the Congo Basin's eastern rim. A half-hour later garden clearings, these more extensive, appear more frequently, and I know we are about to land in Mambasa, a town of some 10,000 people and the administrative seat of the Mambasa Zone. Mambasa is still 130 kilometers from the Ituri's eastern border with the savanna. It is also 70 kilometers east of Epulu, my destination. I will have to backtrack the distance via the road whose passage I have heard has become even more difficult than it was back in May when Deb and I had to walk the last 7 kilometers into Epulu after our friend's Land Rover sunk to its doors in mud. We had stopped then for a ten-day visit on our way to the Ubangi. During that time I made arrangements with the conservation projects based in Epulu for my research and return. This included the use of a motorcycle. Having no means of communicating with people in Epulu over the past three months, I will have to wait until I am on the ground to find out exactly how I will travel on to the little town that will serve as my base for this last leg of the research.

Background: The Ituri Forest

The Ituri Forest, approximately 70,000 square kilometers· has no clear boundaries and so refers to the area roughly outlined by the watershed of the Ituri River, one of the Congo's many tributaries. As part of the largest forest refugia remaining from the Pleistocene epoch, it is particularly noted for its high species endemism and diversity (Wilkie and Finn 1988, 308). The Ituri holds more than thirteen different species of primates as well as an array of large terrestrial mammals including the forest elephant (*Loxodonta africana cyclotis*), forest buffalo (*Syncerus caffer nanus*), giant forest hog (*Hylochoerus meinertzhageni*), and the okapi (*Okapia johnstoni*), a rainforest giraffe endemic to Congo and most abundant in the central Ituri. Other interesting and little known fauna include the Congo clawless otter (*Aonyx* [*Paraonyx*] *congica*) and the water chevrotain (*Hyemoschus aquaticus*). Botanically the Ituri is of

The okapi, a rainforest giraffe endemic to Congo and most abundant in the Ituri, serves as an emblem and keystone species for the RFO.

interest because of its large extent of *mbau* forests, a monodominant forest type in which *Gilbertiodendron dewevrei* (*mbau*) constitutes 90 percent of the canopy, a feature quite rare in the tropical realm (Hart 1990).

Besides its biological richness, the Ituri is also the home of a rich cultural array of forest-dwelling peoples (see Figure 5.1).[1] Various groups of foraging peoples, collectively known as the Mbuti, are very likely to have been the first people to live in the Ituri. For much longer than was once thought, they have been living in a complex, interdependent relationship with various Bantu and Nilotic farming peoples. The relationship is based on a rich configuration of economic, political, social, and religious exchanges that goes beyond the purely material. Besides exchanging meat and other forest products for cultivated starches grown in farmers' gardens, the Mbuti may often play important and necessary roles in various ceremonies the farmers hold. Similarly, various Mbuti ceremonies will incorporate farmers (Peterson 1991, 9–11).

Due in part to this high level of biodiversity, the Ituri has from early on also attracted the attention of the Western conservation community. In the early 1950s Belgians set up the Okapi Capture Station at Epulu for the capture, breeding, and export of okapi to Western zoos. In 1987 this work was taken up by the Gilman

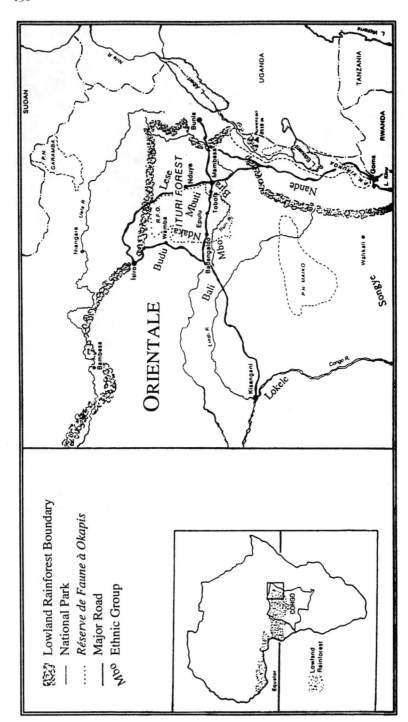

FIGURE 5.1 Northeastern Congo, Orientale Region: Ituri Forest and surrounding area

Investment Company (GIC), a private U.S. conservation firm funded by a paper magnate who as a hobby collects and conserves rare species of animals. Since 1985 the New York Zoological Society (NYZS) has also been involved in a spectrum of ecological research and conservation initiatives in the Ituri through its field research organization, the Wildlife Conservation Society (WCS). Most recently WCS has established the Centre de Formation et Recherche en Conservation Forestière at Epulu, a research center set up to train Congolese students in field ecology and conservation management. From 1987 to 1994, the World Wildlife Fund joined the conservation efforts at Epulu and together with GIC, WCS, and the Institut Congolais pour la Conservation de la Nature worked to have more than 13,000 square kilometers of Ituri rainforest officially "gazetted" in May 1992 as the Réserve de Faune à Okapis.

The RFO was founded as a human-inhabited, multiuse conservation area. Having gotten the area officially gazetted, ICCN, GIC, and WCS, the three groups jointly managing the reserve, are now grappling with what exactly that designation means. All three groups are actively seeking ways to properly manage the protected area through programs of exploration, monitoring, social and ecological research, conservation education, enforcement of park policies, and, for lack of a better term, community relations.

It is this last realm—the RFO's relations with local communities—that serves as the focus of this chapter. As in my conversations with the farmers living near Loko, my dialogues with Ituri farmers and foragers touch on numerous issues in addition to the specific topic of the RFO: the ancestors' ways of protecting the land, the purpose of creation, people's relationships to animals, the causes of current environmental stress, taboos and totemic animals, traditional hunting, trapping, fishing, and farming practices, possible solutions to the difficulties people are facing, and so on. Here in the Ituri, I find people expressing, if not in the exact same words, many of the same ideas as did the farmers in the Ubangi (summarized at the end of Chapter 3).

Rather than reiterate much of that material, this chapter will instead focus on a circumstance facing Ituri villagers that does not exist for the farmers around Loko: the presence of the newly created wildlife reserve and the fact that their lands have come under the increased control and influence of the State and the international conservation community. By recording the sentiments of those living in or near the RFO in regard to the reserve and how it has impacted their lives, I hope to reveal some of the fundamental differences between their ways of viewing the human-environment relationship and those exhibited in the theory and practice of scientific conservation.

From Mambasa to Epulu

There is no motorcycle waiting for me at Mambasa's Catholic mission, but I do receive word that Adrien, one of the logistics people for the projects at Epulu, has gone to Bunia and will pick me up during his return. While waiting for him I spend time reading, planning, and thinking about the research ahead, set up in the same wood-plank guesthouse where numerous times before I have been graciously sheltered by the Italian priests.

Over supper they bring me up to date on the immigration of Nande farmers from the densely populated highlands to the east, the subject I had studied in my master's research six years earlier (Peterson 1991). Now many Nande are moving out of areas of the Ituri due, in part, to the various trumped-up fines and taxes they are being forced to pay by local-level indigenous "chiefs" who seek to gain from hard-working farmers and merchants from the east.[2] Other Nande have gone on to seek elsewhere for the gold that originally brought them here but has now petered out in the boom camps in which they have been working. In the wake of the depopulation resulting from the Nande exodus, many local Bira villagers, especially those east of Mambasa, are reporting increased animal damage to their gardens. In some cases, the damage has been significant enough to lead the Bira themselves to abandon whole villages and move to larger centers such as Komanda, Lolwa, and Mambasa.

In these priests' minds, one of the most disheartening impacts of depopulation has been the devolution of the educational system they helped establish over many decades. As late as 1975, nearly every village along the roads leading from Mambasa had primary schools. Now, heading west, you find a school 25 kilometers down the road at Banana. The next one is at Epulu, 45 kilometers farther. To the north there is a primary school 12 kilometers up the road, but not until Nduye, 60 kilometers from Mambasa, will you find the next. After leaving the school at Mandima, 9 kilometers to the east, you have to travel another 50 kilometers to reach the next school at Lolwa. And the first school you reach heading south from Mambasa is at Maiyowani, 40 kilometers toward Beni. Given such spacing, it becomes impossible for the children of parents who choose to remain in the village to attend school, to ever learn to read or write and get at least a foot in the door toward being able to cope with the bigger, wider, rapidly changing world that yet affects and penetrates their lives. And that is not to mention the affects such scholastic devolution, which is happening in other places besides the Ituri, will have for the future of this country—a whole generation seriously impeded in the skills needed to help Congo manage itself within the complex webs of the global political-economy.

Later, when I reach my old study site of Badengaido (65 kilometers west of Epulu), I will hear about a similar devolution taking place within the health care system. Clinics once stocked with medicines sit idle with empty shelves or, like the one in the nearby village of Bafwakoa, cease to function altogether. Now Badengaido's sick must travel 55 kilometers over horrendous roads to the town of Nia-Nia to receive any medical care.

Tied in to this complex devolutionary process (depopulation, fewer schools, diminished heath care) one also finds dramatic decreases in local agricultural production. In his most recent work with the Mbo people in and around the village of Basiri just south and west of Badengaido, anthropologist Michael Roesler has found agricultural production to be as much as 60 percent less than what he measured in the late 1980s (Roesler, personal communication). This can partly be traced to depopulation, but many farmers also talk about how deteriorating road conditions have made marketing produce nearly impossible. Producing purely for subsistence may decrease pressure on the forest, but it also precludes a major means by which villagers obtain the small amounts of cash they need today to buy clothing, kerosene, soap, and salt, as well as to pay for medicines and schooling when and if they are available.

That this area—marked as it is by such dramatic signs of socioeconomic decay—is also the home of one of the largest rainforest reserves in all of Central Africa is more than a coincidence. Throughout the world one finds that regions rich in biodiversity are often poor socioeconomically. Frequently they suffer from severe infrastructural deterioration and neglect as well (Wilson 1988; Sayer et al. 1992). This juxtaposition of rich biodiversity and poor people raises thorny practical problems for conservation programs, but the ethical and moral dilemmas it poses are of equal import. At one extreme, it is not difficult to imagine a conservation policy that surreptitiously condones socioeconomic deterioration, as it results in less pressure on the forest. At the other extreme is the expectation that conservation organizations should launch full-scale socioeconomic development programs hand in hand with their conservation work. Neither option can be easily justified on both moral and practical grounds. Situated somewhere dialectically between these two extremes is a third position that asks the more fruitful but also more difficult question: How can authentic ecojustice—doing justice on behalf of both the forest and its human inhabitants—be put into practice on the ground in such "real-life" terrains as the RFO?

In order to arrive at any answers to that question, it is necessary to first gain a fuller understanding of how people living in or adjacent to the reserve perceive and relate to the forest, both currently and historically. This chapter aims to enhance that understanding by communicating local people's perceptions of the forest and their concerns about the reserve through the narratives that follow.

* * *

After finishing our simple fare, one of the priests, Père Toma, and I move to the *salon* to continue the discussion begun around the supper table. Framed photos of various Congolese curates decked in red robes and skullcaps stare down at us from high on the pink-painted walls. Books in Italian, French, and Swahili overflow the mahogany bookshelves that line the room. A photocopier is tucked in a corner, completely out of place in a room that I imagine has not changed all that much since the days of Père Longo, one of the first Catholic missionaries to come to the Ituri.

Père Longo, like Patrick Putnam, is somewhat of a legend in this area of the forest. He lived most of his life at Nduye, where he founded a technical school to equip young people with various trades that could help them secure their livelihood amid economic uncertainties. The school continues today but has been moved here, to Mambasa, where it has grown even more impressive and now bears Père Longo's name. Many who knew this giant of a man remark on his humility, openness, good humor, and love for the people of the Ituri, exemplified in long years of service and his refusal to leave his mission at Nduye during the chaos and danger of the 1964 rebellion. Visiting him in the late 1950s, anthropologist Colin Turnbull wrote, "I found myself in the company of the kindest, wisest, and most sincere person I had met in this part of Africa.... He was not concerned simply with adding a lot of names to his roll of converts but primarily with living a life that would be an example that others would want to follow. He made everyone equally welcome at his mission, regardless of religion" (1961, 244). In 1964, Père Longo was speared to death by Simba rebels in Mambasa's central *rond-point* (traffic circle) after being taken captive at his mission in Nduye. My friend Mustafa has told me stories of how, as a young kid, he, along with all the rest of the townspeople, were forced by the rebels to witness the gruesome event, as well as to watch the killings of numerous Congolese.

A faint smell of tobacco lingers in the air as Père Toma shares with me bits of his own story. Toma came to the Ituri in 1993 after ten years of teaching philosophy and theology at the *Grand Seminaire* (seminary) in Bunia. In 1992 he was told by the archbishop that he needed a *repos* (rest) for one year back in Europe, a polite way of removing this teacher from the dangerous work of helping young seminarians to think critically, including being critical of their bishops who love their power and remain committed to preserving the status quo. "How can you teach or do philosophy without the practice of critical thinking?" he asks rhetorically. After a year in Italy, he returned to Congo but was sent to Mambasa, where he now helps out wherever he can—teaching a few courses at the technical school, supervising the carpentry shop, visiting village chapels and catechists, and helping with accounting (or, as he puts it, "plugging the holes").

Taking a puff of his pipe, Toma begins to offer me his perspective on the RFO in response to my inquiry. Although in theory the RFO's policy is to let people living within the reserve continue their traditional use of the forest and thus place no major burdens on the local way of life, Toma feels that in reality the weight of its policies have fallen most heavily on the "small guys" within the surrounding societies. He goes on to explain in the following words, paraphrased from my field notes:

"The reserve is really powerless to stop the big poachers who are supported by the local political authorities such as the *chefs de collectivité* and *commissaires de zone*. The authority of the reserve thus only truly affects those less powerful. Needing to do something, reserve personnel censure small-time hunters, level fines, limit garden clearing, and so forth. In reality, the only ones who are really affected by the reserve are the common villagers and the Mbuti, the *citoyens* and *citoyennes* (citizens), and that in a negative sense. The weight of the reserve falls on them."

Toma explains that another dimension of this burden is the increased damage animals are causing in the gardens of villagers living within the reserve. Controlling such damage with shotguns or wire snares is prohibited, and more traditional trapping techniques using local materials have been forgotten, he says. The end result is decreased agricultural production that in turn leads to a whole array of secondary health, nutritional, and reproductive effects within these forest societies.

The lights in the living room flicker and then come back on, the signal that the generator is about to be turned off. I retire to my room thinking over what Toma has said, yet I realize there are no easy answers to the ethical quandaries posed by doing conservation work among people who have so little and depend on the forest for so much.

Sunday afternoon Adrien pulls into the mission, and by four P.M. we are on our way to Epulu. We are heavily laden with fresh supplies from Bunia, not to mention the two or three extra passengers crammed in between sacks of rice and the packages of fragrant *makayabo* (dried salt fish). But for the extraordinary driving skills of Adrien and the help of the Toyota's electric winch, we would likely never have made it through the three-meter-deep mud holes we encounter time and again soon after leaving Mambasa. With walls like canyons, they have been carved in this part of the trans-Africa highway by the lumbering twelve-ton Mercedes trucks (*les éléphants de la piste*) that somehow find it profitable to continue plying their cargo between the cities in the forest's interior and those to the east. Inching past one of these behemoths, I learn from the driver that this is their fortieth day on the road after leaving Isiro, about 300 kilometers to the west, bound for Nairobi.

At Koki, still 20 kilometers from Epulu, we encounter a scene vividly illustrating Père Toma's assessment that it is the little guys who bear the brunt of the reserve's weight. We pass a contingent of reserve guards, armed with carbines, who have been patrolling the forests east of Epulu for signs of commercial hunting. They are followed by an entourage of Mbuti packing heavy loads of wild meat the guards have confiscated. The official RFO policy on hunting has been understood somewhat ambiguously by reserve inhabitants such as those with whom I am riding. They proceed to offer me a variety of explanations of the infraction the Mbuti may have committed. Some say the Mbuti are allowed to continue their traditional style of net hunting but must first obtain a permit from the reserve warden. Others say that the Mbuti can net hunt but must not be accompanied by, or sell their catch to, food-furnishing villagers deep in the forest. In other areas of the Ituri, villagers, most often from outside the area, have followed the Mbuti into their hunting camps, furnishing them with garden foods and packing out wild meat to sell in area markets. This practice has not only been a major step in the transition from subsistence to market hunting; it has also allowed the Mbuti to stay in the forest for a longer period of time, thus causing an overexploitation of game (Hart 1978). To keep this from happening within the RFO, villagers have been prohibited from following the Mbuti into the forest, a policy I will hear many local villagers criticize vehemently in the days ahead.

In this case, I never receive a full explanation of exactly who has committed the crime—the Mbuti, the villagers accused of bringing in food to Mbuti hunting camps, or both groups. Nevertheless, the wild meat confiscated by the guards represents a loss of protein and revenue to both the Mbuti and villagers, on top of the loss incurred by the fines they may very well have been forced to pay. But the most certain and telling fact is that the Mbuti have been rounded up by the guards to porter the heavy loads of meat back to Epulu.

Later that night they pass us up while we labor to free ourselves from the mud that has swallowed both our axles. This is hard walking even for the experienced Mbuti, laden as they are with heavy packs, feeling their way around perilous mud holes in the dark with no light but that cast by a few torches of *Canarium* sap burning in the night. Adrien mutters something about how with their walking sticks and torches they resemble *BaYuda* (the Jews) trekking through the wilderness. Six kilometers from Epulu we pass the reserve vehicle waiting to pick up the guards. Adrien tells me that when they reach the truck the Mbuti will be relieved of their packs and told to go home with no payment for their work, nothing more than a *merci*.

Within the Ituri's social and political environment, the Mbuti have always to some degree borne the brunt of oppression at the hands of those more powerful, but they have not responded powerlessly. They have become skilled in making strategic choices that allow them to adapt and benefit amid the rapidly changing social and political dynamics of a frontier. A good example of this from my earlier research is the way in which the Mbuti strategically distribute their labor between immigrant farmers and their indigenous farmer exchange partners in order to reap the benefits each group has to offer (Peterson 1991, 83–85). Now, in their encounters with the reserve, their choices can be somewhat curtailed by the threat of a pointed gun. Furthermore, such encounters most frequently involve dealing with carbine-toting, Lingala-speaking guards, many from other regions of the country, who have no understanding of, or appreciation for, the historical and complex patron-client relationships between the Mbuti and local villagers. Under such conditions it remains to be seen just how much flexibility the Mbuti will have in adapting to the new forces affecting their lives, such as those engendered by the RFO.

Several days later in Epulu, after what turned out to be a six-hour trip from Mambasa (average speed, 11.7 km/hr), I am still pondering this enactment of reserve-community relations. It is true that the conservation organizations in Epulu have made serious efforts to educate and improve the guards' abilities to deal sensitively and sensibly with local people. But are such efforts swept away by the ocean of history that surrounds them, a history of established patterns of domination and submission between the powerful and the powerless, patterns to some degree inevitable in a situation where there is so little to go around? Are such efforts fruitless in a situation where guards themselves are underpaid and so, like the military they sometimes imitate, turn to the exploitation of local people to make do? But those are the easy questions of critique: What can contribute to a solution? Do away with the reserve? No. Seek additional funds to pay the guards a living wage while controlling their abuses of power through some means of discipline and punishment? Yes—but that's a full-time job.[3]

In the end any solution will ultimately depend on whether those people who have lived in the Ituri the longest can or will feel any sense of ownership of the RFO. Part of enabling such ownership is to dismantle the patterns of power that have already been established between the RFO and local people, an extremely difficult task. Then again what does "ownership" mean exactly? The Mbuti, too, can be prone to overexploit this forest they consider their home. No one is an angelic environmental steward. One can hardly afford to be when life is so hard and needs are so pressing. The environment doesn't speak back (at least immediately) when people shift the loads they bear directly onto the forest. Therefore, someone has to speak for the trees and the animals, but in a way that is mindful that people too are crying out for justice.

*   *   *

Kitoko, the home of wildlife researchers Roy and Diane Radcliffe, sits perched on a bank looking out over the rocky waters of the Epulu River. Towering, white-barked *Gilbertiodendron* stand like sentinels ringing a cluster of small dwellings including a guesthouse, schoolroom, office, garage, herbarium, and various storage rooms. The Radcliffes have to some degree inherited this compound. When they arrived in Epulu to do their dissertation fieldwork in the early 1980s, local authorities passed Kitoko on to them to serve as their base camp. Earlier it had been the home of other *wazungu* (white people) researchers to the Ituri.

Returning now to this little clearing, I find it has grown more pleasant than even the many idealized images I stored in my memory over the past six years. I settle into a room in the guesthouse I had helped build nine years earlier. Kitoko is especially quiet now, as the Radcliffes have returned to the United States for the time being. I walk down by the river in the golden light of late afternoon, reminiscent of many gone before. The forest shimmers against the backdrop of blue sky and billowy white clouds. The clouds hold no rain but reflect back the sun's rays in a pure, shining white. As always, the river tugs at me to jump in and test my swimming against its current or to let it carry me down around the bend. Even though I used to swim here nearly every day, I resist now, mindful of the crocodiles that have recently killed several of Epulu's children.

"The forest is good," Turnbull quotes a Mbuti man singing to his infant son cradled in his lap (Turnbull 1961, 83). Masalito will later tell me that the forest can also be dangerous, as crocodile attacks reveal. "Even though we know the forest," says Masalito, "we Mbuti also fear things because the forest is difficult and complex. In it we walk in the midst of many dangers of various kinds. There are also many things there that are hard to deal with; many wonders. Snakes—sometime you might meet one, and if you didn't know how to be vigilant in the forest, you would step on it and it would have you—*pah*! An accident!" The forest, complex place of peace and predators, can be both good and dangerous.

The next day I meet with the reserve warden to take care of formalities and inform him of my plans to work in two of the villages where I had conducted research in 1989, Badengaido and Tobola. I also speak with Jacques Dongala, who works for CEFRECOF, the institution through which I had obtained clearance to do research within the reserve back in June. Jacques invites me to his and wife Sophie's home for supper. Over *kuku na wali na sombe* (chicken and rice and manioc greens) there is plenty of time to share our stories and research interests. In response to my explanation of my research approach, Jacques mentions how examining local cultural traditions and environmental perceptions for conservation principles and practices that make sense to local people is "a third level of research that's never really been tried much before here within the RFO [the first two levels being ecological and socioeconomic]."

It is nice to hear that I might be embarking on something at least somewhat new. But, more important, Jacques's point reflects again the vast gap that exists between the world views of Western conservationists and those of local forest peoples. It is another sign that despite all the money, time, and energy the reserve has expended on research the culturally embedded perspectives and understandings of local people are still not being fully heard. In bringing forward the following voices of local people, my point is not to portray community-reserve relations in the Ituri as a struggle between good guys and bad guys. Rather it is to reveal more clearly how environmental conflicts in the Ituri are rooted in the clashing of world views and to plead that in order to resolve these conflicts, in order for ecojustice to take place in the Ituri, these world views need to be carefully and respectfully negotiated.

It is late by the time I ride back to Kitoko on one of the CEFRECOF motorcycles graciously lent to me for my upcoming research sorties. Back in the guesthouse, I drop off to sleep looking forward to the next day's journey to Badengaido—and to the joy of seeing again many of the same people who six years earlier had so graciously received me into their community.

Badengaido: Living Within the RFO

The road west of Epulu proves as poor as that to the east, and I must push the Yamaha through yet another mud hole. Sometimes the only way to move beyond them is to follow the bicycle tracks that lead around or over the mountains of mud heaved up on either side. Approaching a grove of bamboo, I find out that the tracks have been made by a couple of young peddlers. They are resting in the shade next to their bicycles, which are almost hidden by bales of used clothing, cartons of cigarettes, and a plethora of trinkets and Chinese-made sundries ingeniously attached to the bicycles' racks, frames, and handlebars.

The two have come from Butembo, they tell me, about 450 kilometers east, on their way to Wamba, another 200 kilometers up the road. Along the way they've been peddling their wares, and whatever they have left they will unload at their destination. With the money earned, they will buy as many of the twenty-liter jugs of Wamba's rich palm oil as they can load onto their bikes (perhaps a dozen) and begin the journey home. Back in Butembo the cooking oil is scarce and can fetch nearly double Wamba's price.

In days past this trade was mostly done by trucks carrying the oil in 200-liter drums, but with the roads as they are bicycles have become the vehicle of choice—but what a choice! Some of the peddlers, I am told, literally die from exhaustion pushing their overloaded bicycles up the Ituri's steep hills and through the tire-sucking mud. Weakened by toil, many become sick with malaria or other diseases. Having no medicines, they simply endure until the illness passes and then continue on their way. I press on, thinking that only an absolute vacuum of other economic opportunities could persuade people to make such a choice.

After more than three hours of travel, I finally descend the last hill and enter Badengaido midafternoon. My first impressions are of how the village has grown and how much farther back the forest is from the road. I continue past familiar compounds to stop first at the home of Yakobo, the *chef de collectivité*, only to find that he is traveling. But I am told that Kamu, the *chef de localité*, is here. A young boy is sent to fetch him and soon he is welcoming me back with an imploded "*Ndo!*"—a characteristic Ituri exclamation of surprise. I explain why I've come, we settle some logistics, and by nightfall I am set up in the same roadside hotel where I stayed in 1989. Sanu, the hotel proprietor who had been a great help as a research assistant back then, is also away traveling. His wife shows me to a room in the hotel's new tin-roof wing, and here and there I note the improvements Sanu has made. Not everything has gone downhill.

But much has, Matiyé tells me as we catch up after, hearing a quiet knock on the door, I had opened it to find the familiar face of this Budu farmer greeting me with his shy and unassuming grin. He still holds the same air of humility and kindness, untainted by the added burdens of life here following the economic decline that began in earnest in 1991. Matiyé attributes the decline primarily to the deterioration of the road. But later, as we bathe in the clear coolness of the Isayi River and I ask him about the impacts of the reserve, he adds that it, too, has caused them trouble, mostly through being bullied by reserve guards for having wild meat.

Later that night, Kamu stops by my hotel room to visit. He begins to tell me how the reserve is really hurting the local population. Especially difficult is the way in which the guards are burning and closing the gold camps that border the Ituri River, several hours' walk to the south. Whether directly, by digging gold, or indirectly, by selling food to *bouloneurs* (gold prospectors), people in Badengaido have benefited from the camps. Now with *bouloneurs* moving elsewhere to dig for gold, Kamu tells me people are losing a market for their garden produce, thus making it very hard to have the things that can only be gotten with cash: clothes, kerosene, sugar, medicines. Several days later, I will hear a different perspective on the effects the gold camps have had on the local economy, but for now it is undeniable that in these people's minds the reserve is adding burdens to their already weighted-down lives.[4]

Matiyé and Ndjaosiko:
"Our Pantry Is Entirely in This Forest"

After a day of paying visits to many of the villagers whom I'd gotten to know in 1989, I begin going out with people onto their land as I'd done at Loko. Matiyé and his wife, Ndjaosiko, have graciously invited me to stay with them rather than remain at the hotel. My first day out, I accompany them to their current garden to help plant peanuts. Most of the crop will be sold to earn them some cash to help pay for their children's schooling in Nia-Nia.

The sun is especially strong today. Ndjaosiko, with her baby daughter, Anakese, wrapped tightly on her back, works with us for awhile, then finds shade beneath a huge *Canarium* and begins to cook the chicken, *sombe*, and *njelu* (plantain) that will be our midday meal. After a good hour of work, Matiyé, the other children, and I finish seeding a large area of the garden, which also holds manioc, plantains, corn, rice, and a multitude of lesser crops. We find seats on the *Canarium*'s huge roots that spread out from its base like the tentacles of an octopus. After sharing some of Ndjaosiko's tasty chicken and *sombe*, we begin to talk. I ask them whether they think this forest is in danger, if it is being ruined, and if some day it will come to an end.

"It will not end," Matiyé replies. "These days, the problem with the forest, what we see at our level here, according to the way in which we live here, the thing that is causing us difficulties is the way in which now the park [RFO] desires that it should take this forest into which we enter. Because, if that happens, it will add to our problems; for instance, to say that after so many meters a person can no longer farm because that area is a park…. They will no longer agree for us to go on farming further ahead but instead to remain only within the area they give us…. If you want to add on [to your field] there, the State will arrest you…. [But] the forest is vast."

"The forest does not have an end, it is huge!" Ndjaosiko adds.

She continues, telling me about the farming and hunting practices of her people, the Lese. When I ask if her ancestors had any hunting restrictions, both she and her husband say such restrictions did not exist. Matiyé explains why: "God made the animals in order for people to eat…. For you see, you can say that all the things of the forest—"

"They are simply for eating," interrupts Ndjaosiko.

"All the things of the forest, that are in the forest are free," Matiyé continues. "There is no one who can forbid you by telling you not to do such and such. Because it was not a person who created it, but God was the one who told us about the things in the forest, telling us, 'All of you, eat.' You see a tree that has honey there in the forest; okay, you go and remove the honey and eat it. It is not a thing one can say was prepared by a person. God indeed is the one. It is God's work of helping us to live.

"Now, just in these recent times, some of these things that have come to have restrictions placed on them, it is the State that has placed them. However, in the time of the ancestors, when they were here, they did not have any traditions like that. It was that the things of the forest, each thing, God gave."

"But did they begin to lack animals from time to time?" I ask, curious to know if they think this God-given freedom encouraged overexploitation.

The two reply in unison, "No, they didn't lack. They didn't lack animals."

Turning to the specific case of the RFO, I ask them, "Therefore, now what do you think about the park? Is it something you are accustomed to?"

Ndjaosiko responds first: "We are not used to it. We do not know it for it has just discovered us."

Then Matiyé carries on, "These things … our leaders agreed to and then told us the people about them. For they [RFO personnel] discussed it only with the leaders (literally, "big people")."

"Which big people?"

"People like the chiefs."

"The *chefs de collectivité*?"

"Yes, they are the ones."

"They agreed, but you the people … ?"

"Our life, the life of the people is simply to kill an animal. You sell it, your wife gets some cloth along with the children, and you yourself get your own clothes as well."

"But if a large market comes in for the selling of wild meat, won't the animals be finished off?" I ask.

"They will not end. You see, they will keep giving birth," Ndjaosiko points out.

"But they will decrease a bit. They could decrease a little," Matiyé concedes.

"And if they decrease a bit, won't people see that as a problem?"

"Well … there are some places where people have no way of reaching.… The furthest a person goes into the forest is really only 7 or 8 kilometers. There a person can reach. But to go so far, if you don't know that forest, you'll get lost or an animal will kill you there."

I then ask if that is why they have a hard time understanding why the park wants to "close" (*kufunga*) this forest.

"But we really don't know," Ndjaosiko replies, somewhat stupefied. "We don't know the reason why they want to close the forest."

Matiyé is more certain. "This, this is a thing of the State. It is simply oppressing us."

"Do you think that this park will bring lots of difficulties?" I ask.

"In the end, if it gets hard, you will see that people will leave."

"They will leave? Do you also think so mama?" I ask Ndjaosiko.

"Yes, how many would agree to live in a park? 'Oh, don't clear farms; oh, don't do this or that!' And our life is indeed only farming."

Then Matiyé sums up his view metaphorically: "People cannot live … life … like a goat, the way they tether a goat with a rope. The goat only gets to go around in one place. But people at least need to be free, in whichever place they can eat well, they alone choosing what they want. If they want to farm in the forest, they farm there; farm in whichever place they desire. But this being subject to these orders, people cannot hesitate. They must in the end leave."

"And where will people go?" I ask.

Both husband and wife respond, "To a place where there is no park."

"Maybe the Kisangani road or wherever.… To a place where they are free," Matiyé adds.

"Do you think the park has any benefits for a person, for the people?"

Without hesitation Matiyé replies, "Not at all."

"And you mama? Could it have some benefits? If they did things a little differently, could it benefit people?"

"I don't know if it could have any benefits."

"It is not something from within your own traditions?"

"Not in the least," Matiyé replies emphatically. He pauses, choosing his words carefully. "Before among the ancestors, truly they would never sell forest. Something like that did not exist. The ancestors would never sell forest, in the way that today we children of the present have come to sell forest. Never. For see, they are selling this forest."

"To whom?"

"But today, our leaders are selling this forest for the selling. However, before, the ancestors did not—"

"But your leaders, to whom are they selling it?"

"Well! They are selling it to … to those, those people.…[5] For it has become a matter that the president, he himself, has agreed for them to sell the forest. Today we have become a people who have been harmed."

"And you see that as not good?

"Uh-uh."

A bird song from the other side of the garden falls faintly against the silence. The children chatter in the background, but Matiyé's face is quiet and thoughtful. "Therefore, in the traditions of the Budu," I ask him, "what did they think was the meaning of the forest, its purpose? Why did God create the forest?"

"For the purpose of helping us, in the way that humans live … in the way that God placed us in the forest. All food God gives us from the forest. There is no food here in the village, not one bit. Today we are living from the forest. All our things that God has given us are from the forest. If it's the house you sleep in—the sticks come from the forest. Everything, even the *mangongo* [*Marantaceae*, broad-leaved understory plant]. Since we do not have tin roofing to cover our houses, our roofing comes from the forest; it is *mangongo*. There is not one thing—these chords that we use to tie together our houses, they come only from the forest. All our food, and things like mushrooms, whatever, all of it, all food comes from the forest. Our life comes only from the forest; it does not come from the village.

"I think … well, just look a bit at you white people. Your food is already in the house. If you get hungry, you get some sardines there in the house. Everything is there in the pantry. But we black people, our pantry is entirely in this forest."

"Now, it's as if the park has locked the pantry?" I pose.

"Yes, it's as if a person locked the pantry."

"And not just anyone can have a key?"

"No, you will not have one. Today you deserve being able to say, let's have such and such to eat. And they forbid you from going in there. How then will you live?"

"Have they already announced what the limit of the gardens will be?"

"Not yet, we are waiting, and it will happen."

Looking at my watch, I see that we have been talking for only twenty-five minutes, and yet so much has been said. Owing to their eloquence, Matiyé and Ndjaosiko have already given me a good sense of how at least some of Badengaido's inhabitants feel about the reserve. To them, the reserve is a foreign concept lacking roots within their own cultural understandings of the relationship between forests and people. To restrict people's use of this forest that supplies most all their needs they see as a grave injustice as well. They feel oppressed and betrayed not only by the reserve personnel but by their own leaders (including the president) who they see as having put for-sale signs on a forest that was never meant to be sold. It will be important to hear what others have to say, but this couple's story portrays feelings of frustration, anger, and a hint of dejection over the changes the reserve is bringing to their lives and livelihoods. Returning to the road, we part ways as Matiyé, Ndjaosiko, and the children head for home.

Heri and Debo: "If People Die, the Animals Will Also Come to an End"

I continue up the road to find Heri, a Mbo elder and former leader in the village. I had heard that he had moved in with his daughter, who now takes care of him in his old age. Entering the compound, I see that Heri is sitting in his *baraza*. Debo, another Mbo elder, is sitting with him. His daughter is beginning to prepare the *sombe* for the evening meal. She quickly sets out a seat in the *baraza* for me to join her father and Debo. The two elders remember me from my earlier sojourn here, and we spend some time just sitting and talking about all that has happened in the past six years. Then I tell them why I've returned and try to explain the purpose of my research. Debo thanks me for coming, for returning to visit them again. Then, with some formality, he offers his own initiatory remarks, setting the stage for the conversation to follow.

"Riké [Rick] has come to us. We are really amazed because within our language and traditions, we seldom have seen whites visiting us like this—those who come from different places far away to see us, with whom we live and laugh together. So now we rejoice in this our child Riké coming to ask us *wazee* [elders], *Mzee* [singular] Heri Bakana and *Mzee* Debo Bidomé....

"You appear before our eyes having come to ask about our traditions. We will be able to tell you and to give them to you. We are children of the *baraza*, the fathers left us here in the *baraza*. You are a child of the *baraza*, you are not a child who remains outside [*mtoto wa pembeni*]. And you will receive under the *baraza*."[6]

But then Debo also reminds me that they will need some help in turn, as they are old and no longer have the strength they once had. I will need to give them some clothes and, second, "wash their eyes"—give them money with which to buy some dried fish and salt. I am glad I have brought along some gifts for them and assure them that it is only right for researchers like myself to help those who have helped us. "You are helping me and therefore it is necessary that I also help you," I reply.

"I am giving you power," Debo responds. "You know that we receive power here in the *baraza*. When you leave here and arrive there at your place in the West, these stories I will give to you, you will go and tell them—'One day when I was over there, something happened, his name was … '—and I will give you some stories … the stories of long ago."

Heri then joins in with his own words of welcome. But he also wants to let me know his concern over the recent changes taking place in their forest. "This child Riké left America and came here in order to know what? He has come again to the Ituri of today. Because there in America things are breaking down. And here in our home, things are also not what they used to be. These matters of the forest that you are asking about, Riké, you are asking rightfully. Because of what? People are starting to come from far away. When they arrive here in our place, in our forest, they begin talking, saying they will outnumber us here in the forest. Not true! They could never surpass us in the forest. We are people of the forest. Our fathers left us right here. Do you hear?

"Look, some of those who come here enter the forest. They will carry in nylon traps, perhaps sixty of them. Entering the forest here, they will trap and lots of animals will die. And then how will that fix anything? And there in that forest, there is one animal who is really tough. The animal they have said no one should kill. These people still go in and trap and trap and—what are they doing? This thing—it is not good! This thing is not good; they should forbid such a thing. Do you hear? Forbid this thing, it is not good. They come and go into the forest to look for their gain. But now we the people of the forest, we are here waiting and the day we catch a person who has killed this animal, this animal they have forbidden, we will arrest that person, the one who has come to kill us, and we will do away with them completely, completely."

"What animal are you speaking of?" I ask him, assuming that he is referring to the same animal for which the reserve has been named. If so, it appears that, according to Heri, his people also feel strongly about protecting this animal that is endemic to the Ituri.

"That animal—the okapi."

"Even the ancestors of the Mbo, did they also forbid killing that animal?"

"Yes, they forbid it. Even these white men here … who? De Medina [Belgian Captain Jean de Medina], and who? Mr. Putnam [Patrick Putnam, of Harvard], they also forbade it."[7]

"But let me ask if within your own traditions, before Putnam and de Medina arrived, was it also so?"

"You could not kill that animal."

Trying to get more details about Mbo hunting restrictions, I ask Heri what other types of things were traditionally forbidden within this work of hunting.

"There it is forbidden, there in the forest, that is to say, when people go into the forest, what is really truly forbidden is an animal there whose name they call leopard. Don't mess with a leopard! Because a leopard will also attack and wound you. Don't disturb the buffalo! Why? Because a buffalo will hurt you, too. Only kill other animals. Do you hear? Because if you vex a leopard, the leopard will come and attack and hurt you. And troubles enter in. A person dies."

Heri goes on to clarify that people could kill a leopard if it attacked, but most of the time people just passed to the side if they saw a leopard. The same was true with buffalo. When I ask about chimps, Heri says, "Oh, they are indeed people altogether, entirely. Who is going to mess with them?"

"You are allowed to kill them?" I ask.

"How in the world would you kill them when they are people, people who grab hold of you for the grabbing and can really hurt you. How and why are you going to mess with them? For they are really Mbuti [laughs].... Yes, they are Mbuti."

"So you can't kill chimps?"

"No way, you can't kill them."

"Even you, Debo, would you be able to kill a chimp?"

"Me? A chimp? Ah! Those people are strong in the forest; by what means would I be able to kill them? They are smarter than I am when it comes to fighting, even if I have a weapon. Perhaps I throw my weapon at them. They dodge like this, the weapon misses them, and then when I go to recover it they throw it back at me and hurt me. They are really smart."

"And they can fight with this hand. With this hand they can grab your hand firmly. This one here they grab firmly. And with this one over here," Heri exclaims as he slaps his leg, "they can grab too." The old man breaks into rolling laughter as Debo continues extolling the intelligence of this animal they sometimes refer to as *mutu*, the same word for person.

"So, therefore, don't mess with chimps?"

"No way, let them go their way," Heri responds.

"And the same with okapi?"

"Ah, okapi, that is an animal with lots of tradition. That one is an animal with traditional meaning.... They do not attack people nor hurt people, they only go their way."

"But they are food," I argue, playing the devil's advocate.

Debo is firm. "They let okapi go free."

Heri clarifies, "However, if they run into them, I say ... when they run into them, after they have been fighting and the carcass is lying there [okapi are often killed by leopards], then afterwards, they can take it. But to tell people to catch them and kill them, I don't like that. I say there's a problem."

"Because they don't hurt people. And they don't bother people's food," Debo explains.

Heri agrees, "They don't hurt people."

"But a *mboloko* (*Cephalophus monticola*—blue duiker) doesn't hurt people either," I point out, in regard to one of the main species hunted in the area.

"*Mboloko* ... Ah! Those we can eat really well," Heri responds with a smile on his face as if relishing the taste of this little forest antelope.

"But why can you eat a *mboloko* but you can't eat okapi?" I persist.

"No, no. Okapi ... that one is an animal belonging to people [*nyama ya batu*].[8] We do not eat them."

Debo agrees, "No way, they are animals that belong to people. We cannot approve of that."

"Even before de Medina, you could not eat okapi?"

Heri lets out a loud "Ooh!" Then he adds, "Medina will eat his own over there."⁹

"Medina just got here. Medina found us here, when they had already forbid killing that animal," Debo affirms quietly.

"Long ago."

"A real long time ago."

"Who [forbade it]? Your ancestors?" I ask seeking clarification.

"Yes, the ancestors," they echo in unison.

A group of people walks by on the road carrying bedrolls, an assortment of pots and pans, here and there a water jug or a bottle filled with palm oil. Heri's daughter greets them with a loud welcome—*Karibuni*! One of them wears the collar of a clergyman. They must be a contingent of those descending on Badengaido for the regional meeting being held at the local Protestant church this weekend. I have been told that people will be walking in from as far away as Epulu and Nia-Nia. It promises to be a big celebration with lots of choirs, feasting, dancing, and drumming. Not quite the same fare I received in the little Protestant church in which I grew up, but perhaps not too different from what I would experience in America's black churches. Many have encouraged me to be there, and I find myself looking forward to it as the last of this group of pilgrims passes out of sight.

After they've passed, I return to the matter with which Heri had begun: the changes taking place in the forest today. I want to know whether they feel the forest is endangered and, if so, what is causing it to be ruined. In response to my questions, Debo launches into a long history of the early white settlers who lived among them, focusing first on Putnam during his days as a rubber agent for the colonial government. He describes Putnam as one who lived with the Mbuti and learned how to hunt with his own net. He recounts how Putnam captured animals and also collected many of the forest flowers and plants. Then he mentions Putnam taking a picture of an okapi and sending it to Europe. According to Debo, this was the catalyst that initiated the white man's efforts to protect the okapi, beginning with the construction of the Station de Capture des Okapis under de Medina.

"And so Epulu grew up between these two people [Putnam and de Medina]," the elder concludes. "But now, we are amazed! When Medina was here, he did not forbid any group from setting traps. Do you hear? People would set their own traps. And when Medina was here, he did not designate areas of forest, saying that the forest has now become a reserve. Do you hear? The one who has brought the reserve, this name of reserve here like this is you Americans. Eh?"

"Which ones?" I ask.

"The Americans."

"But who do you mean?"

"It is your white people, the ones who are here in Epulu."

"Therefore, you see that the things of the present are different than the way they were before?"

"In the past, Medina did not have this way of prohibiting things.... And so now in this time we have come to be amazed that here within our forest, it has acquired a name that says that we have a reserve! Eh?"

"So the idea of a reserve—in the traditions of the Mbo, was there an idea like that?" I ask.

"It did not exist," both men reply heartily.

I rephrase my question: "The concept of saying, 'This forest—a person cannot use.' Is this a European [*kizungu*] notion or was it also a way of thinking of the Mbo?"

"No way. This is a European concept."

"The Mbo did not have such a tradition?"

"They did not have ideas like that. They did not have them."

"Did they have other traditions that resembled such an idea of the whites?"

"No, no, no."

Debo continues, "We did not know, could not say, that this European concept of closing off the forest [*kufunga pori*] would reach us."

"We did not know such a thing," Heri affirms.

"During the time of the Mbo, the time of traditions, they did not close off the forest?" I ask.

"They did not close the forest. They did not know how to close the forest ... no way," Debo says.

Heri adds, "They didn't want to. No."

"Even during certain periods?"

"No."

"Really?" I ask somewhat incredulously.

"Yes."

"Every day, people would be able to go and kill an animal?"

"They could go and kill their animal."

"[The forest] is not to be closed," Heri declares emphatically.

"It was not forbidden?"

"No, it is not forbidden."

"Why would it be forbidden?" Heri asks me.

Earlier, in response to my question of whether the Mbo had any rules to govern the forest, the two elders had vividly described the joy their people had derived from the forest, from the hunt and the feasting that followed. Perhaps it is this joy that makes Heri so astounded at my questions of forbidding people from entering the forest.

According to Heri, his ancestors "had rules that said that if a guest [arrived, they would] ... go again into the forest. They said, 'Now we are living here and we want to go again into the forest. You, come here, sit here; you, come here, sit here. We want to go again into the forest. Whoever has some plantains, you who have lots of plantains, you are the one who will pull together the nets [*kukokota mukila*, i.e., call the hunt]. All the Mbuti will come and take the plantains to your house. And you will give them tobacco.' Eh?

"And the people are yelling and dancing there around the *baraza* of that person. They would sleep until, time to go, they take up their journey and go into the forest. They are killing animals there—*mboloko*, the red animals [referring to another species of duiker]. They return with them, bringing them all the way into the place of the person who had called the hunt. Who was this person who had called the hunt? None other than the chief! For example, someone like Yakobo who is over there. Do you hear?

"So then [the hunters] arrive. After they've arrived, people start to cry out, to ululate. Once they've returned to the village, [the chief] says, 'Take this piece of meat here, take this piece of meat here, take this piece of meat here. Whoever ... you, you are an elder, here's yours. You there, you are an elder....' He divides it among the entire village. People rejoice and celebrate!"

"No one could go without?" I ask.

"No way!"

"Even the ones who did not go on the hunt?"

"They would not be able to go without. Because, this is none other than our joy and happiness! To cut and share among all the people like this! You see, it is great sport! People are passing the time, people are dancing, people are drinking banana beer.... They are drinking! Then they blow the horn."

Heri tilts his head back and lets out a loud "*Tiyoooooo, tiyooooooooo, tiyooooo.* Eh? Do you hear? Then again there, the one who is the chief, the big man, he wants everyone to rejoice. So he begins to sing as people are starting to fill the place. He sings, '*Heh-oh—*'"

With no inhibition whatsoever, Heri bursts into song, repeating several times a chorus sung in KiMbo. "That's what the people hear. He is full of joy."

"What is the meaning of what he is singing?"

"It means, 'Here I am singing, singing with joy. It is not a day of fighting. I want us all to rejoice! All the women come out, all the elders, let us rejoice!' Do you hear? That's the way it is!"

Not to be left out, Debo adds other details. "A village like this one here, there would be food cooked for everyone."

"Food, everything!" Heri exclaims before Debo continues. "Also then, the elders would be sitting here in the *baraza*. The young men would sit in their place apart, the elders in their place apart. Eh? Dance is falling out throughout the village. Drums!"

Wanting to know if this type of celebration was reserved for special occasions, I ask, "Was this only at the time of the boys' initiation [*Nkumbi*] or at any time?"

"No, the *Nkumbi* has its own special time. *Nkumbi* has its own time."

"This is simply merriment and rejoicing."

"Simply rejoicing and celebration of ... of what?" I ask.

"The joy ... of the forest," Heri says loudly.

Debo adds, "To make the village happy."

Heri nods, "It is of the forest."

* * *

To these elders, the idea of delineating an area of forest and placing it outside the scope of this rich relationship between it and its human inhabitants is clearly very strange. It flies in the face of their idea that, as Heri puts it, "God made all those things saying that we should use them in order to have strength, in order to eat." Furthermore, the reserve's restrictions seem especially perplexing when they have their own restrictions on killing wildlife, as these two elders have outlined above. And for the RFO to forbid even the smaller animals they've depended on for food generation after generation seems unjust given that according to tradition they only take that which satisfies their immediate needs. In addition—similar to what I will hear from villagers in other communities—the two elders clearly distinguish these new restrictions and the new breed of *wazungu* associated with them from the *wazungu* of the colonial era. The latter regulated forest use in a manner much more favorable to local people, whereas current expatriates working at Epulu, referred to simply as the "Americans," are viewed much less approvingly for their new and peculiar style of conservation.

Heri explains how different the RFO's restrictions seem to him from those set by his own tradition: "There [in Epulu], what they are saying is for a person to no longer kill an animal. We may not even kill one animal! If you are caught killing [an animal], they can lock you up. But we, we would kill maybe two or three, and then come back out [of the forest]. But now if you kill here in this time now, you will be beaten, and you will pay fines!"

The limited nature of wildlife exploitation under traditional systems of subsistence use makes it inconceivable to these elders that animals could ever come to an end in the forest. Moreover, if such did happen, it would mean much more than simply a scarcity of food sources. As Debo puts it, "For animals to come to an end in the forest would mean that the entire world has ended." Then Bidomé, Debo's younger brother who has joined us, makes a remark that captures just how integrated in their minds are animals and people, both parts of God's creation.

"Listen here now," Bidomé begins fervently. "You see here, there are animals of the forest and people of the village. But if the people come to an end from death, then in the same way, the animals will also come to an end."

"They will end," Heri affirms.

I am caught off guard by the statement. "Come again? If the people in the village disappear, the animals will also disappear? Explain that one to me because I don't understand it."

"Eh?" Heri is puzzled by my puzzlement.

"*Mzee*, explain it to me. I can't understand that idea, that thought."

"I say, what he said—that if people die out here in the village, the animals will also die out—the meaning of what he said is that God created animals for people to eat. God created villages [for people] to live in and for animals, their own places to live. But that day when God speaks his word and says that people come to an end—animals will end, and people will end [claps his hands for emphasis]."

"They will end together?"

"Together."

"To die, both of them," Bidomé adds.

"But why?" I ask still confused.

"They will die because God is the one who decided ... God created them." *Mzee* Heri pauses a moment and then continues, "I say, eh, Riké, try to understand this: Animals could never disappear because of us. It is God alone who will decide this affair. It is God alone who says whether the animals will end or people will end, not humankind. Because God created animals, God created people."

"Animals and people," Heri's daughter calls out from the courtyard.

"God created the creation of humans together with that of animals," offers Debo.

"Eh-heh-h-h," I respond drawing out the ending to show I now understand. "Therefore, if people die out, the animals will also die out?"

"They will end. Like they said, 'The world has ended; it has broken down.'"

* * *

Bidomé's profession implies that in the minds of these forest dwellers animals and humans are necessary parts of the entire natural system that God has created. The disappearance of animals from the forest would certainly impair human livelihood; but according to these forest peoples, so would the disappearance of humans from the forest (a real possibility in the minds of these villagers given the RFO's restrictions on their livelihood) impair the lives of animals. Both humans and animals need each other; both need to be there if the world is to be kept from coming to an end. Such ideas are common to ideas held by many other indigenous peoples that the natural world to which they belong needs them in order to be sustained.[10] As indigenous inhabitants of particular natural environments, they play a role in the web of life just like any other part of the ecosystem. To remove them or impair the part they play in relation to other parts is to offset the local balance they've helped to create and to distort the web.

Such ideas stand in stark contrast to the views of some Western environmentalists who might find it extremely beneficial to the natural world if humans were to disappear. In such views one finds not an interdependent partnership between humans and animals but a dichotomous antagonism that leaves little room for co-habitation and, thus, few opportunities for humans to learn to live with nature in some degree of balance.

Nonetheless, indigenous views are not always benign. Heri's belief that "animals could never disappear because of us" since "it is God alone who will decide this affair" may represent a fatalism that can blind one to the effects personal actions carry for the natural world. True, the elder's belief portrays a certain humility regarding humans' power over the natural world. But at times such humility can be more of a detriment than an asset to the art of balancing one's own needs with the needs of other forms of life.

Ideology aside, the fact that the RFO is now changing these people's lives in terms of what they can and cannot procure from the forest is definitely having an affect on their relationship and local sentiment toward the reserve. In an essay addressing

African conservation, M.P. Simbotwe, an environmental consultant working in southern Africa, makes an observation that applies well to the case of the Ituri: "Attempts to prevent the use of natural wildlife products, to try to 'preserve' flora and fauna that African people have no alternatives but to use, results only in deep resentment" (1993, 21).

Resentment and a deep sense of injustice are certainly what I hear from Heri and Debo as they continue to describe how they feel about the actions taken by the RFO. "But ... in earlier days we didn't relate to the forest like this," Heri explains. "We were never brought to judgment over the forest [*hukumu ya pori*]. Now you are placing me under these jurisdictions about the forest. You have already set your orders about the forest upon me. Now where will I live while you place on me your supreme power over the forest?"[11]

Being unfamiliar with the phrase, I ask, "What is *hukumu ya pori*?"

"*Hukumu ya pori* means these, my traditions, that you are killing. You are taking hold of your own traditions."

"Of the State," Bidomé says quietly.

"You are taking hold of your traditions of the State," Heri affirms.

"Like saying, 'I should not eat game,'" explains Bidomé.

"Such is not the tradition of the Mbo?"

"Not at all," they all exclaim.

"Therefore, the Mbo manner of taking care of the forest is different and separate from the way in which the people of Epulu are caring for the forest?"

Debo answers, "It is different. It is not according to the laws of my tradition. Another way of explaining it is—say he wants to harm me over what is rightfully mine. For example, take something that belongs to you. Eh? Would I come and take this thing that belongs to you away from you right in front of your very eyes? How then would you reply to me?"

"Ah! I would say that you had cheated me," I reply.

"Ahh! Yes, you would certainly say that I had come to do you wrong. Meaning, you have come ... well, you have oppressed me. For what reason would you come and seize my things, and I see it all plainly? That's the way it is."

"So now, the way you see it, the whites in Epulu have come and cheated you of this forest?"

"They have finished cheating us out of it," Heri replies.

"They have already taken it from us," Debo adds.

"Because it is all under their authority," explains Heri. "It is no longer under our authority."

Debo has more to say. "They forbid us from eating game! You, a person, go to set your traps here like this, and the guards come and discover you with some wild meat in your house, and they lock you up. They beat you; then there is a fine to pay! And you didn't go and steal from someone else. You are only taking what is yours to your home in the village. Do you see? Such fines are what are hurting us."

In his essay, Simbotwe also suggests that the "African approach to solving problems communally is both humane and realistic" (1993, 21). Earlier in our conversation,

Heri had complained to me about the rather inhumane and noncommunal ways in which reserve personnel have tried to resolve conflicts with local people: "They are not saying, 'Now, our brothers, you are all people of the forest....' When we try to speak courteously and slowly to them, they don't like it. They only like it when we say, 'Go ahead! Fine us! Beat us! And arrest us!' Do you see?"

In contrast, when we discuss how to improve the relationship between the RFO and local people, Heri suggests a problem-solving approach more akin to what Simbotwe describes. "How do you think the Mbo, the Ndaka, the Lese, and the Bira will be able to come to a good understanding, a good relationship with the whites of Epulu?" I ask the two elders. "What needs to be done to build good relations?"

Heri responds, "I say, each person, eh? A Ndaka is a Ndaka, a Lese is a Lese.... This is something that ... are you listening? This is the way it is, eh? In regard to preventing these problems and making things right—be it a Bira, be it a Mbo, be it a Ndaka, eh? What needs to happen is for us to set up a tribunal and sit like this. We call the heads of that place of Medina's [Epulu]. We say, 'Sirs, this thing you are doing like this, it is not good. Stop doing it. How are we going to eat now? Where is our forest now? If such is the case, then move us from here and put us in another forest there far away. Because we do not kill a whole colony; our killing is maybe three or four, that's enough for us. Now you are saying that [killing] even one animal is a problem. [Killing] only one or two leads to a beating. Very well then, where shall we live? Whose forest is this? Take us out of here then and take us to another place.'"

"What other place?" I ask.

Heri responds full of emotion, "There, wherever they want, since they do not want us to live here."

"To go to another place, to live in another place, would be better than living here in the reserve?"

"Yes, yes, because they have cheated us. It suffices that we live there, do our living there...."

"Where?"

The old man raises his voice, suggesting a hint of sarcasm, "There where they will choose, telling us all to leave from here."

Bidomé suggests, "Europe or someplace else, perhaps Kinshasa."

"Someplace else, perhaps Kinshasa," Heri agrees.

"That is better than living here in your traditional place?"

As if to imply that I am finally getting the point of his sarcasm, Heri replies to my question, "Good, good." Then he goes on to express where his heart lies. "It is here, the place in which we are living; it is much better because it is indeed what gave birth to us. It indeed is what we have lived with; indeed together with it we have faced difficulties."

"Here? Here among the Mbo?"

"Yes, good."

"To remain here is better than going some place else?"

"Much, much better!"

Bidomé admits, "There we will suffer."

"You will suffer there?"

"Eh," agrees Heri.

"But you say that now you are [also] suffering here? You are experiencing adversity?"

"We are suffering because they have already succeeded in closing off the forest from us. We are no longer in a good state. To leave here and go where they say—let's say it's Kinshasa—one plantain, one thing, do they put it at five francs? Eh? Millions, millions, millions! Where we will get it from? We only want our own [place]."

There is a pause before Debo, looking me in the eye, says, "The thing that makes it difficult for us, *Bwana* Riké, is that we don't have any livestock. The people of the savanna, they have cattle; people of the savanna, they have pigs; people of the savanna, they have goats; people of the savanna, they have chickens. Eh? Well, then, here where you are walking, what kind of livestock do you see? You see that one lone goat that my elder [Heri] has tethered there, eh? If that goat goes in here [to the forest], she doesn't like to eat the forest leaves and grasses. And a goat doesn't like the forest insects. Eh? Let her not meet one of these insects like this, he will sting her there. She comes back here and starts crying, '*Beh beh beh*.' She comes back here to my elder, and soon she is gone.

"Ok, there is another animal, his name is 'Lion, Lion, the Devourer of Chickens' [he likely means a mongoose]. A chicken goes into the forest here and he comes and grabs and squeezes her like a vice in there. '*Bleh bleh bleh*,' right then and there. You see, I myself am missing chickens. Eh? Well then, given these things, what you are doing—to proclaim that this is a reserve—the way I see it, you are killing me."

"But if you see that the people of Epulu are insisting on their reserve, what conditions would you set in order to allow them to have a reserve?"

"We would have some conditions to inform them of, eh? The Mbo, eh? They—the elders among whom I went to sit—according to their traditions, they live from [or "with"—*kwa*] the animals. Eh? One animal they already forbid—the okapi. The second is the elephant. Eh? The third is *kenge*. *Kenge* is the brother of the okapi; he has horns like this.[12] Okay, the fourth, what kind of person? The chimpanzee."

"The fifth is the aardvark. The one who goes in and digs the dirt," adds Heri.

"One can't kill them?"

"Not at all," the two reply in unison.

"Why?"

With some impatience, Heri replies, "Eh! It is forbidden. It is forbidden to kill them."

"That animal was forbidden to kill way before the State forbid killing them," adds Debo.

"Really? The ancestors also forbade killing it?"

"They forbade it."

Debo then makes it clear how killing any of these animals today under the State's prohibitions would also be problematic. "Take the example of you—you are here, eh? We are sitting here with you like this. There is no reason for me to take up a knife while you are watching and say that I am going to stab my elder. If I kill him there, would you not take me in front of the authorities and say, 'This person is a bad man. He killed his brother right in front of me, right while I was watching.'

"So now today, I go into the forest. If you have told me, 'This thing I, the State, do not want to be killed. If I hear news that you have killed this thing, I will arrest you and you will receive a harsh punishment'—would I then be able to kill that thing? Well, that would only be stirring up trouble!

"But those others, *mboloko, mungele* [*Cephalophus callipygus weynsi*, Weyn's duiker], those smaller animals, *lendu* [*Cephalophus dorsalis*, bay duiker]—it suffices that one should leave them for us to eat. Eh? Forbid the killing only of okapi, elephant, chimpanzees, aardvarks—that will do. The killing of those alone you the State [should] forbid.... However, this having enclosed us in a reserve—it is this that has thrown us a lot of sorrow. Because we don't have any livestock. The people of the savanna have livestock."

* * *

Simbotwe concludes his essay with the following advice:

> To realize effective conservation in practice demands the use of African traditions, ideas, language, and terms that communicate with African people effectively, as well as the use of Western technology where it is appropriate to particular problems. In short, a partnership is both possible and desirable as long as Africans' traditional values can be retained (1993, 21).

Perhaps an initial step toward realizing effective conservation in practice in the Ituri is for RFO personnel, both Congolese and foreign, to learn better ways of communicating with and listening to local villagers. As both Heri and Debo convey, local people have a lot to say regarding the reserve and their own indigenous prohibitions against killing wildlife. They also offer a mechanism through which to voice their concerns politely and formally by holding a tribunal in which, according to traditional values, all have a chance to speak and each listens with respect to the words of others, as Heri describes.

Perhaps an initial step by which RFO personnel can improve their communication and listening skills vis-à-vis local people is simply to practice the art of putting themselves in the shoes of people such as Heri and Debo. By so doing they might gain a greater appreciation for the difficulties local people face, as well as feel more inclined to share with them some of their own hardships. Such mutual communication may foster the recognition that finding ways to live sustainably in the Ituri is a challenge local people and reserve personnel face in common, one better met through cooperation than through conflict.

Improving communication skills begins with listening, paying careful attention to the ordinary narratives of local people such as the one below, in which Heri and Debo urge me to speak on their behalf. In what follows the two elders speak for themselves more poignantly than I would ever be able to.

* * *

"Look here where you are sitting, what are they pounding over there? Is that not manioc greens?" Heri asks me after Debo finishes explaining how animal husbandry is not a practical alternative to their traditional dependence on game.

"Every day, only these same leaves," his daughter cries out.

The elder continues, "We will sleep only to find [the same thing] every day, every day, every day. But in our traditions, we didn't have such a thing. Things weren't this way."

"Really?" I respond.

"Yes, indeed."

"What did you eat traditionally?"

"We would simply eat game."

Debo continues, "Game. If I wanted to eat manioc greens, then I would always have a bit of meat and my wife would toss some in there."

"It would make the manioc greens rich and flavorful," Heri adds.

"Then I can eat manioc greens," Debo remarks. "But every day, manioc greens, every day, manioc greens. In addition, these manioc greens make my stomach ill with diarrhea. If the diarrhea continues to pass through me, very well then, I die. Where is there a hospital?[13] These are the things you are forbidding us."

Heri takes up where Debo leaves off: "Here now, we are in a reserve with hospitals. [But] there in Bafwakoa, there is no longer a hospital. Here, in our place, there is no hospital. Well then, what shall we do if a child gets sick, if I get sick? Here now, like we are sitting here—sometimes my stomach starts to rumble, '*Kolo kolo kolo*'. Now, this alone, this forest vine of mine, it's really bitter. We call it *kumiakwaba* [scientific identification n.a.]. It comes out red into the pot, and then I will drink it. The ways of tradition."

"It truly is a difficult situation," I respond feebly after hearing these very real hardships that people are experiencing with regard to the very basics of their lives—food, illness, death.

"You will be able to learn from us about these difficulties—of this reserve."

"I'm simply trying to learn how reserve personnel can come to a better relationship with local people."

"Yes, that's it."

I repeat my purpose, rephrasing it again, explaining that I'm looking to find ways, founded upon local tradition, to conserve the forest and control the overkilling of various forest fauna. Debo then explains that such controls did exist. He refers again to the time of de Medina, telling about how at that time, okapi capture seasons lasted only six months of the year. Hunting was controlled during the other months, but permission was granted to hunt for special holidays, especially those in December. The elder then compares this time to what he is experiencing today.

"Each year, then, we would take care of [the forest and its fauna] continuing up until today. But today, they have succeeded in killing such a pathway. Now [hunting restrictions] apply every day, every day, every day. Before Medina would close the forest for six months. In one year, he would close the forest for six months. Because of what? The animals get pregnant, and give birth ... well, it is clear that animals bear offspring when? October and November. In December, the animals have become again animals open to being hunted. Eh? [This is how it was] during the course of a year.

"However, such things, we no longer see. Eh? Now they close the forest day after day after day. Very well then, what are we to do, we local people? This is really hurting local people. What shall we do now?"

"When are we going to eat meat [again]?" Heri asks.

Debo only reiterates his question, "Regarding this reserve, what shall we do?"

"However, didn't they say that the Mbuti are allowed to hunt?" I ask, trying to elicit their opinions on the means for game procurement the RFO has in theory left open to local people.

Heri replies, "No, the Mbuti, for them to hunt first requires permission. They go there [to the reserve headquarters] and say, 'We have come because we want the Mbuti to go net hunting for us. If they get four or five *mboloko*, they will bring them to us there. And you will also take two of them.' That's how they do it. However, if they don't grant permission, the Mbuti cannot go net hunting."

"These days the Mbuti are also afraid of it ... of going into the forest," Debo adds.

In the background, I hear Heri's daughter yell out, "They are really afraid of it."

"They are afraid?"

"Yes, they fear it."

"Fear who?"

"They fear the guards, and the judgment of what? Of Epulu. Now during the time we are in."

Then Heri clarifies, "What we are saying—Medina here [it appears that in Heri's mind "Medina" has come to be synonymous with all conservation initiatives at Epulu], you will see here, some times when six o'clock comes around, you will see some of them walking around at night in order to check things out, looking here and there. Who will they catch with wild meat? They come, they will come and start walking there, starting around four when there are lots of people around."

"To do their controls!" Debo adds.

"You mean the guards?"

"Yes, the guards."

"During the night?"

"Yes."

"And they also check in bags," Heri's daughter calls out.

"They come from [the patrol post in] Bafwakoa?"

"Yes."

I let out a big sigh and the elders chuckle softly. Faintly in the background, in a voice like a whimper, I hear Heri's daughter say, "And now we are really tired of manioc greens, Papa. Every day pounding them in our mortars."

* * *

As I spend time in other communities, I will hear mention again and again of manioc greens (*sombe*) having assumed an inordinate proportion of people's diets. One elder in the village of Tobola, about 60 kilometers east of Epulu, feels this dietary change that people link to the reserve has real impacts on the health and strength of their children.

Musafiri, a Bira elder, asks me, "Look, plants like these, you feed them with it; [do] they become strong from it?" Turning to his two adult sons, he asks, "Well, you there, where did you get your strength from?"

"Meat," the one replies without hesitation before Musafiri continues. "From animals. Did you get it from manioc greens?" the elder asks.

"No."

Musafiri nods. "Manioc greens are nothing but leaves that make your stomach ache badly. You go and defecate, it is zero. It is worthless. Eh? You see."

Manioc greens certainly have more nutritional value than the tubers that grow beneath them, but if people in the Ituri were really eating only manioc greens, they would not be able to survive. The greens provide little of the protein people need to live. Thus it seems that hyperbolic accounts of eating only *sombe* reflect a larger meaning than that associated simply with diet. The repeated and often colorful narrative people give to it implies that it has become symbolic of the more general intrusions people feel the reserve has made into the intimacies and exigencies of their lives.

Narratives about *sombe* replacing meat also highlight how central a role meat plays, not only in people's diet but also in their culture. Many times I have heard the remark that one has not truly eaten unless one's meal has included some type of meat. Meat defines many forest people's cuisine as much as pasta defines Italian cooking. Imagine the uproar if Italians were prohibited from eating spaghetti or rigatoni.

Procuring meat through hunting also holds numerous cultural meanings for forest peoples. Learning how to hunt is a large part of boys' initiation into manhood. For the Mbuti, net hunting enacts and teaches the delicate balance between community cooperation and individual prowess and property. But for the very young and very old, the entire band takes part in the hunt. As the men set their nets in a semicircle encompassing nearly a square kilometer, the women and children move silently around the perimeter of the arc to the center of the opening. Once the nets are set, a signal starts the women and children whooping and hollering, flushing the animals as they walk toward the center of the semicircle. Clearly the nets nearer the center have the best chances for capturing game, while hunters with nets on the edges often find them empty. To compensate, Mbuti hunters rotate their positions along the arc, giving everyone a chance to occupy the choice places for capturing game. In a similar gesture of egalitarianism, game that happens to fall into your net belongs to you alone, but you are obligated to share it with the rest of the camp. Hunting thus provides for forest people in both a material and an ideological sense.

Actual reserve policy allows people to eat meat obtained through such traditional means (i.e., Mbuti hunting and exchange) so as not to compromise the nutritional status of the RFO's inhabitants and the cultural importance of the hunt. But the practices of reserve personnel in implementing the policy have discouraged people from hunting and having meat. People sincerely fear reprisals such as those Heri and Debo have mentioned. To know the degree to which people are actually lacking protein in their diets due to the reserve's restrictions on hunting and the use of wild meat would require a careful and controlled study. However, in the minds of these villagers, there is little question that their nutritional needs and health status have been compromised by the limited availability of meat. Thus the fact that people fear partaking in the hunt or having wild meat in their kitchens in response to the enactment of RFO policy holds repercussions far beyond people's literal complaint of now having to "eat only *sombe.*"

Later, another group of elders in the village of Bapukele, 6 kilometers east of Epulu, will tell me how their nutritional status is being affected, not only by the lack of meat in their diet but also by the abundance of animals destroying their garden crops. But I still have lots to learn and additional perspectives to gain from other farmers living here in Badengaido.

Mapoli and Maurice:
"Here There Is True Peace"

Not everyone in Badengaido feels this same degree of injustice and anger over the restrictions the RFO is placing on their lives. My conversation the next morning with two other Badengaido farmers stresses again the need to talk with lots of people before drawing any conclusions. But the conversation also highlights the danger inherent in placing too much emphasis on conclusions and generalizations over the particularities of people's experience. There is no orderly, general perception of the reserve and the forest on the part of Badengaido's residents; rather there is a rich and varied tapestry of views, some of which conflict, all of which lend themselves to illumining the complexity of human communities.

* * *

The next day I sit with Mapoli, a Bali elder, and Maurice, a Songye farmer who was born in Orientale Region and has been living in Badengaido for many years. "Look, in comparison to other places, we are living here in peace because the [reserve] guards are taking care of us here," Mapoli tells me. "There are lots of animals here." The shrill whistle of an African gray parrot flying overhead punctuates his words. "If you walk here, [you see] monkey after monkey. Other places it would be impossible for you to see a monkey. Here there is peace, true peace."

In essence, Mapoli is the founder of Badengaido. During his work as an agricultural agent in the Mbo and Ndaka *collectivités*, he had noticed that the villages along the road were far apart, with long stretches of virgin forest in between. "Well then I said, 'Living far apart isn't good. I will go and live in my wife Mariama's village.'" His superiors in Mambasa suggested he live halfway between Epulu and Nia-Nia, so he settled in the middle of what was then nothing but forest. He was slowly joined by other settlers. From these modest beginnings, Badengaido has now grown to be a village of some 700 people representing many different ethnic groups (Peterson 1991, 45). Only within the last ten years has *Chef* Yakobo moved his residence and administrative offices here from Basiri, located on the Ituri River and off the main road. Like many other villages these days, Badengaido is an extremely diverse community.

As we continue our conversation, Mapoli makes it clear that one of the reasons why he approves of the RFO is that its conservation efforts maintain an abundance of game so that people can continue to supply their need for meat from the forest relatively easily, as has been their practice for generations. "Monkeys, to see them like that makes you happy," the elder explains. "You ... eat them. They are not enemies.... You eat your food. It's very good, the way Epulu is defending [the forest], it is good."

"Really?" I ask

"Yes! Very good. In order that animals not all disappear. We are not people of the savanna who have livestock. Our livestock is in the forest."

"However, this matter of the reserve, how do you see it?"

"The reserve is doing the right thing. It is doing its thing justly. Because first of all here, we are in our place here, the Parc Okapi [RFO], our animal, the one that has value by bringing in lots of zaires [local currency].... The people of Epulu and the park warden say they are defending [the forest] since the forest around Epulu has already been degraded. This is good."

"Really?"

"Yes, I agree with it. When you desire, you enter the forest and you see monkeys like that, animals passing by. You get one and you are satisfied. We are used to this because we don't keep any animals like goats and such as livestock."

"But how will you get these animals to eat?"

"Animals for eating have their own people. For example, I don't know how [to hunt], but there are the nets of the Mbuti, the camps of the Mbuti. Today, for example, if someone like Kamu calls the Mbuti, and asks permission from the park warden, they then go into the forest to look for that food. The Mbuti will come then and bring us animals and [in return receive] manioc. They come out to buy. They come with a *mboloko*, they give it to us. You then give them some manioc...."

"That system is working well?"

"Very well!"

"Do you agree Maurice? This system for getting animals to eat?"

"It isn't bad," the younger farmer agrees. "It's much better than making a mess of the forest. Because they were disturbing a lot of forest, even killing animals like okapi.[14] To do such things is not good. Every single thing needs to have some security in order for you to live in peace. You see?"

"But some others here in Badengaido have told me that they are really hungry from the lack of eating meat. Is this true or how do you see it?"

Mapoli replies, "We only want the animals of the Mbuti. When the Mbuti go into the forest, we will come and give them food in the village. No one setting traps and all those things. That is ruining the forest."

"Such is ruining the forest; it is wrong," Maurice affirms.

"However, why did others tell me that they hunger for meat? Is it true that they lack meat?"

"They could be hungry because they [the reserve guards?] do arrest people and prohibit some things," exclaims Maurice. He laughs while Mapoli explains how for a long time they have had some restrictions on hunting here in the Ituri.

"Before they would close off the animals beginning the 1st of August ... because that is when the animals become pregnant; fish, everything becomes pregnant. They will give birth in December.... During those ... months, animals can't be killed.

"But then, they take up their shotguns each hunting season until they begin causing problems and strife. What are they doing? They want to harm the forest! Eh? The law is the law. We are here in a reserve, we must do things according to the law. We want only the food that comes from the Mbuti, the Mbuti who go to the forest and hunt with nets."

Mapoli carries on for some time, describing to me how among his people the Bali hunting was mostly done collectively. According to him, hunts were concentrated around the time of the boy's initiation period (*Mambela*). "They would go into the forest and stay for one month. After finishing [their work in] the forest, they would leave the forest to remain in peace."

"They wouldn't go in again?" I ask.

"No, no! They would return all the way to the village. There they would remain. They would hunt only those animals nearby since they were working in their fields."

During the agricultural season the Bali may have undertaken minor net hunts, but they primarily used traps with chord made from raffia palm (*sakasa*) to ward off garden-ravaging animals such as baboons, pigs, and mangabeys. Such practices recall the periodicity of hunting mentioned in Chapter 4, one among other factors that limited the extent of environmental damage. Mapoli, like some of the farmers whose stories I recounted in Chapter 4, recounts another factor, namely, that his ancestors did not hunt for the market but for "their subsistence. To preserve their lives.... They would go hunting and kill an animal or two or three. These they would then share among them all, and they would eat.... They didn't practice [market hunting]; only for exchanging for food. You have some food, you give it to your relatives. They share it with their relatives."

When I tell them about the soldier in the Ubangi who killed more than sixty monkeys in one day, and ask them what they think, Mapoli and Maurice are amazed. "It is bad, it is wrong," Mapoli responds. I ask him why.

"Because he is finishing them off! If he finishes them off, then what are we going to eat?"

"Sixty monkeys? In one day?" Maurice asks incredulously.

"So you think it is wrong, but why is it wrong?"

Local Voices on Conservation in the Ituri

"Because he's destroying, to kill so many like that," Maurice exclaims. "Even in one year he couldn't eat all those. Should we then kill sixty?"

Mapoli adds, "Sixty. If so, then the animals would all be gone."

"Then tomorrow—where will you hunt tomorrow?" Maurice asks rhetorically.

"Yes, where then will you eat, if it's all gone like that?" Mapoli asks. "Kill one or two—to eat that's all. If you have a gun, you may kill one, but sixty?"

"That's no good?" I ask.

"No good at all!" Mapoli says fervently. "You are destroying the animals. What the people of Epulu are saying, that they forbid [the killing of] animals here in our place—good! Very good!"

"It's forbidden by law," adds Maurice.

But Mapoli clarifies that not all of the animals should be prohibited. "Animals … the ones they call baboons—[those are] bad animals!"

Maurice then points out another aspect of reserve policy that I had not heard from Heri and Debo the previous day. "They [reserve personnel] even told us in regard to our livelihood that if it's baboons, we are free to kill them, to trap them with *sakasa* around our fields, because they are animals that bring hunger. Animals like pigs, you can trap them with *sakasa* because animals like these bring hunger. Because they are animals that destroy. The mangabey—that is one animal who comes to your garden and leads you into hunger."

It is clear that the actual things prohibited by the reserve have been understood differently by different people. Maurice continues, "It is only these animals that you are allowed to trap with *sakasa* around the garden in order to make your garden secure."

"They gave permission [for that]?" I ask.

"Yes, that's what they said."

"Therefore, your view is that what the Epulu people are trying to do is good?"

"It is good," Mapoli responds without hesitation.

"But others have said that it is really oppressing them?"

"Uh-uh [no]. Oppress how? [They only want to] finish off all the animals. Let them finish them off then," Mapoli says with a tinge of sarcasm.

"Those who say that, it's up to them," Maurice remarks, not wanting to speak for others who might see things differently. "Long ago when we arrived here in Badengaido, we ate the animals [provided by] the Mbuti. They go to the forest, come back with some animals, their *mboloko*. You trade some manioc or plantains, and you have your meat. We didn't eat these animals coming from these nylon traps. They are forbidden by law."

Seeking more knowledge of their own people's laws governing the forest, I ask, "Now then, according to the traditions of the Bali during the time prior to the State, were there also some forest animals whose killing was forbidden by the ancestors?"

"There were some."

"Which animals did they forbid people to kill?"

Mapoli replies, "Animals—up there in our homeland, in Bomili, there are okapi."

"And the ancestors forbid people to kill okapi?" I ask.

"Yes, they forbid their killing. You see, it's an animal that if you kill ... if you kill an okapi and eat it yourself, you will die."

"Really, that's what they said?"

"Yes! You will die, if it's you alone. If you go and kill and eat one, you alone, you will die."

"And if you kill one as a group?"

"If you kill one and bring him or her out, they would say, 'Ah! Why in the world did you kill this animal? This animal—don't kill them anymore.' It wasn't that they would go out and kill them with arrows like that! Okapi would sometimes get caught in a *sakasa*. Only rarely and by chance would they meet their death that way. [More often they would break the snare] '*Kpa*!' and they would take off. Only the leopard would kill them. They struggle and the leopard kills the okapi, and you come upon it. Then you could eat it. It is not an animal to get through hunting with the bow, because it is strong."

We have been speaking for quite a long time, and the morning sun is beginning to bake the open spaces of the village. Mapoli rises, saying it's time for him to return home, but before he goes he briefly reiterates the benefits of the Mbuti-villager exchange system as a means of getting meat. However, when I ask him what time I should come by that afternoon to speak some more with him, his response seems to indicate that at least for the moment the exchange system has not come through. Apparently concerned about what he might have to offer me during my visit, he asks, "But how will we really see some meat with our own eyes?" The old founder of Badengaido gets up and walks back to his house, farther up the road, without explaining to me why exactly his larders are empty.

I continue talking with Maurice and ask him if he agrees with Mapoli that the people of Epulu are doing good things. He feels they are because their conservation actions are "guarding the security of the forest." According to him, there have always been times, even for the ancestors, when they went without meat. And then there are always the cows butchered now and then by the cattle herders from the east who periodically drive their animals through Badengaido on their way to market in Kisangani and Isiro. If the reserve had prohibited killing cattle, then that would cause suffering, Maurice feels, but in his mind it is good that they are keeping the forest secure.

When I remind him that only yesterday he had described to me how the reserve was causing lots of suffering, he seeks to clarify his point of view. "[The reserve] brings suffering because of what? Because of the menace [of the guards]. You see, it doesn't mean that it brings troubles over issues concerning the forest. For example, we were just in the gold quarry here. [The guards] are going into the gold quarries, threatening people, washing out their diggings. They stay for a week, a week at a time, beating people in the forest without any motive. They are not arresting people with wild meat, so what are they doing? Other times they confiscate your gold diggings, or insist you pay a fine over your identity card, stupid things like that. Doesn't that cause us to be troubled in our hearts?

"If they are going about under your laws, then let them follow your laws only. Real policing is when you find someone who has truly committed an infraction and you condemn only that person and take him away. They should not say that all are at fault. So you see … this is one thing that is really causing us to suffer."

He goes on to mention how the people of Badengaido had a talk with the guards and assured them that the population wants to safeguard the forest, to work hand in hand with the reserve personnel to take care of the forest. But such abuses break down trust and need to be curtailed. In his mind, the guards need more discipline. "At least then, let all of us, we the people and they [the guards], let us take care of this forest because it holds advantages and benefits for us. Quit going in and threatening the people. How then will the people have any peace again in their hearts? They won't be able to have peace."

Later that night, the words I am hearing from people like Maurice and Mapoli, Heri and Debo, continue to burn like a deep orange ember within me. Beneath the pleasing aspects of this life, which I at times tend to focus on, there lies a hidden and subtle violence—children with reddish-tinted hair and swollen bellies looking at me with listless eyes; young men wasting hard-earned money and shortening their lives on caustic, home-brewed corn whiskey; no schools for children who at fifteen years of age still can't read; the day-after-day, year-after-year routine between garden and stomach, forest and field. That's the dark side of the dialectic. Its counter is found in the still-abundant hospitality and generosity of these people; the importance of being, of presence, of talk, sharing, and helping. It is these I experience as I join Matiyé, Ndjaosiko and all their children—Joli, Venance, Idey, and little Anakese—gathered close around the fire. This is a time of being together, a time for telling stories, for giving advice, or simply a time to sit. I sit with them, quietly looking into the flames thinking about all that I have heard today. It is not long before my eyes grow heavy and I excuse myself from this circle of family. Returning their wishes of *lala salama* (sleep in peace), I enter the room that earlier they had vacated so as to give me a place to lay my head.

Nkomo: "The Only Solution Is That There Be Some Understanding and Agreement"

Early in the morning while it is still dark I am awakened by a choir of troubadours passing through the village calling one and all to attend the day's "big Sunday." Later, over the course of the marathon twelve-hour service, nearly all of Badengaido makes its way at some point to the huge palm-branch structure that has been set up outside the Protestant church to shelter all those who have come. I stay for as long as I can hold out, maybe an hour or two. Choir after choir gets up to sing, each one unique in composition and style. One of the most impressive is the Choral Mutchatcha, made up of young gold prospectors who have come in from the gold mines for the special event. They are dressed to kill and sing a home-penned song entitled "*Nkombo Ezali Pamba* (One's Name is Worthless)"—a fitting number for these young adventure-seekers who are often frowned upon by many of the older farming generation. At every special number, three or four of them hold up gargantuan boom boxes to record the singing as well as to display the wealth they have nonetheless been able to earn from what I have heard older farmers describe as "worthless labor."

I am accompanied to the service by Nkomo, a Lokele fisherman and elder who has spent a long time living in this community. Like most Lokele, Nkomo has been a fisherman ever since his father began teaching him to throw the nets soon after he began walking. Twenty-four years ago he left his home on the Congo River and came to Basiri to fish the Ituri. Kandoko, the Mbo chief at the time, gave him a fishing concession and he began a small community of Lokele fisherfolk that in its prime numbered as many as seventy people. Realizing that living away from the main road made it difficult for his children to attend school, Nkomo asked Kandoko for some land. The chief told him to settle in Badengaido, which he did in 1982.

Nkomo continued to fish the Ituri while living in Badengaido but also opened a small store in the village. One day while checking his fishing lines, he nearly drowned. Swept beneath the Ituri's strong currents, he struggled to free himself from the huge fish hooks in which he'd become embroiled. Eventually he was rescued but for some time remained quite ill. As a healer, Chief Kandoko helped him to recover but told him he could no longer fish and instead should stick to farming and running his store. Nkomo heeded the chief, and today it has been nine years since he has gone to the forest to fish. But then, a few years after his fishing accident, his store along with everything in it burned to the ground. Following this mishap he decided to just live by farming, "in peace."

Nkomo's story reminds me of a particular genre of popular painting one can find hanging in homes, bars, hotels, and restaurants throughout Congo. Though many variations exist, commonly the paintings depict a man up in a tree seeking refuge from the leopard that has been chasing him. A rather vicious-looking snake coils its way up the trunk while the branch overhanging the river on which the man has been standing is beginning to break. Face aghast, the man clambers for support to prevent him from falling into the open jaws of the crocodile in the river below. Although we are left to imagine what life's caprice brings to the man in the picture, the hardships and dangers Nkomo has weathered have not succeeded in drowning his warmth and hospitality. As it was six years before, his house remains filled with *wageni* (guests) I notice, as we make plans following the service to meet together the next day.

As a fisherman, Nkomo's views about exploiting the forest's rivers echo those of the farmers and hunters whose harvests come from the forest's soils and fauna. Just as many feel the forests and animals can never end, so does Nkomo regard the fish in Congo's rivers. Seated together in his tiny living room the next day, the elder tells me, "Among the Lokele, to say that the fish will not end is not just a laughing matter but a thing of awe. Not without reason since within one fish, even a small fish, once it becomes pregnant, that fish will hatch so-o-o many eggs! All those small eggs are fish. So, inside of her how many eggs do you think she has? In one year how many fish will be born? Millions and millions.... All those fish, they in turn bear offspring more than ever; each one of them bears abundantly. So in the old times they weren't afraid because they knew that fish were plentiful."

However, Nkomo also acknowledges the changes in fishing that have been wrought by the introduction of Western technologies. "But look around today," the elder continues. "The number of fish has gone down a bit. Why? During the time of the ancestors, they didn't have the know-how of using lots of nets. These big nets of the white men, they weren't around. They only had one kind of net ... [made from] the same chord the Ndaka have been using to kill animals [chord made from the *kusa* vine].... But in the water, the nets do not last long. However, ever since nylon string came into Zaire [Congo], when they opened up the nylon factory, from then on you saw that many people started having nets and all kinds of fish started dying."

"So what would you say is the main cause of this current decrease in the number of fish?"

"The main reason for it is that there are more nets all the time, of every kind. During the time of the ancestors, there weren't very many nets.... Sometimes a net would kill one fish, two fish, with luck maybe three. You would not kill many and the fish would remain nevertheless. The fish would go on bearing lots and lots of offspring."

"And did they sell fish for money and then use the money to buy food, or did they exchange fish for food directly with those who had gardens?"

"In the old days, among our true ancestors, when the Lokele would come in contact with other ethnic groups, they engaged in exchange: I have something like this that you want; you will give me a few fish and you take the thing and I take the fish. But beginning with the entry of the white man, they started saying that people should not walk around naked, for in that time people wore the clothes of the ancestors. For instance if my wife had a cloth, she might tear off a piece for me and I would wear [it]. But then when that was seen, they said, 'No, we have become civilized, such clothing is no good. A man wears pants or shorts and a shirt.' So then, things started being sold for money. The desire for money stemmed from the time the white man came. The white men are the ones who came with money."

Bringing up another item introduced by the white man, I ask, "Have you heard that instead of using the forest poison of the ancestors, some Lokele are now using the poison of the white man, the one called Aldrin?"

"Aldrin? The real strong stuff?... Fish [caught] like that are not good fish.... Once [Aldrin] enters the body of the fish, it is no longer any good. If you eat that fish, you had better cook it with lots of palm oil, because if you eat it—kill it just like that and go and cook it without oil—that poison will enter your own body and stay there."

"But do they use it?"

"Yes, it happens. They do it. We hear that ... near the plantations ... many use it in the smaller rivers and streams.... But such a thing is not practiced in large rivers.

It wouldn't work in the Ituri. Even if they poured a bunch in, it would work only at first but once the current took it, it wouldn't do much."

"How do you view using such products?"

"It is bad."

"Why?"

Nkomo laughs at the naïveté of my question before he responds. "I see it as wrong because—if you have your garden, would you want elephants to come in and in one day polish it off? In the same way—fish, rivers, streams—for us Lokele, these are our gardens. Even though I am here like this, if I go to the river today, tomorrow I will eat fish. We will eat fish here in this house. What is it? It is a garden. Well, then, if you go and use it all up in one day, is that good? No, that poison is bad because it's as if those who use it are saying, 'Even if [fish] belong to the ancestors, and even if it means that generations of children will not eat them, we want to finish them off in one day.' That is wrong."

"It is wrong because—"

"They want to finish it all in one day."

"But if they kill all the fish in one day, why is that bad?"

"My goodness! You rejoice to be able to eat each day, you alone. Well then, the ones who are born later who will do their work after you, what will they have to live on? Okay, when they killed Jesus, God became angry and wanted to end the world in one day. What did Jesus say to God? Christ said, 'Father, let them be. They know not what they do. Forgive them.' Because he saw and said, 'They have already killed me, but if they are wiped from the earth immediately, how then will I find a way, how will I be able to rescue these people that they may live?' God gave fish so that people can go on living from it forever. If you finish it off all in one time, do you think that God will come back and work to put more in? You have done wrong."

The elder clicks his tongue as if my questions both puzzle and perturb him. "Ah, young child, you have come with such big questions, so difficult like this?" Then to my relief he breaks into a big grin and laughs. I laugh with him.

The chatter of conversation between Nkomo's family and his ever-present guests filters in through the open window above our heads. Somewhere far away the boom of a drum makes me wonder if yesterday's "big Sunday" has pushed over into Monday. Taking a sip of the sweet coffee Nkomo's wife had brought in earlier, I realize that although we have been talking for some time the issue of the RFO has not yet come up. I decide to raise it with this elder who, although having lived here a long time, is not indigenous to the area. He may have a different perspective on the reserve than do people like Heri and Debo.

"I wanted to ask a bit now, to hear your point of view—I have heard that the people of Epulu have now created a sort of reserve here to protect the forest and the waters and the fish. How do you see this, the way they have set up these laws? Is it good, or is it different from the ways that you Lokele have been taking care of your waters? How do you see it vis-à-vis your methods that date back a long time?"

Nkomo measures his response carefully. It appears that he has given the issue a lot of thought. "In comparison to our traditional ways, I have to say it is wrong in one way. Take the Ituri River: According to me, a Lokele, if they told me I could no longer fish the river, I would have troubles [literally, "I would see pain"]. It is the same way with these people of the forest here. They are habituated to their work with animals because it is the custom with which they were born. If you forbid them—for each and everybody has their work that they are used to doing; this is what they are used to, it is their custom—they have good reason to see—they are seeing—pain, and many of them are saying that things are not good with them. It is [not] good for them to close off their forest and say it is a reserve, a park.... They have already suffered a lot. I see that they are still suffering a lot.

"They have not yet really come to a final solution with which [local people] are happy. They asked [the reserve personnel] to build them schools, or to build them dispensaries, or to build them houses with tin roofs. For how many years have [the projects in Epulu] been coming and taking okapi, sending them out, sending them away? And [local people] do not have even one souvenir from this interest in their forest. And now to also prohibit them from hunting in their forest, they very well cannot be happy. That is the thing that I see is wrong."

"You also see it as wrong?"

"I see it as wrong because if they get—Eh! As I told you, today I don't go to the river [anymore]. If [local people] come to sell me some game, well then I will buy it and eat. If they don't get anything, what will I eat? This is their thing that they are accustomed to."

"So are the means and methods of the people in Epulu quite different from your ways or those of the people who have been here for a long time, their methods of taking care of their forests and their animals? Do the two differ a lot?"

"For these people, the way they guard and take care of their land and their forests?" Nkomo asks seeking clarification.

"Yes, so that they will not break down."

"The people of Epulu have already been hard on them because they are binding them forcefully. If they eased up a bit on them and said, 'Okay, what is forbidden is to kill this and this and this. Since according to your customs [you kill] that which you want for eating, if you want to do that, then do it.' Then they wouldn't have a problem. But the things they've said they don't want to see—even if you come out with one *mboloko* and they meet you, that is one big infraction. But such are things to which these people have been born. They are used to such ancient practices, just as I told you regarding myself and the river. If I go, I am able to get one or two fish. If I can no longer go to the river, I will not be happy."

"So if someone were to ask you for your ideas as to a solution—you had said that they, the people of Epulu, have not yet really reached a good solution with the local people—if a person comes and asks you, what would you say would be a solution?"

Nkomo breaks into a laugh and then continues. "For a real solution—Ah! All the things that the people of Epulu are doing are simply the work of raising up our Zaire [Congo]. Eh? So then, the bad thing is that here in Zaire [Congo], our leaders do not know the way of work. Things like those animals, the okapi … when they are exported, you know that they carry one huge credit. A huge credit comes from them. Okay, what work could that money do? If it comes here to Zaire [Congo], it is right for it to do work that gladdens people's hearts.[15] So then, you will not make my heart happy. Will I be happy if you leave me ... what ... empty-handed?

"The only solution is that there be some understanding and agreement. They should try hard to come to an agreement with people in the villages. Because a person is only a person. Eh? If they were to fool them, to provide them with dispensaries, schools, or to build people like Chiefs Basendua and Yakobo big houses of brick and whatever—is it not true that one receives a good heart because of things to laugh at and things that make one happy?

"If they only know how to fool themselves, a solution will not be found. Because the leaders, if they see and say, 'Eh! These people have already given, have already done for us what I [couldn't do].' Like me here, I don't have bricks. If you were to build me a brick house, put in your hand and help me build a brick house here, I will not forget you. 'Ah! Were it not for my child, I would not have this nice house like this.... And then if a person wants to harm you, I will say, 'Ah! Why are you tormenting that child? Leave him be to do his work.'

"That then is the solution; if they had the wisdom to first please the chiefs, the *localités* [village-level political administrative division] and the people of the village. They need to give them some real things, things they don't have the strength to get. It will amaze them. You will see that those people will leave the forest free."

I am a little surprised that Nkomo suggests a solution based on "fooling" local people. But then I remember that he himself is not indigenous to the area and came here with his own plans for "development" among a people he considers "backward" and "resistant to change." His words remind me that pejorative views of the Ituri's forest peoples can often be found locally. Nevertheless, I find Nkomo quite empathetic to the plight of the indigenous forest dwellers among whom he has settled. This comes out clearly when I raise the idea of people moving out of the RFO on their own accord.

"Others," I recount, "told me that if ... the people of Epulu ... put a limit on gardening—telling people to go this far but not beyond it—and the way in which they are placing fees on fishing, and the way in which they are prohibiting people from having meat, some have said, 'Well if this is the way it is, I will take my leave.' They say they will leave here and go someplace else. Have you also heard this, that people will move, move from here and go someplace else?"

The elder is quick to respond. "There is no way for them to move." With his finger Nkomo draws a map on the table's mahogany surface. "It is like this: The forest is here, and here is the road. The forest comes like this, the forest extends here, the forest extends there. They realize that the forest extends all the way to the Ituri [River], but how many kilometers will they leave for the people of the village in order for them to cut their gardens and set their traps? They have set the boundary so that the park begins right here behind the houses and goes all the way to the river there. What then will you do? If they want to come to an agreement, they should leave them a little reserve of 2 or 3 kilometers and then tell them they can walk [i.e., hunt or farm] within that. But they don't want to do such a thing. [Instead they] see you have an animal in your hand that you have killed. Even if you tell them, 'It is for me to eat,' they will still arrest you. This is what is giving [local people] bad hearts. They should leave them a little bit freer to garden, to set their traps, whatever.

"Sometimes they will find me with the meat of a *mungele* [Weyn's duiker], the meat we had with our manioc greens yesterday. If they catch me with it in my hands, they will assume I was the one who killed it. And following that—fines that exceed what I paid for that animal! That kind of thing will not give a person a good heart. You have done wrong. And so now they are angry. They need to leave them free—a little bit.

"This is not to say there should be no restrictions. For instance, their animal, the okapi—if you find somebody with it, you can very well arrest them for an infraction of another sort. But others, smaller animals, the ones given to them by their ancestors, the ones they live from, to prohibit these and say that killing them must no longer happen—they will not agree.

"The only thing I am seeing is that it is good and necessary that they come to agree, they and the leaders and those below, those of the village. If the reserve personnel are really smart, they should trick them, eh? They should trick them by giving them something; by doing for them some work that gives the impression that ... because you know that an okapi, eh, if they give you the authorization to capture and export okapi for one entire year, even if you capture ten, or even five or six, then how much capital will you gain? A lot! So then, if you give me something, if you do some work for me that amounts to half the value of an okapi—say you build me a house, or buy me a bicycle or a Yamaha or a small pick-up truck—well then, I will be happy and say, 'You have done me well.' I will be cornered into saying that you have done me well. Surprise! You are benefiting more than I. I have already been co-opted into saying that what you have done for me is already enough for me. Is it not like that?"

"It is true."

"Yes. So that's why I said they should trick them with some little thing to soothe their hearts. Then you will see that they will leave them free [to have their reserve]. But this continuous hounding them with words, words, words! If they deprive them so severely, if it aggravates so as to really become a problem, truly, once they take on a bad heart according to their custom, they will cause some trouble. It will be a reserve on paper and be known as a reserve but as for getting animals, they won't get any."[16]

* * *

There are good reasons for giving these four stories from Badengaido much of the space of this chapter. These narratives illustrate well the range of concerns held by some Ituri villagers in regard to the conservation initiatives impacting their lives. However, I want to present a few additional perspectives that are important. They come from some of the other people I encountered as I moved on from Badengaido, back through Epulu and the nearby village of Bapukele, to Tobola, a village situated outside the reserve where I had also conducted research in 1989. Although my research in both the Ubangi and the Ituri focuses on villagers, I also want to discuss some of the perspectives of the Mbuti, the foraging people of the Ituri. Part of the reason I do not focus on the Mbuti is because so many other people have.[17] I also want to avoid the false impression that they are somehow paragons of ecological virtue that undue focus on them might represent in a work of this nature. As mentioned earlier in this chapter, it is neither fair nor accurate to consider the Mbuti, any more than any other people, as angelic environmental stewards.

But these reasons certainly do not justify leaving Mbuti voices out of this study. Thus, following are the narratives of several Mbuti who live in Epulu. As with the villager voices I have presented, I make no claim that these narratives are in any way representative of all the Mbuti people of the Ituri. Furthermore, the Mbuti I spoke with had all worked for the conservation projects at Epulu at some time. To what degree this colored their views on the RFO is hard to tell, but it likely has had some effect on the way they perceive the conservation initiatives affecting their lives.

After presenting and discussing these Mbuti narratives, I offer a few additional stories from villagers in Bapukele before summarizing the different perspectives on conservation to be gained from listening to local people. Finally, I conclude the chapter with one last story that, more than any other, poignantly reveals how dangerous the situation in the Ituri might become unless things change soon.

Mbuti Perspectives
on the Forest and the RFO

As has happened to me so many times before, having Masalito, one of the local Epulu Mbuti, as a guide and walking companion proves to be a wonderful education on the ways of the forest. A few days after returning to Epulu from Badengaido, I head out with him into the forest. Leaving Kitoko about 9:30 A.M., we travel across the Népusé and Nembongo streams through stunning stands of *mbau* (*Gilbertiodendron dewevrei*) trees. We are going out to check the termite-baited fish traps Masalito has set a ways downstream in the Epulu River. Lingering low clouds darken the forest interior and lend a smattering of rain, but there is still plenty of light to view the troops of monkeys we encounter only several hundred meters into our trip. At one point, the canopy is literally exploding with noise and movement, perhaps as many as 100 monkeys leaping and calling out in a variety of chirps, grunts, and barks. Several troops of different species—Mona (*Cercopithecus mona wolfidenti*), red colobus (*Colobus badius*), agile mangabeys (*Cercocebus galeritus agilis*), gray-cheeked mangabeys (*Cercocebus albigena*), and red-tailed guenons (*Cercopithecus ascanius schmidti*)—have converged, perhaps drawn by the ripe figs overhead. While some fly through the air from one branch to the next, others prefer to sit still and stare back at the strange, binoculared eyes looking up at them.

We continue down the trail, and I listen as Masalito tells me about all the things the forest gives: the food and light that come from the *Canarium* tree's olivelike fruit and its slow-burning sap; *pusiya* (*Treculia africana*), the seeds of which can be roasted like peanuts; *etaba* (muliple species of genus *Dioscorea*), the wild yams that grow deep beneath the soil; *asali* and *apiso*, the two types of wild honey the forest provides—the list goes on. After another fifteen minutes of walking we come alongside the Epulu. The river is already quite high from the rains but not nearly as high as it will get come the daily downpours of November and early December.

Masalito carefully lowers himself into the fast-flowing current and pulls up his first trap to find a collection of fish bunched together at the bottom. He takes them out one by one and puts them in the woven rattan shoulder bag he carries—all except the little ones, which he tosses back in the Epulu "in order that they may *komea* [become fully grown]," he tells me. After harvesting and rebaiting all the traps, we press on to the Amabanzoki stream, where we startle a crowned eagle (*Stephanoaetus coronatus*) holding stealth on a troop of Monas. The monkeys holler in alarm as the giant bird disappears through the folds of the forest's green tapestry.

We decide to head back. On the way we find a small clearing in the *mbau* forest on the Epulu's right bank. The river is wide here. Several small islands split its flow, anchored against the current like huge riverboats, abandoned and overgrown with trees. It is a perfect spot to share the coffee and bread I have brought and to talk for a bit about the forest, the reserve, and the ways in which Masalito's forebears practiced their own sort of conservation.

Masalito: "The Forest Is Our Garden, It Is Our Life"

Handing Masalito some coffee, I ask him about the ways in which the Mbuti protect the forest. "What do you do or what do you forbid being done to keep the forest from becoming quickly degraded?"

"Today," Masalito replies, "the Mbuti forbid [certain things regarding] the forest in order that in these days the forest not be ruined. We do not like to garden in the forest; we do not like to cut trees in the forest; we do not like to disturb it by having too many people enter it in a disorderly fashion."

"Why?" I ask.

"Because ... where [then] will the Mbuti live? Our strength is in the forest, we the Mbuti."

"A Mbuti cannot live in the village?"

"A Mbuti cannot live in the village. The life of a Mbuti is in the forest."

"Since a long time ago?"

"A long time ago; from the time the ancestors began."

In the background I can hear the familiar sound of cicadas screeching in the early afternoon heat. From early childhood, the sound has evoked in me a certain languor, a stillness both soothing and oppressive. Perhaps this is because I heard it most clearly during the siesta hour when things were always quiet except for the cicadas. Shaking off my reverie, I return to our conversation and ask Masalito about Mbuti hunting restrictions. "When it comes to the matter of the Mbuti and animals, how many animals are you accustomed to kill in one day or in one hunt? Are there any limits?"

"Limits exist. The limits of hunting, of killing animals are simply the limits [dictated by things] like the nets. We cast our nets, and then close them. *Ama fumafuma, ama babilo* [KiMbuti words Masalito leaves untranslated but that refer to specific sequences of casting the nets]. If we kill maybe two [animals] there, we then try to exercise our skill again. We leave there and cast a second time. If we don't get anything we decide to leave that place. So then, limits pertain to such castings that we call in KiMbuti *amakpola. Amakpola lalakuya.*"

"What does that mean?"

"It means that we will only cast in this forest here; we will not pass by over there. You cast here, the end of the research is here. *Amakpola.*"

Masalito loves to spice up his talk with choice words that echo the delight and play on his face. Often he will borrow a word from French or lapse into long phrases of KiMbuti when the Swahili just doesn't muster an equivalent meaning. Like now, he often follows his speech with laughter. Talking and listening to him, one is infused with good feeling.

"And why do you place limits there?"

"We place limits there in order not to disturb that forest. Let us not disturb it for now. It should just remain there. Later we will know more."

"But why would you let some forest be?"

"The reason why we leave some forests be is so that we will be able to work in them another day. Let us first work nearby here. That forest over there, let it just remain there for now. Because for us Mbuti, it is our garden."

Like so many of the other Central African forest dwellers whose voices I've already shared, Masalito articulates the strong sense of restraint governing people's traditional use of the forest. In contrast, I raise the example of the soldier in the Ubangi who killed sixty monkeys in one day. Masalito is amazed. "Oh la la! To kill sixty monkeys like that, the Mbuti in their hearts would see that such is killing the entire world. We Mbuti do not kill lots of animals like that, like the villager who killed so many monkeys like that. We kill according to dates, according to a little luck, what God gives us. One should not wipe it all out in one single instance. You take a little bit, the rest should remain there. Let it also live awhile in the world. Later then, you also have a little something to eat."

"In the same way that you returned those fish?"

"Yes, similar to that."

"You do the same thing with animals?"

"Yes, that's the way it is."

I then ask if there are certain animals the Mbuti forbid people to kill. Masalito replies, "Those animals that according to their traditions the Mbuti forbid people to kill—do not kill okapi; do not kill elephant; do not kill buffalo; do not kill other animals like *sibili* [*Atherurus africanus*, African brush-tailed porcupine], and *boro* [*Colobus guereza*, black and white colobus]—do not kill them."

"Why?"

"Those were the arrangements of the State in the past."

"However, prior to the State, before the time of the State, did the Mbuti themselves forbid people from killing certain animals?"

"No, it was a restriction of the State alone, according to the traditions of the State of long ago. For the State is also part of the traditions of the past regarding the ways of living, the ways in which we Mbuti live."

* * *

Masalito's words capture very well the sense in which tradition is never an exclusive or homogenous entity. In the minds of present-day forest dwellers, "tradition" is a mixture of influences from both within and without. What comes from outside the society is absorbed, changed, and reworked to fit in with local meanings. What comes from within gets shifted around and amalgamated with new ideas as people adapt to new forces affecting their lives. "The State is also part of the traditions of the past": Masalito's remark makes it very clear that what I might like to separate out as existing before the colonial State cannot be so easily delineated. It highlights the important fact that for the Mbuti, as for any group of people, tradition is fluid, flexible, and ongoing, not brittle, monolithic, and relegated only to some point in the distant past. Were it not so, the Mbuti as a people and forest-dwelling culture may very well have died out long ago. As mentioned in Chapter 2, traditions must change in order to remain alive. One way they change is to incorporate, bend, and mold to their liking even the policies of the colonial State. Tradition is never "pure."

Nor is it ever interpreted in a singular fashion. One forest "tradition" whose meaning remains especially ambiguous to me is the tradition of "closing the forest" (*kufunga pori*). Various immigrants who have settled in the Ituri have told me that indigenous groups such as the Mbuti and the Mbo have certain ritual practices that make it impossible for people to obtain animals from an area of forest, something they refer to as "closing the forest." It appears that one use of the practice is as a defense against the State's history of intervening in the forests of local people through the capture of okapi and other animals.

One immigrant farmer describes the practice this way: "They [various forest peoples] have what they call in KiSwahili *siri* [secret; hidden thing; mystery]. They have their ancestral secrets. [For example], in response to reserve policies, local people can say that okapi will not be captured, and it is possible such a thing will happen. If they say that in that forest people will not get even one thing, whether it be chimpanzees, or monkeys, or whatever animal they choose, they are able to close it if they do it according to their customs, with one heart like they did during ancestral times.... If they really get angry, they have the power to close the forest."[18]

In contrast, when I had asked the Mbo elders Heri and Debo about the practice, they said the Mbo did not know how to close the forest. When I ask Masalito about whether the Mbuti have a practice of closing the forest he replies that he hasn't really heard of it or understand why the Mbuti would want to do such a thing. "Within the tradition, when they talk about the Mbuti closing the forest, I don't really see the reason or the meaning why the Mbuti would close the forest. Because if the Mbuti close the forest, how then will they eat? Where will they live? Such things I have not seen. In the past as well, the ancestors did not close the forest in the sense of prohibiting it. However, it is true that all such ways have come from the State." It is clear that "closing the forest" holds disputed meanings, linked by some to ancestral traditions and by others to contemporary actions of the State.

"Is the reserve [the RFO] that has been set up here, is it a way of closing the forest? How do you see it?"

"Well, this way of setting up a reserve, that is another thing. That is a way that comes from the side of the State. But it is not a way that comes from the side of the Mbuti, to say that they close the forest."

"The Mbuti are not accustomed to closing the forest?"

"No, the Mbuti are not accustomed to closing the forest."

"It is only the State?"

"It is only the State, or the government. However, the Mbuti, to close off the forest from the Mbuti, or to say they cannot go into the forest to walk and hunt for even three days, that is what it means to close off the forest from the Mbuti."

"And how do you see that? If the State does such a thing?"

"Closes the forest? Well, then, I would not really be able to say if it's bad.... Because if it were a bad thing, we would have heard about it from our ancestors. We the children of these years, in our lives we would have heard about how it is wrong, about how at the time when our ancestors were born, they would say, 'Regarding the forest, its closure is like this or like this.' But without this [knowledge], I am unable to say that it is bad."

Tungana and Cephu: "You Need to Eat While at the Same Time You Protect"

Masalito may be hesitant to offer his perspective on the RFO, but other Mbuti I speak with share their feelings about the reserve more openly. Several days later I talk with Tungana, Masalito's brother-in-law, and Cephu, another Mbuti, about how they perceive the reserve. "In the way that the State is making a reserve here, it is something to help us as well," Tungana replies. "Meaning, the reserve is a way of saying that you need to eat while at the same time you protect, you conserve a part. What that means is that you don't eat—in the same way I said before—in one day you don't eat lots and lots. It is necessary to eat a little, while conserving some of the rest. Because, for example, if today I get ten [animals] and eat all ten of them at one time, then in the morning I will cry from hunger! Therefore, the State has decided it is necessary for us to create a reserve for the purpose of forbidding people to eat a lot all at once. Because the way the Mbuti ate was one at a time, in order to ease their hunger."

Cephu then adds his own perspective. "The reserve, this way of creating a reserve here, they also did a good thing in light of the fact that *bakpala*[19] have started going into the forest in order to kill animals. And they don't differentiate; that is, for them an animal is simply an animal and they kill all of them, way more than the Mbuti do."

"How do they kill them?" I ask.

"They just kill them. Sometimes they might kill by using that wooden-handled metal piece [*mpini*] belonging to you Europeans [Cephu most likely means a gun]. Then they begin killing animals like crazy. It is no longer a matter of choosing, saying, 'This animal, I will not kill.' Off they go, for they no longer know about that. They kill them all wherever, whatever type they want to kill. Thus the reserve has decided to come here because of such things. It did not get created on account of the Mbuti because the Mbuti have not been doing things like that. The Mbuti, they only concern themselves with ... *mungele*.... We hunt with them in the forest, killing maybe one animal, or two, that's it. All the rest remain."

Unlike Masalito, who is somewhat ambiguous in regard to the RFO, Tungana and Cephu view the reserve as a positive means to curb the overhunting of animals, especially by *bakpala* who kill with guns and without differentiation. Even so, Masalito implies at one point that the reserve has actually encouraged more *bakpala* to enter the forest. "Regarding what the reserve is doing," Masalito explains, "I can see that there is a little something wrong. Because *bakpala* are again entering in, coming into the forest to disturb me, a Mbuti, in little ways.[20] And as for continuing to walk there, I will not walk there any longer. Because a *mukpala* [singular] does not mix with a Mbuti when it comes to walking in the forest in the same place."

Masalito pauses, reluctant to say anymore. "Thus I will stop there for now since—" He clicks his tongue against his teeth in expression of the delicate nature of the issue. Then he concludes, "A Mbuti does not walk in the same place with a *mukpala*."

Earlier Masalito had mentioned that one of the ways in which the Mbuti took care of their forest was to make sure that "a *mukpala* not enter the forest to the place where the Mbuti are staying. Why? Today *bakpala* are trying to find out the places where the Mbuti are living; but this, according to our ancestors, is very bad. Why? Our traditions of the ancestors—before they did not remain with *bakpala* together in one place. *Bakpala* would stay in the village, in the sun; their residence was in the village. Thus a *mukpala* is not someone to know the forest. The forest belongs to the Mbuti according to the manner in which our ancestors dwelled; according to the Mbuti way of life. A Mbuti and a *mukpala* should not dwell in one place. Therefore, we Mbuti are keeping those laws even now. A Mbuti does not live with a person of the village in the same place. A Mbuti remains in the forest; a *mukpala* is one who stays in the village. *Bakpala* should not bring their life to the place where the Mbuti are living."

And at the end of our talk, Masalito reiterates that the forest is being ruined due to villagers bringing their lives to the places where the Mbuti are living. "The thing that ruins the forest is the *mukpala*. Because a *mukpala* will go and want to bring pain to the Mbuti right in their own forest. They say that they will garden in the same place where the Mbuti are living well, the very place where the Mbuti sleep not with sheets … [but] with *kasadui* [*Marantaceae*?] leaves."

* * *

I am curious why Masalito singles out *bakpala*, the traditional exchange partners of the Mbuti, as the ones who are ruining the forest: What might explain why he is so adamantly opposed to sharing the same area of forest with them? Indeed, historical research has shown that complex and to some degree mutually beneficial interactions between the Mbuti and various groups of farming people (*bakpala*) have been taking place in these forests for more than 1,000 years (Vansina 1984; 1986, 432; Wilkie 1988).

Certainly one still hears the view that the Mbuti and their farming exchange partners relate primarily along antagonistic lines. This notion was especially reified in the writing and speech of anthropologist Colin Turnbull, who, despite his long-term residence in the Ituri, continued to paint *bakpala* in somewhat pejorative terms, as people afraid of the forest and as bumbling neophytes when it comes to forest skills (Turnbull 1961; 1965). However, my own research, as well as that of others, has indicated that the Ituri's farming peoples are in no way strangers to the forest who possess little forest acumen. Like the Mbuti, they are very much forest peoples who have adapted quite well to the forest environment and hold much local ecological knowledge.

Neither does research confirm that the relationship between *bakpala* and the Mbuti has been primarily antagonistic. Although like most patron-client relationships this one is not egalitarian, both groups acknowledge dependence on the other for survival. In the Ituri especially, the relationship has maintained a greater degree of symbiosis than in other places in Central Africa where external influences such as plantation agriculture, mining, and logging have contributed to a greater commercialization of the exchange system (Peterson 1991, 11). Thus it seems a bit surprising to me that Masalito expresses so strongly the separation that needs to be maintained between the Mbuti and *bakpala*.

Perhaps Masalito is glossing the term *bakpala* to include some of the Ituri's newest immigrants who come from areas outside the forest. Earlier during our walk, Masalito was telling me how some newly arrived immigrant farmers had begun clearing forest directly behind his camp located near Kitoko. Excitedly, he recounted how he and others had quickly fed the word to the directors of one of the conservation projects at Epulu. The immigrants' attempts to turn the forest into field were soon thwarted. Research has shown that these newcomers do farm both more intensively and extensively than do the Ituri's indigenous farmers (Peterson 1991). Therefore, they very well could be a cause for the concern Masalito expresses so emphatically.

Other Mbuti clearly use the term *bakpala* to refer to newcomers, especially those immigrants who have come to pan for gold in the Ituri's rivers and streams. For instance, when I ask Tungana who he considers to be the people ruining the forest, he replies, "Now, the ways of ruining the forest today, [it is] especially [because] the *bakpala* are now walking in the forest. It is no longer mostly the Mbuti. They are walking in the forest looking for their gold. They make a big village in the forest. They cut the trees, they do whatever, and it begins to cause the trees of the forest to dry up. Or perhaps, they have their mine there and it ruins that piece of the forest because they will dirty the forest in that area entirely."

Or perhaps Masalito's sentiments about *bakpala* are a reflection of the fact that he has been an employee of the WCS projects at Epulu (including CEFRECOF) over the course of many years. As a CEFRECOF employee, he may well be obligated to abide by certain terms of work such as those I saw posted on the sentinels' *baraza* at *Kitoko* one day (see Box 5.1). I translate the document in full since in addition to illustrating rather pejorative views of *bakpala* it also reveals the many ways in which the conservation initiatives at Epulu penetrate deeply into the livelihood choices of local people without taking the history of human-land interactions in the Ituri much into account.

Of significance is the way in which the notice indiscriminately links farming to destruction of the environment when in fact farming has been carried out for centuries in these forests, which remain some of the most intact in all of Africa. Also interesting is the way in which the notice limits farming to secondary forests even though long-term experience has taught these forest farmers that plantains, one of their main staples, require primary forest soil in order to grow well.

The notice also makes it clear that project workers are not to follow the Mbuti into the forest but should wait to obtain meat in the village. Some have mentioned to me that this restriction applies not only to project workers but to villagers in general. This is the explanation that comes up in further conversation with Tungana and Cephu over whether the reserve is causing them any suffering. Interestingly, unlike Masalito, the problems Tungana has with the reserve are not over *bakpala* infringing on the Mbuti way of life. Rather they stem from the fact that some of the Mbuti have started to take up farming like *bakpala*. When I ask him, "Therefore, you Mbuti do not see the reserve as bringing you suffering?" he replies, "Well, the reserve causes us suffering now in what way? We also are adopting the idea of farming, and the reserve doesn't like farming. That has become a little problem within it. Okay, the reserve now forbids a person to just travel into the forest at random. For us Mbuti, the forest, indeed the biggest part of our life is in the forest. Thus there is the difficulty concerning farming."

"What does it mean that you are now forbidden to go into the forest at random? What is required now?" I ask him.

"There the State is following its routine. On our side, for us Mbuti, it has not forbid us. It mostly forbids people of the village from walking in the forest. Because what they do, they go into the forest and ruin it recklessly. They do not know the ways of walking in the forest. But for us Mbuti, even though there is a reserve here, it concerns us only in regard to ways of walking. Meaning, we walk about according to the ways of the forest. We know how to tell the things that are bad and the things that are good. We cannot do such and such here, and here we must do like this.

BOX 5.1 Notice to Employees of CEFRECOF

CENTRE DE FORMATION ET DE RECHERCHE
en Conservation Forestière (CEFRECOF)
Réserve de Faune à Okapis
Epulu, Zaire

Terms of Work
In order to advance its work, CEFRECOF is posting the following terms regarding the protection of the forests of the RFO. The employees of CEFRECOF, as well as those of the IZCN and the other projects, are asked to exemplify respect for the fundamental laws the RFO has put forth to protect the Ituri Forest.

[Four] terms have been decided on by the directors of the project:

1. To refrain from cutting the forest.
2. To refrain from marketing the meat of wild game.
3. To forbid newcomers to build and live among us.
4. To refrain from marketing gold.

1. How can an employee respect the law regarding the cutting of the forest?
To farm requires the cutting of trees and to cut trees is to destroy the forest and to destroy the forest is to destroy the environment. For example, one destroys a honey-producing tree by burning it to make charcoal or by burning a garden. This year we want each and every employee to only farm in secondary forest, not in primary forest. And crops should only include corn, leafy vegetables (manioc and other types of greens), and peanuts grown alongside the village.

Agriculture for the purpose of trade, buying, and selling will no longer be permitted by CEFRECOF. Corn must not be grown for the purpose of making whiskey. Each and every employee of the project is strictly forbidden to brew or engage in the retailing of whiskey or to house someone whose occupation is selling whiskey.

2. How shall an employee refrain from marketing wild game?
The project does not prohibit the eating of forest animals but it wants employees to respect the fundamental rules of protection. To keep oneself from marketing wild meat or to refuse to lend a hand in the business of marketing wild meat is obligatory. The wife, child, or family member of an employee should not follow the Mbuti into the forest in order to trade liquor, plantains, and so forth for meat. People must stay in the village in order to be able to buy meat and only meat that is for eating.

> **3. How shall an employee restrain newcomers from settling and living in the reserve?**
> Employees of CEFRECOF will refrain from calling their entire family to build and to live with them here (especially in Epulu). Because they too then call their own families, clear trees for building and farming, and engage in meat marketing in order to satisfy all their needs (food, clothing, medicine …). Thus such newcomers add to the destruction of the environment. Employees of the project are asked to live with their wives and children; they are also asked to remove all those who don't have work from their homes.
>
> **4. To refrain from buying and selling gold**
> Buying and selling gold or prospecting for gold by one's wife, children, or other family members really causes the work of CEFRECOF to regress, especially the work of protecting the environment. By engaging in such practices, one is lending a hand to those who are destroying the forest and instilling a huge loss to both animals and people.

"So, even though they have put a reserve here, the warden has authorized us to walk in the forest without needing to go to ask him permission. However, he does lay down some rules. According to the regulations of the reserve, what is forbidden is for us to degrade the forest. Thus, now, he has given us the liberty to walk in the forest. What is wrong is to degrade or ruin it."

"Given liberty to the Mbuti or to everyone?" I ask.

"Only to the Mbuti, especially to the Mbuti. The people of the village, those who live here in the village can follow me into the forest for one day only. If they say, 'Ah, we're hungry again; we will follow our brothers,' what does it require? That they go before the head of the village or the warden and tell him, 'We are feeling hungry. We are going to follow after our brothers here.' And he grants them leave. But us Mbuti now, we walk about according to our own ways. But if villagers want to follow us to the place we go, they need permission. If they don't ask permission, the warden can arrest them and lock them up."

"In other words, a villager cannot follow the Mbuti into the forest without permission?" I ask seeking clarification.

"Yes, that's wrong now."

* * *

Such a policy elicits quite a different response on the part of the forest farmers I speak with. Several days after talking with Masalito, Tungana, and Cephu, I travel over to the nearby village of Bapukele, 6 kilometers east of Epulu. I speak with Mama Amanzala, a Bapukele native and one of the village's many female farmers. She eloquently relates how the policy of preventing *bakpala* from following the

Mbuti into the forest flies in the face of their traditions. According to Mama Amanzala, the Mbuti and the Bira have always been one, living together in the same place. Pascal, a Budu farmer who has joined our conversation, puts it this way: "Long ago here—take first the example of the Bira. They labeled this a 'Pygmy Zone' for what reason? Because they knew that the Bira and their Mbuti were what kind of people? One. You see? Ever since long ago. An animal killed by a Mbuti helps a Bira. An animal a Bira kills helps a Mbuti.... Because the Bira have been supported by the Mbuti and the Mbuti have been supported by the Bira. That's the way life was. Why then are they trying to split them, saying only one person should work when they have always been working together as one? The Bira and their Mbuti have worked in the same space. If they are going to split them, then both of them should quit that line of work. That's the way it is!"

When I mention Turnbull's views of villagers, Mama Amanzala tells me flatly that he is wrong, that *bakpala* did and still do go into the forest and that it is a lie that villagers are afraid of the forest. According to her, exchanges between the Mbuti and farmers did not just take place in the village but also in the forest. When I ask why *bakpala* didn't just wait in the village for the Mbuti since carrying cultivated foods into the forest and carrying out meat is a lot of work, she replies, "We rejoice to go into the forest, for us and our Mbuti to see each other." At other times she says, the Mbuti would invite their *bakpala* into the forest, especially during honey season, a time of gaiety and sweetness.[21]

According to Mama Amanzala, Pascal, and many others I talk to, the view that the Mbuti and their *bakpala* occupy separate worlds that should not mix is off base. Therefore, to try to separate them as they interpret the reserve is doing will only serve to enhance conflict. Similar views are shared by many contemporary anthropologists who insist that Central African foragers and indigenous forest farmers must be considered as parts of one whole, together making up an interdependent system of food procurement, ritual meaning, and way of life (Bailey, Bahuchet, and Hewlett 1990, 6).

Mama Amanzala astutely makes the point that it was only after the Belgians built the road through the forest that this matter of "following ones Mbuti into the forest" had any significance. Prior to the road, foragers and farmers all lived in the forest; according to Mama Amanzala, often but not always, they lived next to each other, the *mukpala*'s house surrounded by Mbuti huts, similar to the settlement patterns found today in villagers' gardens. When the Belgians forcibly moved the Bira and many other forest farming peoples to the road, they moved them *out* of the forest, so to speak, out of where they had been living in relative proximity to the Mbuti with whom they had been interacting for generations. Only then did the idea of *bakpala* following the Mbuti into the forest acquire any meaning.

Thus long ago, the road and the settlement patterns it initiated lent a certain artificiality to the Mbuti-villager exchange system. Spatially, they rearranged the Mbuti-villager exchange quite differently from the way it had been carried out for centuries. Therefore, when RFO personnel now prohibit local villagers from going into the forest after the Mbuti, they are not only creating ill will among many but also basing policy on faulty historical and anthropological foundations.

* * *

I cannot hope to know all the reasons why Masalito, Tungana, and Cephu feel the way they do about *bakpala*. What Masalito does make clear to me is that the Mbuti are not the ones to blame for forest degradation. To my question of how the Mbuti might be damaging the forest, he replies, "For a Mbuti to damage the forest? For a Mbuti to really … in truth, the Mbuti will not damage the forest. The Mbuti will not ruin the forest because if the Mbuti ruin the forest, their eating place, it will be bad. It is their garden, it is their life.… They don't know how to cause damage. They don't know how to ruin it. Because if they destroy it, they will not eat well and their wives and children will not have enough to eat; they will not be satisfied. Because they know that the forest has lots of wealth."

As Masalito begins a long litany of forest foods, I spot what looks like a person on the distant shore of one of the islands in the Epulu River. I point out the rocks upon which I had seen something move and hand the binoculars to Masalito. After a bit of focusing, he exclaims, "Ah, it is an animal!" As we continue on our way back to Kitoko, I am thinking how "handing the binoculars" to local people—taking seriously their perceptions of the forest and its conservation—can often clarify, or at least alter, what it is we think we see when we look at people's place in the rainforest.

Moke and Masina: "You Must Coax and Entice the Forest's Favor as a Lover"

When I had asked Masalito, Tungana, and Cephu about Mbuti stories, they all referred me to Moke. Indeed, Moke was no stranger to me, as I had spent many hours with him tracking okapi when I had worked for WCS in the late 1980s. I had already visited with him some since being back and looked forward to visiting with him some more.

Back in Epulu after spending some time in the village of Tobola, I head down to CEFRECOF one evening to take my meal with the other researchers at the center. On the way I stop to visit Moke and his wife Masina. They are sitting together outside their house, the pale light of a kerosene lantern casting their shadows against the house's mud walls. Moke is in a good mood to talk but, as I will find out, unlike what Masalito and Tungana told me, it is Masina, not Moke, who really knows the old stories. I sit down on the low stool they offer as we carry out the ritual exchange of greetings. The night is filled with the music of crickets. A lone tree hyrax begins to call, the screech slowly crescendoing to its blood-chilling climax before it abruptly ends, leaving the forest unseemly still. Soft conversation emanates from outside the houses of other members of the band, occasionally punctuated by loud exclamations and laughter. I can hear the lilting rhythms of a *likembe* (thumb piano) being played around the fire of the central *baraza*.

We begin to talk about Mbuti ways of taking care of the forest, their traditional hunting restrictions, and the mobility and periodicity of Mbuti hunting and gathering. I ask if these represent ways in which the Mbuti preserved the integrity of the forest. Like many of the voices already recorded here, Moke's reply reflects the perception that people protected both forests and their own livelihoods by living in and with nature rather than by placing nature outside the realm of human influence.

"They were [living] there, however, not necessarily in the sense of preserving [the forest] ...," Moke explains. "What this means is that the forest has been the strength of the Mbuti ever since ancestral times, there when the world began.... So the fathers, the ancestors lived well in the forest without any disorder or tumult. The work of the fathers and the ancestors was simply the joy of the forest. There wasn't this gossiping and accusation, this apprehending the Mbuti in order to protect the forest.... Our work from the time of long ago, [the work] our fathers taught us was simply to stay in the forest, to build shelters in which to sleep.... We laugh well, simply celebrating and enjoying [life], with drums, no disturbing noise; to eat, to give birth.... Thus long ago, that's how the fathers did things. They lived very, very well."

A rather idealized picture of harmony between humans and the forest perhaps? I wonder what Moke sees as its opposite. "According to your thoughts," I ask him, "if a person were to ask you 'What does it mean to harm the forest?' how would you respond?"

"To harm the forest, if people say they want to ruin the forest, they start killing all the animals. They start to kill the animals without discernment, without any discrimination. They do not choose. They kill even the large animals. Thus that is ruining the forest ... [to kill] elephants, buffalo, pigs, *moimbo* [*Cephalophus sylvicultor*, yellow-backed duiker]. Then what are you going to do? That is indeed ruining the forest. However, such things the ancestors did not desire!"

"Why?" I ask.

"They did not desire them because they said, 'If we destroy our forest, where will we eat? We will tarry; we will be too late.... We might want to return to this place over here but there we find that all the animals are gone, the whole forest has dried up! We can no longer find any animals, or maybe only one or two.' Then the worries really abound and they exclaim, 'Ah! We have finished ruining this whole forest!' Thus our thinking is simply to go on following what the ancestors did—to coax, soothe, and entice the forest's favor as a lover, to treat it a bit kindly; to coax it, along with caring for it."

Curious about this vivid description of intimacy between the Mbuti and their forest environment, I ask Moke and Masina, "What would be an example of how they would entice its favor?"

"To entice it? It means to allow it to give birth again," Moke responds, eloquently encapsulating a principle the science of ecology has also discovered. To the Westerner, Moke's descriptions may appear to be simply poetics, but to the Mbuti coaxing and soothing the forest, allowing its rebirth, are vital aspects of their relationship to the natural environment on which their lives depend. Turnbull records the Mbuti band with whom he lived enacting a variety of ceremonies to maintain a harmonious relation to the forest. Music and singing often play major roles in these ceremonies. One Mbuti elder tells him,

> The forest is a father and mother to us, and like a father or mother it gives us everything we need—food, clothing, shelter, warmth ... and affection.... Normally everything goes well in our world. But at night when we are sleeping, sometimes things go wrong, because we are not awake to stop them from going wrong.... So when something big goes wrong, like illness or bad hunting or death, it must be because the forest is sleeping and not looking after its children. So what do we do? We wake it up. We wake it up by singing to it, and we do this because we want it to awaken happy. Then everything will be well and good again. So when our world is going well then also we sing to the forest because we want it to share our happiness (1961, 92).

"Did they sing as a means of enticing the forest's favor?" I ask Moke, curious to know his thoughts on the role singing and music play in relations between the Mbuti and the forest. "They would sing, yes," he answers. "For the nets—songs of the hunting nets, songs of honey, songs of *Esumba* [key ritual celebrating the forest, restoring peace and order, or calling on the ancestors for help in times of stress], songs of *Elima* [girls' initiation rite], and wedding songs. Five things. Meaning, these were our things that were absolutely necessary. The fathers that died here, they have already taught us the traditions."

I am especially intrigued by the songs sung for the hunting nets. Moke expounds, "Yes, in those hunting net songs, they would say, 'May the forest be very good to us. May the forest not cause us difficulties. Our fathers, we and you, it is our forest within which you birthed us. When we call to you for something, bring it to us in order that we may rejoice.' Thus within these songs of the hunting nets, that's the way it was. 'That we may eat, and live well, having honey and so on. From out of all the honey, all the forest, may good come so that we may not tarry and be left behind. We have lots of family. Move toward us and encourage us. Encourage us our ancestors.' The ancestors of long ago, those who have died, they called them again, with songs."

"And in the way the forest or the ancestors would grant you those things, whether animals or honey, was it also necessary for you to give something back to them?"

"Yes, indeed, there were other things to give back to them so that they would give to you again another time."

"What would you give back to them?" I ask.

"[You would] make a small offering of some food to give them in order to call them. You would make a little house. Thus, if it were some meat, you would cut some; if it were plantains you would pound some in a mortar and pestle. You would take the meat and place it there outside: ... 'Here, here is yours. Let goodness be with us. Make us happy. Let us walk well in the forest. Measure us out lots of food.' Such was the way of our fathers."

"And would they petition the ancestors or the forest?" I ask, curious to know exactly who they believed was doing the providing.

"They would make requests of the ancestors, not of the forest," Moke replies; "[of] those who had died there in the forest, in the camp."

"But they did not ask things of the forest itself?"

"No, not of the forest itself, in the camp; to call the ancestors, those who had died."

I am also wondering what Moke and Masina might have to say regarding the personifications of the forest implied in some of Turnbull's and other observers' accounts of the Mbuti (Turnbull 1961, Zuesse 1979). "Were the Mbuti able to converse with the forest? Did they think that the forest could speak with a person?" I ask.

"The forest does not speak," Masina replies emphatically. Moke echoes her words.

"It was the ancestors alone within the forest?" I ask.

"Yes, they would call on the ancestors alone, the ones who had died. They and their leaders would listen and hear."

Moke continues, recounting how the ancestors would also speak to them in dreams, telling them what to do, or letting them know they would get a lot of animals or find a lot of honey the next day. The ancestors would grant them what they had dreamed about—whether animals, honey, or mushrooms—and they would rejoice that their fathers had given them and their family some food. Then, similar to what others have told me, Moke underlines the importance of sharing the fruits of the forest with others, in the same way that the ancestors share the forest's gifts with those who succeed them.

"The whole family would receive some food. You would start sharing it with all of the family. Not just one person! No! Everybody!" he remarks.

"Each person received their portion?" I ask.

"Every person would get a little, even just a little bundle if it were honey. If it was an animal, a little here, a little here, a little here."

"Let's say somebody gets an animal or some honey and keeps it all to themselves—how would you view such a thing?

"If I get an animal or some honey and don't give some to my family, they will call me a bad person. They will feel toward me as if I am a bad person.... Thus if you fail to give all of the family even a little bit each, it would be a very bad thing."

In the background, I hear Masina say something about such people being the ones who kill others. Moke continues, "The fathers, the ancestors did not like doing things like that. It is good that we all eat a little bit, until all have some no matter how many houses there are. No matter the number of houses, we all eat the same. Thus that's how it was."

I respond simply, stating how much we whites have to learn from the Mbuti since we have exchanged much of this ethic of sharing for a rabid individualism upon which both our society and natural environment now stumble and teeter.

The lantern's flame starts to sputter, making our silhouettes dance sporadically on the walls. Either the wick needs trimming or the small dose of kerosene measured out at sunset is coming to an end. Moke reaches over and turns the wick down to make the little kerosene left go a bit further. We continue talking in the feeble light but never get to the subject of the Réserve de Faune à Okapis. Instead, in response to my question about Mbuti stories, Masina begins a long, long tale—a myth, actually— that explains the origin of the exchange marriage that the Mbuti and many other forest peoples have practiced for generations. It is the story of two twin girls who are born attached back to back. Here and there in the telling, Masina pauses and sings the twin's sad lament, her voice carrying softly through the night.

> *We share only one bed.*
> *Me and my sister love each other, the two of us.*
> *There are not three near us, it is only the two of us who*
> *came out together from our mother's womb.*
> *We love each other.*
> *Our bed is one.*

Through trickery, two young men succeed in separating the twins. In response to this amazing feat, the twins' father gives his daughters to the two young men to marry, but only on the condition that they in turn offer their sisters in marriage to his sons.

It is some time before Masina comes to that phrase used by storytellers the world over—"and that is the end of the story." Thanking her for her delightful telling, I rise, realizing just how late it is. Talk of the RFO will have to wait until another day.

Summary

Certainly these few narratives do not speak for the Mbuti in general with regard to the RFO and its effects on their lives. Yet several points can be drawn from these conversations that add to the picture of how conservation initiatives grounded in Western science and administered by the State are affecting local people, especially, in this case, those most indigenous to the forest.

To some degree the RFO favors the Mbuti over indigenous farming peoples in regard to its policies on forest use. Whereas it restricts village farmers from entering the forest to hunt (with or without the Mbuti), theoretically, at least, it permits the Mbuti to hunt and gather in the same way they have been doing for generations. Tungana, for instance, expresses this understanding above when he recounts how the warden has granted the Mbuti authorization to walk freely in the forest, whereas if villagers follow the Mbuti into the forest without his permission he can arrest and lock them up.

However, as mentioned earlier, local people hold a variety of understandings of the exact nature of RFO policies on Mbuti forest use. Some report that the Mbuti *are* required to obtain a permit from the reserve warden prior to spending long periods of time in the forest. Such an understanding represents a radical adjustment for many Mbuti who in the past have been completely free to choose, according to their own culturally grounded criteria, when to go into the forest, where to set up their hunting camps, and how long to stay there.

Beyond the confusion over official reserve policy on Mbuti forest use are the ways in which the Mbuti choose to respond to RFO policies as they are actually being implemented by reserve personnel. Even though Mbuti hunting may be permitted by the reserve, many Mbuti and villagers express that the Mbuti are afraid to enter the forest and procure wild game due to the repercussions they anticipate from reserve guards. As the above narratives make clear, reports abound of interrogations, beatings, and confiscations of Mbuti-procured wild meat by reserve guards. I also witnessed first-hand some of the oppressions the Mbuti are suffering at the hands of reserve personnel as recounted in the opening of this chapter. Thus reserve policies designed to favor the Mbuti, the Ituri's earliest human inhabitants, may not in reality have any such affect. The Mbuti share with their Bantu and Nilotic farming neighbors similar perceptions and experiences of their livelihoods being seriously altered by the presence of the reserve.

One of the most poignant testimonies of this comes from a couple of Mbuti I meet while spending the day in the garden of a Bira farmer from Tobola, Meka Musa. The Mbuti have built their shelter next to Meka's as is common among traditional exchange partners. At one point, Meka speaks to them in KiBira, explaining what he and I have been talking about. They respond at some length. I ask Meka what they are saying, and he replies, "They are saying, 'This forest, there is no place for us to speak. The forest has become hard because the *wazungu* have taken it. The Americans there in Epulu have finished taking this forest.... Well then, what are we to do? There is no place to walk. They have finished taking the forest from us.'"

Barely audible, the two Mbuti, speaking now in KiSwahili, continue to talk about how they live from the forest and how the forest has now been taken from them. The reserve, they say, is definitely making their lives difficult. "They want the world to return to be like it was before," Meka adds.

In addition to the fear that now accompanies the prospect of Mbuti hunting, other Mbuti express that the reserve is also thwarting the relatively new ventures some of them are making into agriculture. As quoted above, Tungana mentions how the reserve is causing them some difficulty over the farming they have begun to adopt. Thus the RFO's preferential treatment of foragers over farmers may become ambiguous as the Mbuti adopt new ways of making a living from the land and, like their indigenous farming neighbors, begin to feel the limitations the RFO has placed on agriculture-based livelihoods.

Meanwhile, the reserve's attempt to differentiate between the Mbuti and indigenous farming groups in regard to certain policies has not gone unnoticed by local people. Many villagers, for instance, express that the RFO's attempts to implement preferential policies for the Mbuti is really unfair to the Ituri's other indigenous inhabitants. Furthermore, they feel it fails to appreciate the unity under which farmers and foragers have been interacting in the Ituri for centuries. To reiterate how one farmer articulates it: "In the same way they've prohibited the Bira from killing game, they should also forbid the Mbuti from killing game. That's it. Because the Bira have been supported by the Mbuti and the Mbuti have been supported by the Bira. That's the way life was. Why then are they trying to split them saying only one person should work when they have always been working together as one? The Bira and their Mbuti have worked in the same space. If they are going to split them, then both of them should quit that line of work. That's the way it is!"

Indeed, RFO polices that differentiate between the Mbuti and village farmers are affecting the long-standing exchange relationship between the two groups in significant ways. The Mbuti narratives recorded here attest that a key issue for the Mbuti is their relationship to village farmers, or *bakpala*. Whether constructing the relationship as beneficial or antagonistic, these narratives confirm that the relationship is changing in part due to the presence of the reserve. Masalito, for instance, links the reserve to increased *bakpala* activity in the forest (see Note 19), a perception that seems to have heightened his feelings of antagonism toward both *bakpala* and the reserve. Tungana and Cephu, in contrast, welcome the reserve's restrictions on the mining and hunting activities of newcomers to the forest, people they identify, in a more general sense, as *bakpala*. All three Mbuti strongly criticize these newcomers, or *bakpala* in general, insisting that they are the ones who are really ruining the forest.

It is difficult to gauge the exact degree to which reserve rhetoric toward newcomers and reserve restrictions on *bakpala* entering the forest serve to heighten the antagonism some Mbuti feel toward *bakpala*.[22] But the reserve clearly represents a new player entering into this complex and sometimes tense exchange relationship, bringing policies that do not always reflect an appreciation of the history, fullness, and complexity of the long-standing association between forest farmers and foragers. Continuing to act without that appreciation can only exacerbate the ill feelings the reserve has already generated among both the Mbuti and their farming neighbors.

Finally, these Mbuti narratives make clear that conservation as preservation—separating people from the forest they depend on—is a completely foreign idea to the Mbuti with no local cultural cognates. For Masalito, closing off the forest from the Mbuti raises the questions "How then will we eat? Where will we live?" He goes on to say that his ancestors "did not close the forest in the sense of prohibiting it" and that a reserve is something "that comes from ... the State, but it is not a way that comes from ... the Mbuti." Moke confirms that "apprehending the Mbuti in order to protect the forest," this forest that has been "the strength of the Mbuti ever since ancestral times," plays no part in their cultural heritage.

Such perspectives on how the current conservation initiatives are affecting Mbuti culture raise serious questions, not only about the survival of the forest but also about the survival of the Mbuti themselves:

- If the Mbuti begin to significantly forego their forest hunting, what will be the affect on their culture?
- Will the forest they depend on fall asleep for good, unable to be roused even by their singing?
- If the forest is not allowed to care for them, will they continue to see it as father and mother, as provider, as goodness itself?

- Will things have gone so wrong that the Mbuti lose their faith in the forest to look after its children? Lose their faith in their culture's ability to keep the forest happy? Lose their faith in themselves?
- And without such faith, how indeed can a culture, much less a forest, ever survive?

Regardless of its official policies, in reality the RFO is cutting off the Mbuti from certain long-held forest practices due to the great fear repressive actions by reserve guards have instilled in the Mbuti. Unless the actions of RFO personnel begin to more closely reflect official RFO policy toward the Mbuti, as well as toward local people in general, the reserve runs the danger not only of fueling local resentment but also—and this holds much more serious import—of contributing to the demise of Mbuti culture.

The Bapukele Elders: "The Animals Have Received Their Independence and Are Destroying Our Food"

I never manage to get back to hear Moke's views on the RFO, but I hear plenty the next day from a group of elders in Bapukele, a village 6 kilometers east of Epulu. About a dozen men are waiting for me in the central *baraza* when I arrive on the motorcycle. I apologize for not meeting with them the day before as originally planned. Upon setting out, I had noticed that there was enough wobble in the Yamaha's rear tire to make a duck's waddle straight. What I thought would be a simple tightening of a bolt turned into an entire morning's labor. Taking everything apart, I had discovered that the rear wheel's bearings were shot, likely from their continual submersion in all the mud holes I had been traveling through. New bearings were hundreds if not thousands of kilometers away. So, as they say in Congo, *debrouillez-vous* (fend for yourself). After some serious cleaning, heavily endowing the bearings with grease, adjusting the chain, and tightening the axle, I had the bike roadworthy by noon. But by then another huge tropical storm was about to break. The trip to Bapukele had had to be postponed.

Despite the delay the maintenance had caused, the work of combining material objects that have only one way of fitting together into a tangible and unequivocal whole had offered a much appreciated change from the omnipresent ambiguities and obscurities of research. If only the puzzles of people and parks—the process of doing research with complex human communities—might be so straightforward. Then again, I suppose we would miss out on the fascination, wonder, and, often, the delight that real-live human beings in all their intricacies and convolutions afford us when we try to understand why and how they do what they do, believe what they believe, and say what they say.

* * *

Under the Bapukele *baraza*, the elders and I are soon engrossed in a fascinating conversation. Many of their statements concerning local means of conservation and those of the RFO replicate what others have been telling me. But then they raise a concern regarding the reserve that is especially troubling them. Others have emphasized the costs of not being able to eat wild game; these Bapukele elders are concerned about game now eating them—eating them out of their gardens, that is—due to the restrictions on their traditional means of controlling garden-ravaging animals.

The issue of animals ravaging gardens comes up when I throw out the question of what these elders think it means to ruin the forest. In response, one of them indicates just what exactly he thinks is being ruined. "To ruin the forest means that—" (the elder pauses as if reconsidering the question, then continues) "because now the animals are oppressing us. If it was still like it was in the past, we would kill them. Now, the animals—elephants, every type of animal, baboons—they have made their way even all the way into our houses!"

"Baboons, monkeys, all of them, they are destroying our food," another elder adds. "Baboons, monkeys, and sometimes pigs, which also ravage at night, … elephants! Here I am, I am together with him and him. Elephants enter [the gardens] at night. Sometimes they come about nine P.M. We are amazed and try to chase them away—'*Wooah wooah wooah*!' They come again and then they run away. But when they run off they don't go far; they just circle around close by like that."

"Thus what you are saying is that now it is not people who are destroying the forest but it is animals who are destroying people's food?" I ask.

"It is only the animals," the group choruses.

"It wasn't like that before?"

"No way! It wasn't like that."

"Before it wasn't like that," one of the Bira elders explains. "Elephants were people who would go way off into the forest. The Mbuti would go after them there, a long distance away. But now, at this time, elephants are right here in the village. In the village!"

"Why do you think the elephants have come close to the village now?" I inquire.

"Because they are being protected; meaning, here people are forbidden to kill them—" He is cut off by a rousing discussion that eventually posits another theory: "Since they [elephants] are being killed [elsewhere], they come close" (i.e., elephants come from more remote areas of forest where they are being poached to the protected areas of the reserve, a theory I have heard others in Epulu mention).

Then one of the Bira elders who has been entertaining us all morning with his ironic sense of humor adds his own twist on things: "You see, the animals have gotten their independence."

"They've received their independence?" I ask, surprised by his choice of words.

"Yes, they've received their independence."

The elder sitting next to him expounds: "Now the animals are simply destroying our food. And you can look at them, but you had better not kill them. Now the animals are killing us. And the children, the ones we are giving birth to, whose children are they? Are they not children of the State? They are dying of hunger! And the guns with which to chase away the ravagers exist![23] What then are we going to do here? What will we do there? It is destroying people because at the same time as we eat, so are baboons, elephants, pigs, monkeys.... So then what are we supposed to do here? You see it is destroying the children themselves!"

Mulimaji, an immigrant from the eastern savannas who has lived in the forest long enough to be considered a local by these Bira elders, then offers his own novel explanation of how the relationship between animals and people has changed: "Before during the time of the ancestors of this place, none of the animals recognized the foods that we planted." Exclamations of agreement ripple around the group. "The work of all the animals in the forest was simply to eat the seeds of the forest."

"And their leaves," others chime in.

Mulimaji continues. "After the animals had developed a bit, they began to recognize the other foods that people plant near the village, and they began to eat them, continuing up to today. Manioc, rice—the monkeys are eating them! Plantains! And now even beans, they are starting to eat. All of this stems from the animals being more developed than they were before. Before animals did not know these things. They began to try a little here and there, and soon they had penetrated the whole garden. Before, there were no animals that would come to the gardens; only those that they killed a long ways off; or perhaps [those they killed] near the garden there, but it wasn't because they were entering the gardens. Now, they have successfully developed. Just like the people in the village, they have begun eating all the foods that people are planting in addition to their foods of the forest...."

"Is he speaking the truth, *wazee* (elders)?" I ask the others.

"The truth! He speaks the truth!" several chorus.

"What he is saying is exactly right!" another elder affirms. "And what he is saying we have also seen with our own eyes! For is he not one of us here? Eh? He speaks the truth. What he is saying is also taking place in his own garden. All of us speak as one the same words he is saying. After the animals try a little bit of food like that—'Ooh, ah! I say, food is good stuff! Well then let's go after it in order to eat more!' Thus these animals have entered into our gardens. Later they say, 'If they come to kill me here, me an animal, if they come to kill me, they will simply come and take a count of me.' Because of what? Because they have forbidden killing them, they the people of the State. Thus the animals become strong and more daring. They then take the whole forest to pieces, even coming all the way here into the gardens! Thus in that way they have received their independence, their freedom from being killed."

"According to you, what can be done to solve this problem?" I ask. "What should we do to end the damage animals are causing your gardens, the strife they are causing you?"

Mulimaji is the first to respond: "In thinking about it, I also am concerned about this. The way it was before, as we have just mentioned here, animals were not seen nearby because people would go after them. If they saw one, they would quickly go after it and it would flee. Now, frankly, we need to figure out how we can bring back such ways to kill ravagers. People should be able to kill them if they see them coming here. If they are seen [in the gardens], people should kill them and they will see that all of them will return [into the forest]."

"They will scatter," another in the group says.

"According to our tradition of the past—chasing them away from around the gardens," adds another elder.

I then attempt to summarize what I have heard: "So what you mean is that according to tradition, it was like this: When animals came near the gardens, your ancestors would either chase them away or kill them, and they would return into the forest, to their place. And now the reserve's custom is not like that?"

"It will finish us all off here," one elder replies.

"Is that what you are saying?" I ask the others.

Another elder responds, as others have done, by drawing a distinction between current RFO policy and that of Jean de Medina in the 1950s and 1960s. "Now, yes. Here during the time when the park [i.e., the Station de Capture des Okapis, not the RFO] became established, it was good with Mr. de Medina. Then, when the elephants would come to our gardens, he would send his hunters and they would come and shoot the elephant. We would see baboons, and he would say, 'Okay, if you see baboons and mangabeys, hunters will come and kill them.' Bit by bit they carried out such a program of killing ravagers throughout the area, and thus the animals, the baboons, would get up and flee and go a long ways away.

"Now we cry out, but to no avail. We are crying but for what reason? It doesn't do a thing. Their work [that of reserve personnel] is simply to come and look at the elephant footprints. They say things like, 'Today an elephant defecated here.' And then they write it down. Or after seeing that an elephant is in the garden, it beats me what kind of bullet they shoot her with that tells her, 'Come, come all the way here right to the middle of the village!'" The others roar with laughter, but the elder's face is serious as he goes on: "So, then, we are crying out. We are begging for mercy. How are we supposed to live?"

* * *

It is clear that CEFRECOF's research of counting and making careful observations to determine the "actual" extent of garden damage caused by animals carries little value or holds little meaning in the world of this elder, a world where animals and humans have always coexisted—but each in its own domain. Now the lines between domains are becoming blurred, leaving this elder confused, frustrated, and, most of all, hungry. In the past his subsistence trapping kept his family fed, his garden secure, and the animals at bay, yet it also precluded his need to invade their territory, overhunt their populations, and bring ruin to their habitats. Like the components of any ecosystem, he and his people and the animals they have lived with have developed complex means of balancing each other's needs for space and food. Now such complex means of keeping one group of animals from invading another group's territory and overriding them are being disrupted. Simply researching that process makes little sense to him. He would like it changed. The RFO has upset the balance in favor of the animals, and he simply wants to be able to play his part in helping that balance be restored.

I point out that one of the reasons the RFO has established such hunting restrictions is to prevent people from upsetting the balance in *their* favor, as has happened in other African rainforests. Reserve personnel want to keep the Ituri from experiencing the same depletion of wildlife characteristic of many other parts of Congo. As an example, I recount my experience of tracking okapi back in Équateur Region where I walked for three days through primary forest and didn't see one animal. "Couldn't something similar happen here if no laws are established?" I pose.

In the conversation that follows, the elders make clear the need to distinguish among the many different ways in which different people interact with animals, some of which upset the balance in people's favor and some of which don't. Sanguma, a Bira elder, is the first to respond to my question: "Here we don't have any guns and we don't kill as do the people over there. We hunt and kill only a portion. But those people over there [in Équateur], do their killing so as to finish them all off.... [We kill] for eating only. But the people over there finish theirs off because they want to get a lot of them in order to sell. However, we kill our animals only according to what we need to enable us to eat."

"Thus, if I understand you well, the thing that will finish off the animals is simply the market?" I ask.

"Indeed, it is the market that finishes them off."

"However, to kill animals only for eating will not cause them to disappear?" I question.

I hear a chorus of "No!" from around the circle. I then go on to give the example of the soldier in Équateur who killed sixty monkeys in one day and ask them how they would view such a thing. First there are expressions of disbelief and disapproval around the *baraza* and then, one after another, the elders respond, "It is bad, it is wrong."

"It is wrong?" I ask.

"It is altogether wrong!" one of the elders exclaims, setting off another boisterous exchange from which I catch only intermittent phrases: "That is the style of the white man!"; "Among us a person doesn't kill like that!"; "No way!"; "Even to kill four or even ten or five, you could not kill that many; two or three, that's all!"

Then, raising his voice, Mulimaji breaks through the din. "As I said before, here there are no livestock. Even though there are no livestock, this area's livestock has always been forest animals. But they also would not kill animals in large quantities. They would kill them in moderation. Because, as with the livestock that you keep, you will take one or two chickens and eat them. You would never finish off all your chickens in one day!" The others voice their agreement. "And when guests arrive, you butcher one of your goats but leave the others.

"Now ... [suppose] a person sets a trap and ... it kills an elephant. That one elephant will feed the whole village for an entire month; they will even be eating that one animal for two to three months!"

"For two months!?" I exclaim, a bit incredulous.

"Or even longer!" Pascal responds.

Mulimaji continues. "Yes, they are eating from that one elephant there. They continue eating only that one elephant in order to conserve for another day. They would never be able to kill ten elephants all at once in one day. Who would eat it? It would rot and spoil."

"Most of it would rot!" one of the elders confirms.

"That one elephant would even feed the people of Epulu!" another adds, prompting additional comments from the others.

"It would suffice for all of the people."

"If they killed one here, all of the people of Epulu would come running!"

"They kill one here, all of Epulu and Bandisende would come to eat it!"

"And there would not be one person who would return empty-handed because of the meat having been for sale. [It would feed] all of them!"

"In the past that was the way in which they conserved the animals," Mulimaji sums up. "One could not inflict ruin on the whole forest."

I ask the others for confirmation, and they all say that Mulimaji is telling it like it is. "Therefore, in your view the only thing capable of finishing off all the animals in the forest is the market?" I conclude.

"The market alone."

"And guns," someone else adds.

"So if they forbid the market and forbid guns, such restrictions would not bring suffering?" I ask.

Some time earlier, Mama Amanzala, a native of Bapukele, had quietly joined our circle. She speaks for the first time: "But people would still be troubled by hunger."

I continue probing, "However, if you were allowed to kill animals for subsistence, that would be a bit better?"

"Absolutely! It would certainly be better," an elder responds.

Mama Amanzala replies more emphatically: "The only thing we want is to eat!"

"To eat, yes, for where is the market to sell it in?" someone asks rhetorically.

"But people are passing by here all the time. They will buy it," I counter.

"We don't know anything about marketing," one of the Bira elders rejoins. Mama Amanzala agrees: "We don't know about it. There is not one Bira who would go and sell game in some faraway city."

"All of those things have stemmed from those who live in the cities" Pascal affirms. "See, guns come from the city—do you hear? In the first place, not even one monkey has been killed here in the past twenty years. Even though they set traps, they don't kill any monkeys. Anybody who is from here, from the day of their birth, they have not killed a monkey. How would they go to sell it then? These are problems that stem from the city. A person sees those markets that eat a lot of money. He comes back with a gun, and that's when you get the sixty you told us of. For those are bullets! Even if you're talking about arrows—how many would that kill? Often with arrows you return with nothing! Such problems exist over there, in those cities over there."

"However, for now the law of the reserve forbids all killing?" I ask the group seeking to confirm if I am hearing them correctly.

"All of it!"

"Even for subsistence?"

"Everything! They forbid it all!"

Our talk under the baraza continues for another hour or so before we say our goodbyes. As I get up to leave, I promise all of them that what I have heard I will pass on. I will write about it and try to bring their voices into places where they may have a hard time being heard. But riding back to Epulu, the questions knock on my heart: What good is this research really going to do them, these people who have so openly shared with me their struggles and their lives? How can it serve them? How can I apply what I've learned in these *baraza* exchanges so that they truly see some improvement of their lives? I ride on, sheepishly aware that these questions have no clear answers.

The Upshot

Despite the richness of the preceding narratives, it is difficult to derive any conclusions and generalizations. To attempt to do so assumes that conclusions and generalizations, the standard products of our Western science, surpass all other investigatory insights. Following a strict "scientific" approach, we might be able to come away from a research essay such as that described in this chapter with a "generalized" picture of what the people of the Ituri perceive regarding the RFO. Such pictures have value, but they are also fraught with problems. First, they result in gross simplification of complexity. Second, they lose the advantages of what particularity can offer: a sense of the more true-to-life disorder and contradiction that exists within any community. In addition, they lose the points of view of the minority, each of which is legitimate. Too often these dissenting voices are silenced by the generalizing grids we so fetishly place over any research experience out of our fear of disorder, uncertainty, and complexity.

In contrast, the narrative approach I have used in this chapter holds a different sort of epistemological value: It offers insights from the particular stories of actual people, each one bearing a unique way of seeing and acting upon the world. It is a method that allows a little of the real-life person to actually come through and have a chance to be heard above the roar of percentages and the din of statistics. But let me re-emphasize that the narrative approach is meant simply as a complement to, not a replacement of, other more conventional methods of social science.

So what follows is less a general summary of this chapter's conclusions and more a testimony of what can be learned by listening carefully to the different perspectives on forest conservation local people have to offer. As voiced by the people whose narratives have filled this chapter, such perspectives teach us that in regard to the *meaning of the forest*:

- All the things of the forest are God's gifts to human beings to help them live. It is local people's pantry and hardware store. Since God alone created the forest, no person has the authority to forbid another these gifts.
- Wild meat, one of those gifts, is a vital part of forest peoples' diet and culture. To subsist on manioc greens and other vegetables brings illness and diminishes one's strength.
- Hunting in the forest fulfills not only material needs but also needs for community celebration, ritual, and emotional well-being. Failure to hunt divests forest cultures of important meanings and educational opportunities while depriving forest peoples of a significant aspect of their identity.

In regard to *people's interaction with the forest and its animals*, we learn that:

- Forests and rivers are considered to be people's gardens; forest animals, people's livestock. Implied in such images is the idea and practice of careful stewardship of natural resources in order to sustain them for future use (a striking contrast to the perspective more prevalent in the West that views humans' relation to nature as a zero-sum game, that is, nature is either to be raped or restricted from all human interference).
- Animals and people were created together and co-exist in an interdependent relationship. Without animals people will not be able to survive but without people, the animals will also come to an end. Both need each other.

Local perspectives teach us that in regard to *ethics and restrictions pertaining to the forest*:

- An ethic of the future and of future generations discourages people from overexploiting forest and fauna.
- The killing of animals was limited, and should still be limited, to supplying people's subsistence needs. Killing for commercial markets is considered wrong and seen as stemming from urban and modern technological influences.
- Traditional taboos already offer protection to key species of forest fauna and coincide with reserve restrictions. However, traditional restrictions part company from those of the reserve in regard to species (mainly duikers and smaller animals) used for subsistence. It is not understood why the reserve restricts even the killing of these animals.

And finally, in regard to the *current conservation policies of the reserve*, these perspectives teach us that:

- There is some appreciation for the reserve because its protection of wildlife guarantees that game for subsistence will remain close at hand. Also, some see the reserve as an effective way of keeping immigrants from settling in the Ituri.
- However, protecting the forest by closing it off from human use is considered not only a very foreign concept but also a grave injustice.
- People feel that they have been cheated out of their forest by the State, the international conservation organizations (often referred to simply as the "Americans"), and their local leaders who, without the consent of local people, have transgressed ancestral rules by "selling" the forest to outsiders.
- Reserves are thus considered to be yet another extension of an oppressive State that pries into peoples' lives, another means by which local State officials gain authority to exploit people. However, people are able to differentiate between the goal of State conservation programs (to guarantee a perpetuation of forest and fauna) and the abusive way in which those programs are carried out (e.g., "the guards need more discipline").
- The abuse of power and authority on the part of reserve guards is a primary complaint people hold against the reserve. Despite official policy allowing the Mbuti subsistence use of the forest, such abuse is also causing many Mbuti to refrain from hunting due to the repercussions they anticipate from the guards.
- Gardens are suffering increased animal damage due to the RFO's wildlife policies. This, in turn, is seriously limiting people's health and nutrition. In the light of such hardships, research on animal damage by reserve staff makes little sense and carries little value. Instead villagers would like to resume their ancient practice of trapping ravagers or in other ways moving them away from around their gardens.
- People do not clearly understand the reserve policy on killing garden-ravaging animals. Some villagers say they will suffer fines and beatings if found with any wild game, whereas others say that trapping garden ravagers such as pigs, baboons, and mangabeys with traditional traps is permitted.
- Under existing animal husbandry practices, domesticated animals do not provide an adequate alternative to wild game as a source of meat since they often die from predation, disease, or insect bites in the forest.
- People prefer the game restrictions of the colonial era (during the time of de Medina) to those of the present-day reserve, as the former did allow for subsistence hunting and operated according to hunting seasons rather than by restricting territory.
- Many villagers view the RFO's preferential policies for the Mbuti as being unfair to the Ituri's other indigenous inhabitants. Especially troubling are restrictions on villagers following the Mbuti into the forest. Furthermore,

such policies are seen as failing to respect the unity under which the Ituri's farmers and foragers have been interacting for centuries.
- People may elect to move outside of the RFO given the difficulties they attribute to the reserve's restrictions. However, this is not their preference, and many see it as a severe hardship if not an impossibility.
- Ways of resolving the conflicts between the reserve and local people include: (1) holding a tribunal between reserve officials and local leaders to work out their differences with regard to the use of the forest; and (2) using proceeds from the export of okapi to offer local people some compensation for the restrictions placed on their livelihoods. Suggested compensations include providing communities with schools, dispensaries, houses made from durable materials for chiefs, bicycles, motorcycles, or a small vehicle.

In Chapter 7 I will discuss how differences in perspective between local people and Western-trained conservationists are rooted in fundamentally different metaphysical understandings of people's relationship to the natural world. I will also discuss the implications such differences hold for both local conservation practices and environmental theory. Prior to that, I present in Chapter 6 the voices of university-educated project staff and local academics, another group of actors in this drama of forests and people with their own perspectives on what an Afrocentric view of ecological sustainability might look like. Their perspectives offer interesting contrasts to the voices heard so far, as well as to perspectives found within Western environmental science.

Before that, however, I want to close this chapter with one last story. Of all the many narratives I recorded, it impressed upon me more than ever how important it is for those in power within the RFO to hear the voices and grievances of local people.

Paulin Mboya: "If Local People Don't Support the Reserve, It Will Fail"

The day following my visit to Bapukele, I am graced by one of those spontaneous research experiences that can often provide more insights than many of the prescribed sessions we so carefully conduct. It is my last day in Epulu, and I head down to the Okapi Capture Station to take in the beauty of the river one last time. The sky above the forest's ever-near horizon is intense, towering clouds broiling and turning, pregnant with another approaching storm. Streaks of lightning begin to flash across the sky, and I decide to seek some shelter. I hit on the idea to head up to town for a parting beer with Kasereka, one of the Nande storekeepers who had become a friend years earlier during my master's research. Finding his little shop and hotel locked up tight, I decide to head across Epulu's main drag to bid *au revoir* to other friends, Paulin Mboya and his daughter Marie Claire. Their hotel and restaurant, the Kata Bei, has been a longtime Epulu landmark, a most popular overnight for trans-Africa truckers as well as the choice locale for swinging off a little of life's frustrations to the bubbly beat of Congolese *soukous*.

Rain-soaked and mud-splattered, I knock on Paulin's door and am quickly ushered in by Marie Claire, who along with her father has been visiting with friends in the small sitting room of their mud-wall, tin-roof house. They offer me a beer and without prompting soon begin to share with me their perspective on the new reserve and the suffering it is causing. What follows is their testimony, re-created from my field notes, of how out of step the reserve is with the practices, needs, and understandings of local people.

* * *

"Who is tranquil? Who is at ease?" Paulin begins. "How can we be when we face animal damage to our gardens and can't feed our children? When we have no means to remedy the situation by hunting these predators—not protected species, but baboons and mangabeys? When guards abuse and mistreat us, confiscating meat we've legally purchased from the Mbuti and then going and eating it themselves? When we've seen none of the benefits we were promised? How can we support the work of these projects?

"Do you think we have no *conservation coutoumière* (traditional means of conservation)? I, Paulin Mboya, have taken on *commissaires de zone* who are big poachers. We don't want that here; nor do we want commercial meat marketers from Kisangani who come to fill their sacks. But let us have our *mboloko* for food, and kill the baboons and monkeys that ravage our corn; not to sell or eliminate them, but to move them off deeper into the forest.

"We are in a 'Pygmy Zone' where we have little or no livestock but live off hunting. How are we supposed to live if you take these things away from us, cut off our source of protein, confiscate meat we've bought legally from the Mbuti and then fine us? Take the gardening policy that's just been announced: You plan to cut a new corridor 750 meters away from the station forest in order to let that forest recover. Old gardens people can still harvest but after six months to a year, they have to leave them. How can you justify doing this to someone who went to a lot of work to plant that garden or who has property there such as oil palms? How can you tell them to just leave it? If you do these kinds of things, how can we support you?

"These policies of the RFO are very different than how we've lived here for years and years. I would even call this some new type of colonialism, but worse than that of the Belgians. At least they respected local chiefs and gave them a means to control animal damage to people's gardens by providing them with guns. But today the *chefs de collectivité* are shown no respect. For example, after one of the conservation meetings here at the station, the Ndaka chief asked to be transported back down the road 20 kilometers or so—and this was when the road was still good. His request was met by an offer to buck him on the back of a motorcycle, he carrying his backpack and all. His response was telling: 'Who do you think I am, an animal?'

"Lack of respect is also shown by your failing to notify government officials of expatriate visitors. Some even come and land their planes here as they please. You could never get by with this in your own country but here you speak of *Le Zaire* [Congo] *spontané* and come and do as you will.

Even me, how can I offer advice to the projects when they see me as worthless, old, and poor and never come here to talk with me; never try to establish good relations with me? Or when it is clear that some of them really don't want black people living next to them?

"In February of 1996 things are going to get really hot! There's going to be a real shelling. They are calling a meeting of all seven *chefs de collectivité* in the zone of Mambasa with their two *conseilleurs*. They will hear what people have to say, hear their grievances, whether they are for or against the reserve. And they will take our message all the way to the top and, unless things change, demand that things here return to how they were in the past under the Station des Okapis.

"If you want this to remain a reserve, then you have to fulfill the things we've asked for and need—each collectivity having a dispensary, school, and good office; each chief a good house. Don't come with gifts of jars of wild honey as one of you gave the chief of the Ndaka—wild honey you've gotten from the Mbuti in our forest and then come and given back to us! Treat us with respect! We have yet to see one benefit coming from your projects. In fact, these projects are ruining Epulu. Since they've begun bringing in supplies and building canteens for their workers, store owners in town who cater to the rest of us speak of getting up and leaving because they do not have enough customers.

"If you keep on this way, you are going to see some real problems. People will get fed up and, tired of seeing they no longer have any benefits and that benefits are only being given to the animals, take matters into their own hands. Perhaps they will just say no and go directly to the minister of the environment and the president and tell them, 'We don't want your reserve if this is what it means for us, this added suffering!'

"You have to go slowly with people, building agreement and good relations if the reserve is going to work. If the local people don't support it, it will fail. But local people feel they are getting no clear message from you. The World Bank delegation comes and declares there will be a 10-kilometer buffer zone along the road where we can pursue our lives as in the past. No one has ever surpassed 10 kilometers. But then we hear 5 kilometers, or that this is now forbidden or that is now forbidden. This is not normal! People are starting to wonder why in the world their leaders agreed to the reserve in the first place.

"Further, you think you can push everything through by going directly to Kinshasa, but what power can you find there when ministers change overnight? And given our country's recent decentralization policies that give greater autonomy to the regions and governors, you will not be able to go on bypassing local or even regional officials. You can't just go to Kin!

"We are not dumb people. We hear and see and feel and know what is going on and we are not happy!"

Conclusion

My son Benjamin recently visited NYZS's Bronx Zoo with his mother and grandparents. He returned home with the following photographs (taken by his grandfather) of a placard that sits outside the new okapi exhibit (see Photos 5.2 and 5.3). The placard presents the public with a picture of WCS's work and the RFO that is partly accurate but mostly ironic. Although the text recognizes that the support of local people is vital in order for a reserve to survive, it is striking that it offers not a hint of actual local sentiments such as those expressed so eloquently by Paulin Mboya or the other voices in this chapter. Instead the placard makes it clear that according to WCS the solution to "securing the reserve" is not, as Paulin Mboya points out, treating local people with respect, going slowly, building agreement and good relations, and providing local benefits; rather it lies primarily in training Congolese for the task of "careful ongoing protection and scientific management."

Rather than starting from local people's perceptions, concerns, and knowledge, WCS, like many other Western conservation organizations, strives to change people into adherents of their own Western scientific ideology in order to help carry out their own goals and values, all under the guise that through such training programs local people are being integrally involved in establishing and managing the reserve. Such claims of local involvement are dubious, as the narratives in this chapter make clear. Furthermore, although some of the people trained at Epulu are indeed local, the vast majority has come from areas of Congo hundreds if not thousands of kilometers away. Moreover, the placard clearly reveals that despite such training programs it is the international personnel who remain the "experts and scientists"; local Congolese are the "assistants."

Yes—as the placard correctly affirms—"Ultimately, people everywhere must care." But care about what? Missing from such official representations of the grand work of preserving wilderness is the admonition to also care about the local people

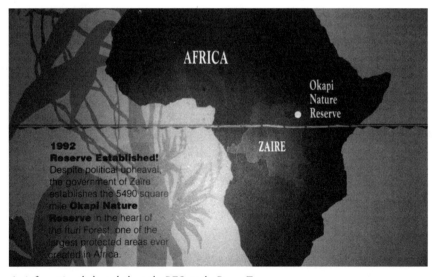

An informational placard about the RFO at the Bronx Zoo.

Local Voices on Conservation in the Ituri

An informational placard about the RFO at the Bronx Zoo.

deeply affected by such efforts. No matter how much training takes place, in the end, the words of Paulin Mboya still ring truest: "If the local people don't support it, the reserve will fail."

This chapter has made it clear that for a wide variety of reasons, local people do *not* fully support the RFO. Testimony such as that of Paulin Mboya also indicates that when conservation projects lack such support they sit on a time bomb and run the risk of having all their work blown away unless they make serious efforts to negotiate their differences with local people. Before proceeding any further, it is vital, both for their own interests and for the good of the forest they are striving to protect, that conservationists sit down with local people—chiefs and peasant farmers, government officials and Mbuti hunters—in an atmosphere of respect and strive to create the good relations and agreement so desperately needed to resolve the growing conflicts that endanger not only local people and local projects but the forest itself.

Notes

1. The names of several of Congo's eastern Great Lakes are still in a state of flux as far as I was able to determine. For the two northernmost lakes, the names Lake Mobutu and Lake Amin are out, and it is not clear what the new names are, will become, or will remain. On this map I have resorted to using the old names, Lakes Albert and Edouard, although I do so with some uneasiness.

2. I place the word "chiefs" in quotes to signify the somewhat artificial basis, meanings, and power bequeathed the position during the colonial period. Vansina gives an excellent description of this process that often produced "chiefs" with no indigenous claims to power: "During the first generation of conquest the conquerors expended considerable effort to 'un-

derstand' local societies and cultures. Reams of ethnographic and legal questionnaires provided a stream of information. But from the outset these incorporated an abstract view of what such societies and their laws should be like. Such efforts soon resulted in the creation of a conscious model of what societies in equatorial Africa were and should be. As early as 1909, the ethnography of de Calonne-Beaufaict, a disciple of Durkheim, provided the blueprint of spurious, segmentary, hierarchized, patrilineal society for use in the Belgian Congo. This allowed the government to build up a practical administrative line of command. By 1920 agents in the colony were enjoined to 'discover' this model wherever states did not exist, and the population was physically regrouped to conform to it. Dependent status did not exist in the model, so all dependent statuses from client or pawn to slave were gradually abolished. Leaders became mere 'patriarchs,' bereft of any source of power that did not directly derive from the colonial administration" (1990, 246).

Contemporary political administration in Congo continues to reflect this structure to some degree. Moving from the local village level to that of the region one finds the following hierarchical organization of leaders and political administrative divisions: *chef de localité*, *chef de groupement* (a collection of several villages), *chef de collectivité* (the division corresponding most closely to ethnic and clan distinctions), *commissaire de zone* (the *commissaire* is often nonindigenous to the zone), *commissaire de sous-region*, and *gouverneur de region*.

3. Since this research was carried out, the ICCN and chief warden at Epulu have taken additional and more effective measures to improve the quality and sensitivity of reserve guards. Such actions are commendable, and one hopes they will help to prevent the types of abuses documented in this chapter.

4. In contrast to Kamu's viewpoint, other villagers will later complain to me that marketing garden produce to *bouloneurs* has brought hunger to their households since the flush of money causes them to oversell their food stocks. As Maurice, a Songye farmer, explains, "We farmers are few in number and those who eat have outgrown the number of us who are farming. You see? For example, the dried manioc that I've placed outside there—if it's just those of us who live and work here in my household, I can eat it for six [months]. But others who eat come; a person comes and says to me, 'Give me that manioc and I will give you fifty thousand [zaires] for that one.' I start to shake over the money and I give him the manioc. Lo and behold, I end up starving, me and my household. That fifty thousand that I've taken, what am I going to do with it? ... It won't help one bit! I take it to the boutique, even this one shirt ... and I will no longer see that fifty thousand!"

5. Matiyé never specifies who exactly he means by "those people," but I imagine he is referring to those in charge of the reserve.

6. In most all forest societies, the *baraza*, a meeting or reception shelter, is the place where children receive instruction from their elders. A "child of the *baraza*" is someone who heeds such teachings whereas those who have displeased the elders by not following their advice remain outside.

7. See Chapter 1, note 2.

8. By designating the okapi as "belonging to people," Heri may be a referring to the State's protection of the animal or to the long-standing role okapi have played in forest cultures as symbols of power and status.

9. It is unclear what Heri means here, but aside from the literal meaning, he may be implying that the okapi captured by de Medina served him well through the money earned from exporting them to zoos. In many African societies, "eating" has multiple meanings beside its literal one. Most often it implies benefiting oneself at the expense of someone or something else.

10. See, for example, works on the Koyukon people of central Alaska, especially Richard Nelson's *Make Prayers to the Raven* (1983).

11. This passage proved difficult to translate. Heri's actual words are, "*Ninakujibu kusema kama mbele sisi hakukuwa na pori hivi. Hatukukuwa na hukumu ya pori. Sasa wewe unanibeka* [from other contexts Heri means -*weka*, to place or put] *na hukumu ya pori. Wewe unakwisha kunibeka na hukumu ya pori. Sasa nitakaa wapi na we unanibeka na hukumu ya pori?*" The crucial words here are *hukumu ya pori*. According to *A Standard Swahili-English Dictionary*, *hukumu* can mean: (1)judgment, jurisdiction, authority, supreme power; (2) legal process, trial; (3) sentence, verdict, decision, order, either in a civil or criminal case. *Pori* is the word used for forest in this area of Congo. Hence in the sentences immediately preceding this footnote, I have translated *hukumu ya pori* in various ways, all of which capture the sense that, according to Heri, the forest has now come under the authority of the State rather than that of local people. When I ask Heri what *hukumu ya pori* means, he links it distinctly to the realm of tradition by describing it as a process of exchanging his own traditions of forest regulation for those of the State.

12. The local meaning of *kenge* is ambiguous. I did not hear it being used often, but Moke, a Mbuti elder, once told me it was the word for "rhinoceros." Turnbull translates *kenge* as "an antelope, sometimes an okapi" (1961, 285). Here it is unlikely that Debo means either rhino or okapi, since the former are not found in the forest and the latter he has already specifically listed as protected by the elders. Perhaps he means bongo, a large, horned antelope found on the fringes of the forest whose markings can sometimes be confused with those of an okapi.

13. The French word *hopîtal* is often used to refer to small village dispensaries or clinics.

14. Maurice does not signify exactly who he means by "they."

15. Nkomo's comment "If it comes here to Zaire [Congo] ... " is interesting. It implies some doubt as to whether the significant monies garnered from the export of okapi ever make it to the people of the country (instead winding up in the pockets of politicians), much less to the people of the Ituri in whose forest the okapi originates.

16. Nkomo is most likely referring here to the okapi capture project at Epulu not getting any okapi in their pit traps.

17. See the following works, representing only a small portion of the research on the Ituri's hunter-gatherer peoples: Abruzzi 1979; J. Hart 1978, 1979; J. Hart and T.B. Hart, 1984, 1986; Putnam 1948; Schebesta 1933, 1936a, 1936b; Turnbull 1961, 1965, 1983; Wilkie 1987, 1988; Zuesse 1979.

18. Similar terms, meanings, and practices are found in other parts of Africa. In their fascinating reappraisal of Guinean forest loss based on both local accounts and remote imagery, James Fairhead and Melissa Leach report that the Kuranko people of Guinea speak of the forest or the waters being "tied" when hunting and fishing fail. The same word is used to describe a woman who fails to become pregnant. Furthermore, as they importantly point out, "Certain activities can provoke a state of tying across all of these domains; if, for example, a menstruating woman enters a fishing pool, both she and the water may simultaneously become tied. Treating tying merely as a metaphor would obscure this, as well as hiding common aspects of fishing and hunting ecology in local representation: for example, when a senior hunter dies, both fishing and hunting immediately become tied. Importantly, attention to such links throws into relief particular ways that representations of ecology might be socialised.... [It also reveals] how agro-ecological reasoning helps to constitute and reproduce the socio-political conditions within which it gains its sense, as well as the social implications of differing 'ecological' opinions and assertions" (1996, 8).

In the Ituri case, "closing the forest" thus takes on a rich meaning bearing ecological and political significance at the same time, the latter primarily in terms of local people protesting the "social implications" stemming from the State's ecological perspective that is very different from their own. "Closing the forest" is a metaphor demanding serious attention in order, as Fairhead and Leach put it, to "show up relationships which are considered to be stronger than mere likeness; where social and ecological processes are linked together intercausally" (1996, 8).

19. The term *bakpala* refers to the Bantu or Sudanic village-based farmers who share an exchange relationship with the Mbuti.

20. Masalito might be referring here to the way in which the reserve's research program brings the Mbuti and villagers into the forest together. Perhaps he might also be implicating the increased number of villagers that have settled in Epulu, drawn by employment opportunities the reserve has to offer.

21. Both the Mbuti and *bakpala* refer to each other using possessive pronouns indicating the depth of interaction between them. Indeed, the interdependent relationship between *bakpala* and Mbuti families is often passed on from one generation to the next. Although the relationship is certainly not egalitarian, the reader should not interpret the use of such possessives to mean ownership or slavery. Like other patron-client relationships found around the world, the relationship is more complex than simple ownership (see Peterson 1991; Grinker 1994).

22. See Box 5.1. This CEFRECOF work notice contains statements strongly disparaging the forest activities of newcomers and villagers in general, implying some degree of a pejorative attitude toward those groups.

23. He is most likely referring to the guns used by the reserve guards, as few villagers living within the reserve have guns.

6

One Step Removed: The Voices of University-Educated Project Staff and Local Academics

If the environment—the soil, the forest—has life, then farmers have life as well. If it doesn't have life, they don't have life.

—**Mbio ya Kotake**

A benefit commonly attributed to a university education is the opportunity it provides to step back from a phenomenon and attempt to view it more "objectively." One step removed from the nitty-gritty of a situation, the university-trained are theoretically freer from the emotions that may cloud the vision of those more directly involved, freer from the danger of missing the forest for the trees. They can therefore see things more clearly, or at least view any phenomenon as situated within a broader context. Or so it is said.

Such assumptions are, of course, fraught with problems; as many postmodern theorists have eloquently argued, none of us can ever achieve a purely objective point of view. All points of view are situated somewhere, and all of us are actors deeply involved in what we choose to see. The perspectives of the university-educated, no less than those of people physically engaged in an activity, are also selective according to their own set of value judgments, ideals, and inherited biases (Escobar 1995; Flax 1990; Haraway 1988; Harding 1986).

Thus the voices of the university-educated recorded in this chapter are not to be privileged, but neither are they to be dismissed. Examining land and culture from a distance *can* provide insights one might miss by looking only at close range. Furthermore, although the following perspectives on land and culture may not be as firmly grounded in direct experience as are the perspectives of local villagers, such experience is certainly not absent. Project staff and local academics may be one step removed from the experience of life out on the land, but that step is a very short one. They, like many of the farmers, foragers, and hunters who have already spoken, also gain at least a portion of their sustenance directly from their gardens and the forest. Moreover, despite living in modern homes situated within institutional compounds, project staff and local academics are not so thoroughly separated from the culture of the village as to have forgotten ancestral traditions. In short, their perspectives on land and tradition are in no way purely abstractions. They are also grounded in experiences of farming the land and stem from people who have had significant experience with the beliefs and practices of their respective cultures.

Nevertheless, project staff and local academics differ significantly from villagers in their degree of being influenced by Western education. Whereas many of the farmers and foragers I worked with had attended primary and in some cases secondary school, all of the staff, university students, and professors who speak in this chapter have received some postsecondary education. Their views on the same issues I broached with farmers and foragers are more extensively influenced by both the knowledge and the blinders that Westernized education affords. Although in general project staff and local academics hold a broader understanding of the causes of and responsibilities for environmental problems, their narratives also exemplify how the Westernized education they have received can blind them to the possibilities for finding solutions to such problems from within their own traditions.

Abbé Mosolo, one of the professors with whom I speak, eloquently expresses how such blinding can happen. When I ask him why he thinks African world views have been studied less than others, he replies: "I have the impression that there are very few Africans who promote their own culture. Schooling and studies are all being done in the United States or in the West in general. Furthermore, in regard to subjects of study, one prefers to work on subjects from over there. There are very few people who are really interested in Africa. It doesn't attract the majority."

Mosolo goes on to say that the same Western focus students find abroad pervades their own universities and schools back home. Congolese students study a curriculum that tends to "uproot" them (*déraciner*), since the textbooks available are often written from a Western perspective. Thus they are taught the geography of America while very few courses on African land and traditions are ever offered. And the lessons that do incorporate African traditions are little used. Teachers lack the instructional materials to teach about African traditions, whereas books coming from the West abound. Thus they teach about the geography of France and America's large rivers because those are the countries that send the books. But as for their very own homes and regions and cultures, they teach very little.

"The problem for me is the very orientation that is given to education in Africa," Mosolo concludes. "It uproots us and opens us up more to that which is alien than to ourselves.... [It] has made extroverts of us. We are open to the foreign but less interested in our very own matters."

In this chapter, I discuss the major themes that emerged from two focus groups I conducted with the university-trained staffs of the projects at both Loko and Epulu. With both groups I sought to elicit how staff perceive the environmental challenges of their work and what, if anything, they think their ancestral traditions have to offer in the way of solutions. I also include the perspectives of two professors teaching at postsecondary educational institutions in the city of Bunia, capital of the Ituri Sub-region, with whom I held in-depth conversations. A principal reason I have included the voices of the university-trained in this work is because they, especially the project staff, will have the most contact with local people in trying to improve the level of social and environmental sustainability within each region. Arranged thematically, this chapter draws out the ways in which the views of university-educated staff and university educators are both similar to and different from those of local land-users. Environmental perceptions shared by the university-educated and local people provide common ground upon which to further the work of ecological sustainability already begun. Pointing to the differences in perspectives between project staff and local people draws out what each group can learn from the other and clarifies the areas in which further negotiation needs to take place.

Voices of Contrast

The environmental projects at both Loko and Epulu benefit greatly from the dedication, knowledge, and authentic concern for both people and the environment that is shown by their Congolese staff. These young administrators, wardens, agronomists, community development educators, and natural and social scientists persevere in their work and research against incredible odds. Their salaries, although higher than those of many of the people around them, are still meager in comparison to world standards. Funding crises frequently frustrate their ability to carry out the important work they have envisioned. With uncanny coolness and efficiency they maneuver through logistical challenges that would leave many an administrator in more developed parts of the world completely at a loss. They remain committed to trying to make a difference for the human and natural communities with whom they work despite the numerous constraints posed by Congo's economic and political malaise. I have been truly humbled by their example.

However, in spite of such dedication, the distance between their world views and those of the local people with whom they work can be significant. Such distance may be understandable but it can also hinder the effectiveness of their efforts to build sustainable livelihoods hand in hand with local people. Discussing with these university-trained practitioners and professors many of the same issues that marked my conversations out in the villages, I am struck by the contrasts between their perspectives and those of local farmers, hunters, and foragers. Following are some of the main themes of contrast existing between these two groups of actors.

Ancestral Traditions:
Skepticism, Complexity, and Critique

In general, the project staff with whom I speak hold a much more nuanced assessment of their ancestral traditions and how efficacious they might be for addressing the current environmental challenges facing Central African forest communities. Whereas many of the voices recorded in the previous chapters speak to the positive role ancestral thought and practice played—and may still play—in preserving the integrity of the natural world, project staff are much more skeptical and critical. A major source of this skepticism is that they see the contemporary world so differently in comparison to the world of the ancestors. According to project staff, too much has changed for the ancestral ways to be of much use in combating contemporary social and environmental ills.

University-educated staff also highlight how any attempt to access ancestral values and apply them to today's environmental problems faces numerous complexities. Several themes reflecting this complexity surface in the focus group I hold with the staff at Epulu:

- Environmental problems and their prospective solutions need to be situated in the specifics of time and space, that is, what worked then may not work now; what works there may not work here.
- Traditions, societies, and cultures are dynamic; they are not static but evolve. Therefore, any attempt to bring conservation efforts more into line with indigenous traditional values must take into account the dynamism of people's ways of thinking, that is, one cannot try to restore value to nature through means that people no longer believe in or understand.
- Not all ancestral values and ways of thinking are positive. Some are based on lies, threats, or accusations of sorcery. Others are linked to practices that are difficult to condone, such as the prohibition to visit an area of forest, hill, or mountain that had been a place of human sacrifice.
- Natural resources are specific to certain physical locations. Traditional practices and beliefs that may serve to conserve those resources are also tied to natural specificities and are not uniform across space. Therefore, any attempts to incorporate these traditions into environmental programs must be selective and not exploit things out of context.

These are all excellent considerations that draw forth just how difficult it is to ground environmental work more firmly in cultural understandings and practices inherited from the ancestors. In fact, some staff members feel attempting to do so is a waste of time and that their work should instead focus on the more practical and pressing needs of today. "Some of the rules and thoughts of the ancestors are really no good for the current conditions we face," says Ngalina, one of Loko's staff. "If we try to apply them today, it will only be a waste of time.... In fact, they will put a brake on the work of development or on the matter of [conserving] the environment." He uses the example of ancestral taboos that forbid people from planting native fruit trees in the village and warn adults not to eat sweet fruits for fear of developing hernias. Such ways of thinking can frustrate Loko's present-day efforts of reforestation using native fruit-bearing species. Contrary to such teachings, he feels the important work is to "discover these and other [native] seeds and plant them in the village so that even though such species may be disappearing in the forest [due to garden clearing], they might remain and be conserved here in the village." Thus he sees that "it is necessary that we remove such ways of thinking, the ones they had long ago ... because they are not true."

Dieu-Donné, another staff member, has a bit different perspective. He acknowledges that many of the ancestral practices and ways of thinking such as the belief in spirit-inhabited springs and totemic animals, taboos against planting and eating certain species, benign technologies, and rotational patterns of settlement indirectly did conserve nature. The latter especially went a long way in providing some equilibrium between people and the natural world, as he explains: "In the old times, even though they removed things [from nature] they had a way of replacing them through fallowing. After a number of years, they would leave this place and go to another. It was a technology for replacing the resources that surrounded the village. Once these had decreased, the soil had grown tired, the animals had become a bit rare near the village ... people got up and went someplace else. It was therefore a technology by which they tried to replace the things in the environment that had become exhausted ... [a way for] animals and soils to restore themselves."

In further discussion, it becomes clear that ancestral taboos against planting native fruit trees, which Ngalina decries, and rotational settlement patterns, which Dieu-Donné affirms as being key to sustainability in the past, are in fact linked. But it takes an elder who understands what might have been the reasons for establishing taboos to make this relationship clear. Bill Lundeen, one of Loko's American personnel, recounts the explanation an elder living near Loko once gave him. "I asked the elder what was the reason for such taboos against planting native fruit trees in the village. And he told me that he didn't know everything but that he thought it was because if people plant forest [fruit] trees in their compound, ... they develop a need for that place since, through planting a tree, they have become psychologically attached to it and invested a lot of time in it. So then when the leaders of the village agree, 'No, the gardens have become distant, the children have lots of illness; it is necessary that we move our village,' people will resist since they have planted a lot of things in that place. So that is why it was necessary for them to prohibit people from growing such things in the village ... in order that they remain more flexible and compliant to move their habitation. By not investing too much in a place, ... they could more easily see that another place was indeed better, and that the place they were had grown old and not one thing of value remained there for them."

Mobility and "A Sense of Place"

Listening to this discussion of mobility and sedentariness, I am intrigued not only by how the ideas expressed relate to those of the villagers with whom I spoke but also by how they diverge from certain intellectual ideas in the West concerning sustainability; in particular, the current popular discussion within Western environmental literature concerning what has been called a "sense of place" (Berry 1972; Jackson 1996; Jackson and Vitek 1996; Sale 1991; Schoepf 1988; Stegner 1986).

Such works, and even fields such as restoration ecology that favors planting species "native to a place" rather than exotics, all seem to stress that environmental sustainability, as well as sustainable human communities, must come about through staying put, sinking roots, planting things, and stopping the addiction to mobility.

The elder mentioned above, however, explains that his people had taboos against such sedentarizing actions—and for good reasons. By planting a forest tree in one's individual place, one could become too attached to it to the point of resisting movement of the collective place so that it could be restored and the community could find new soils and hunting lands with which to sustain itself. Dieu-Donné confirms that ancestral patterns of mobility—moving on once surrounding resources began to be depleted—were a means of maintaining sustainability. Similar to the shifting agriculture by which people fallowed and rotated their gardens in the short term, shifting the entire village once gardens grew distant or hunting grew poor was also a means, a "technology," used to renew the earth and forests over the long term. The fact that African peoples have now become more sedentary and lost the ability to move for one reason or another has contributed to soil depletion and environmental stress.

Such perspectives on mobility and settlement contrast markedly with ideas expounded in current Western environmental writings on a "sense of place." They point out that for some peoples it is mobility, not staying put and developing a "sense of place," that is important for ecological sustainability. For many African peoples, mobility was the environmentally sound behavior, not "becoming native to this place." To invest too much in any one place would diminish that place's ability to restore itself, as well as lessen social cohesion within the community. And so it was taboo to plant trees from the forest in one's compound, to settle in and resist moving the village, and by so doing to place one's own perceived good above that of the entire community.

This is not to argue that one group's perspective is right and the other wrong. Rather the value of listening to others, such as project staff members and village elders whose voices are perhaps less well represented, is to remind us that all theories are bound to particularity. A different place, a different people, a different path to sustainability. What may be beneficial to contemporary America may be the source of the problem in places like African rainforests. I am not implying that those in the West who write about the need for people to develop a "sense of place" believe their message holds relevancy for all peoples and all places. My point is only to reinforce that there are many different windows through which to look and gain insights on our environmental problems and the paths leading to environmental sustainability. These problems and paths are manifested globally, yet each maintains a crucial particularity.

Environmental Problems and Responsibilities: Greater Scope and Specificity

Despite the ways in which certain taboos and settlement patterns may have combined to enhance sustainability in the past, Dieu-Donné goes on to stress how for a variety of reasons people today have become sedentary and the option of rotating one's habitation is no longer possible. In fact he feels that most all of the ancestral means of promoting sustainability, whether directly or indirectly, have collapsed under the weight of modern influences. The despiritualization of the natural world under Christianity, sedentarization, population pressure, and the growth of urban markets—all such changes lead him to conclude that "we are now ruining the environment quite rapidly" and that it is very difficult to return to any of the values and ethics of the ancestors for solutions. "There are traditions, but for us to return to many of them for the advice they offer will be very difficult today because technology and our hopes and desires have changed completely. The peasants have also completely changed."

Dieu-Donné's assessment reflects a second way in which project staff view these issues differently than do those I spoke with out in the village. In general project staff and local academics view the causes of their environmental problems and those who bear responsibility for them with greater scope and greater specificity. They attribute social and environmental breakdown not to such general matters as young people no longer obeying their elders or the coming of "civilization" but to specific historical events and certain ongoing phenomena such as Christian evangelism, population growth, poverty and economic injustice, the market and the creation of consumption, and the role of the State.

The Role of Christian Evangelism. The historical role played by Christian missionaries in significantly diminishing the vitality of ancestral traditions is one such historical event. It receives much stronger criticism from these university-educated staff and professors than it does from local farmers. During the Loko focus group, Dieu-Donné offers his point of view that both Protestant and Catholic Christianity succeeded in despiritualizing the natural world of the ancestors so as to remove certain restraints on people's uses of nature. He explains how evangelism, the coming of the "word of God ... changed the traditional structures a great deal. For example, in times of long ago, if you encountered certain animals, you would surely be afraid because they were animals to be feared.... Even if they caught them in their traps, they would simply let them go. They said that this is an animal but it is also our ancestor who has changed shape. In this way people would respect certain animals and certain birds ... since they believed spirits were inside of them. Due to such beliefs, animals gained some degree of equilibrium [vis-à-vis people].

"The same thing happened with water sources. Long ago it was rare to find someone cutting [a garden] next to a water source because they believed that bad spirits inhabited such places. And even though they did not forbid it, it was taken for granted that a person would not go [cut?] alongside a stream or near a spring. Indirectly, such beliefs protected these water sources. But today, due to eyes having been opened ... such things are no longer [believed to be] true. There are no spirits there. And so people will continue cutting and cutting and cutting until they reach right up to the spring."

Similar sentiments are voiced by staff at Epulu. Malengo, one of CEFRECOF's researchers, gives several concrete examples of how the Christianity he experienced in the Catholic Church served to degrade ancestral values that had some conservation merit. He remembers the church specifically teaching people that they were being fooled by the taboo against killing their totem, in this case sparrows. Consequently, people began attacking the birds. He also mentions the example of his bishop declaring that anyone caught using ancestral herbal medicine could not be baptized or partake in any of the sacraments. Previously, access to areas where medicinal plants grew was restricted as a means to protect them. But with the Catholic teaching against the use of ancestral medicines and against other ancestral values, people have now destroyed all the plants.

"It is true," he concludes, "that within the ethnic groups of Central Africa and particularly those of Zaire [Congo], there were environmental ethics in the societies of our grandparents that surely contributed to the protection of nature. But what do we ascertain at the moment? What we ascertain for the present is that there has been a degradation of all those ancestral values due to the development of Christianity. That is one factor. The development of Christianity has almost completely banished all the ancestral values. Due to the influence of religion, today's young people think that all the things our ancestors deemed as values are no longer worth paying attention to. That is one aspect that is truly against the maintenance of ancestral values."

Among the villagers I spoke with, the role of Christianity in destroying ancestral traditions did not receive such explicit attention. Some villagers such as Duateya, the elder whose story I recounted in Chapter 4, spoke positively about what their conversion to Christianity has meant. "I am happy with all of it," he told me. "Our ancestor and your ancestor is the same. For this reason, Jesus died on the cross." And rather than decry the loss of belief in ancestral totems, others such as Tata Fulani (see Chapter 4) have found such beliefs nonsensical due to their Christian faith. Speaking about his people's leopard totem, Fulani asked, "Since we have already come to believe in God [*Nzambe*], how can it be that an ancestor's spirit has become a leopard? Then what happens to the spirit that is God's? If he does become a leopard, then it is because his very own body has become a leopard. But then did his spirit disappear? ... So it is really kind of crazy.... Even in the midst of such beliefs, we are able to escape danger. And we God's children pray fervently to him that he will show us the way and put us on the path for the praise of his name, and that people will know that such things do not exist. There is nothing that can come before God." Tambwe, one of the Bogofo elders, did allude to Christianity's role in ending some of the *Gaza* initiation rites that he feels were so important for teaching children respect (see Chapter 4). However, he also remarked, "Now *Gaza* is the word of God [*Nzambe*] that has come out, and we teach [our children] from it so that they will know these teachings and obey some of their mothers and some of their fathers."

The Role of Population Growth. If not so much Christianity, a factor that villagers did blame for the deterioration of their natural environments is a burgeoning human population, as I recounted in the stories of Mama Malabo and Mama Kagu in Chapter 4. However, in my conversations with project staff, population growth assumes even greater importance as a cause of contemporary environmental stress. Staff at both Loko and Epulu are quick to point to it as a primary issue, a problem demanding a solution. As one of Loko's staff puts it, "Since environmental destruction has continued all the way up until today, perhaps it is a demographic problem in the sense that lots of people are coming here and there is a great need to work in order to get things…. Thus I see that environmental problems are really due to demography. The population is growing rapidly increasing the pressures on the natural world since now each person is looking to fulfill their needs to the point of saying 'I must do this, I must do that.' So when they begin exploiting the forest, it soon becomes impoverished and they must again move on, only to advance like that until finally the forest will be finished."

At Epulu, one of the CEFRECOF scientists argues that although traditional cultural restrictions may have helped preserve or promote the rational use of natural resources, at some moment one also encounters population growth that dramatically increases the level of needs. Therefore, the problem cannot be situated within only one context. In the past lots of space and low population densities kept some areas intact. But with high population densities needs increase, and people are forced to use the natural resources for their survival. "Therefore," he says, "in my opinion the solution is to seek one way or another of slowing population growth so as to prevent population densities from becoming too high. Look … because the people continually tell us, 'We have been living in these forests for a long time without any of the theory that you have come here to give us. And the forest exists today because of this type of conservation that we have practiced without any scientific theory. It is the very reason you are able to have this resource and the opportunity to study it at all.' Therefore, in my opinion, the problem resides at the level of population growth. It is necessary to try to manage it in one way or another."

The Role of Poverty and Economic Injustice. However, unlike some discourse on the role population growth plays in the deterioration of the natural environment, the views of these staff members also incorporate other factors that work in tandem with population growth to produce ecological stress. It is not simply a matter of there being more people but also the fact that many of these people suffer from a poverty for which they alone are not to blame. During the Loko focus group, Ngalina discusses how Congo's impoverished economy provides few options for poor people to avoid using the environment unsustainably. Further, he eloquently argues that Congo's economic malaise and its people's poverty can also be traced in part to injustices within the international economic system: "There is another aspect that I would like to add," he begins. "It is true, as we have been saying that the State bears a large part of the responsibility, but people of the village must also bear their portion of responsibility. All of us have the responsibility to understand issues of the environment and to conserve it. This is true but as we said regarding the problems of the economy, these are also very important.

"And I am adding another aspect—that the developed countries also bear a large responsibility when it comes to economic matters. For as we have said, if the economy is doing poorly, the burden also falls on the natural environment. For example, we know that the developed countries and the donors such as the World Bank and others give funds for popularization [*vulgarisation* in French, akin to agricultural extension] programs. They want us to go to the village and teach coffee-growers new techniques in order to produce lots of coffee. So then farmers put these techniques to work and grow a lot of coffee. However, it is over there [in Western financial centers] where they fix the price. This year was a good example: Coffee prices were quite good but then later when farmers went to sell it, they were told, 'Oh! We are no longer buying it at such a price.' You sell at their price even though it is falling. Thus the burden falls on the people of the village who work hard to get some coffee but then, when it comes time to sell, receive only a worthless amount of money that doesn't go very far. And so they are obliged to leave it and must either make charcoal or cut firewood to sell in order to get a little extra revenue.

"This then is what I want to say—that the developed countries also bear some responsibility. They want to help, but they must also help with ways to encourage our economies here in underdeveloped countries like Congo. If they would only take some part of the responsibility and help out in such domains by agreeing to buy our export goods at a fair price in order to encourage people here and improve the economy. Even with this, it will still be difficult to overcome all these environmental problems."

Ngalina also emphasizes that beyond the issue of population growth lies that of industrialization and the destruction it has wrought within Central African environments. Again he implicates ways in which the world external to the forest, and not just forest people, bear responsibility for the forest's demise. "Long ago [modern] industries did not exist in Africa. With the coming of industry and the high demand for lumber on the part of large companies, Africa's ecology and natural environments began to be disturbed. Companies went in and cut the forest, felling the trees. We have an example today in companies like SIFORZAL [a subsidiary of DANZER, a German logging firm]. Starting up in Haut-Zaire [Orientale], on past the city of Bumba, they are cutting tons of trees for export, disrupting the environment.... Large companies are cutting trees like crazy but as for replacing them through programs of reforestation, there is nothing."

Staff at Epulu also extend the scope of responsibility for Central Africa's environmental problems wider than do local people. Alphonse, an environmental educator, makes a long statement about how the external world plays a very significant role in changing indigenous people's use of the forest. Therefore, he feels that in addition to local efforts to solve environmental problems, controls must also be placed at the

A local villager hired by an Italian commercial logging company wields a chain saw on mahogany logs waiting to be hauled to the river port of Kisangani, 130 kilometers to the west. Note the worker's bare feet only inches from the saw's blade.

source of these problems. "I want to know if there is not a way to establish controls at the level of the source," he asks, "among the people who come and impose on local people disorderly uses of the forest; that is to say, at the level of Europe, the West, let us say, of Europe and the United States.... For we have found that due to external influences from Westerners and Americans—when they favor the commercial exploitation of a natural resource then indigenous people lose their knowledge of how to use it rationally." He raises the example of the ivory market and describes how ivory buyers are fostering the irrational use of natural resources. "Thus, concerning this point, I would like to know what must be done in order to prevent the exterior world from continuing to subdue these local people for whom the elemental use of the forest is innate."

The Role of the Market and the Creation of Consumption. Although such narratives reflect a greater scope and specificity than do the narratives of local landusers, both sets of narratives share a general critique of Western influences and modern capitalist markets. As in my conversation with the Bapukele elders (Chapter 5), the problem of the market receives primary emphasis during the focus group with

Two men produce pit-sawn mahogany lumber in the Ituri. The tree, located directly behind a village homestead, had died while standing and was subsequently felled. Such hand-sawn operations have long served as a sustainable method of harvesting timber for local needs.

Epulu staff. At one point Jacques Dongala, one of Malengo's CEFRECOF colleagues, counters the latter's stringent critique of Christianity with the assessment that more than religion, it is the culture of the market and commercial exploitation that has caused the most environmental damage. And similar to how Tata Fulani repeatedly linked cultural and environmental decline with the coming of "civilization" (see Chapter 4), staff at Loko mention numerous ways in which modernization and Western influence in general have undermined the dynamic harmony that existed between people and the natural world in times past.

Abbé Mosolo, a local academic with whom I speak, reflects a similar position in response to my question of whether he thinks culturally sanctioned limits to exploitation actually existed among his ancestors. He expresses this criticism of the West shared by project staff and local land-users even more forcefully: "Yes, I think that the fact is, Westerners have contaminated us. Contaminated us in the sense of going beyond, beyond those immediate needs. You showed us that we could surpass what we had. But I think that in Africa you can encounter people who are content with what they and their children have, that which enables them to live. Thus in the beginning I don't think that they regarded nature as a means to produce results for themselves. Nature existed to enable survival.

"Thus it is only today that one hears of such things as somebody killing sixty monkeys because they see that by doing such they will be able to obtain more money. But the African of yesterday did not do such a thing.... In the past we sought to find well-being. We never sought overconsumption. For the most part, everyone was content with what they produced for subsistence, to meet their needs but without keeping stocks. In the house, there were no reserves."

In such ways, project staff, local academics, and local land-users share a common perception: They all realize that the environmental crises they face are linked to larger-scale forces, nonlocal in time and space, that nevertheless have dramatic impacts on their lives. But these narrative critiques of the West also make clear how much local staff people have to offer the farmers and foragers with whom they work in terms of educating them more about those forces. Project staff have more knowledge and greater access to information about how "industrialization" and "civilization" have entered deeply into Central Africa's natural and social communities; about the complex ways in which Western influences have contributed and continue to contribute to environmental decline. They link local problems more clearly to global phenomena rather than view them solely as local issues, or leave them situated only in the colonial past.

Such knowledge and understanding can help to strengthen attempts to resist such forces and the environmental decline they engender. By more fully comprehending the particular ways in which global political and economic forces affect their local environments, local land-users, in partnership with project staff and local academics, can enhance their capacities for improving the conditions of their lives and the conditions of the forests that are their home.

The Role of the State. Another issue on which project staff and local land-users both agree and disagree is the role the State plays in both the degradation and the protection of the environment. Both groups are critical of the State for its complicity in environmental destruction through such means as corruption; its failure to follow, much less enforce, what environmental laws exist; and its officials explicitly taking part in illegal activities such as commercial meat trading. Paluku, who works with Loko's animal husbandry program, gives the example of sawmill owners going to pay their environmental taxes only to be bribed by government tax officials to cut more trees and share the proceeds with them. One of the researchers at Epulu mentions that even State officials working in conservation can become involved in the illegal trafficking of wild meat.

Yet unlike local land-users, staff members, especially those at Loko, feel that the State can and must become even more involved in managing people's relationship to the environment. Although they realize the State is part of the problem, they also feel it can be a major part of the solution. Increased government environmental regulation backed up with severe penalties in cases of violation is absolutely necessary, they feel, and needs to accompany any attempt to appeal to people's culturally grounded ethics.

Mamuya, a community development organizer at Loko, puts it this way: "And those laws need to be really strong for if they are only like it was during the time of ancestors—" He pauses for a moment and then, without elaborating on the comparison he has drawn, continues: "But there needs to be pressure coming from a tough superior level. If they only come with a system of trying to restore people's conscience, to persuade people, they will have a lot of difficulty.... So, I see that regulation is also good. Of course, conscience still exists but in these times most everybody's conscience is not good; it has really been weakened. Take a look at the State law against killing okapi—if people kill one, they will really try to hide it well.... If not, the State will arrest them. Because of [this threat of arrest], people are becoming more conscientious and such laws have become like a protective barrier to keep such [killing] from happening."

Yet as the previous chapter shows, in situations such as Congo's parks and reserves where the State has been given greater mandate to enforce environmental regulations, local people often suffer unfairly, heightening local resentment and mistrust of government officials and programs. Although Loko staff are not aware of the problems engendered by increased government environmental regulation within the RFO, they seem to anticipate them. Perhaps that is why they qualify their recommendation of increased government environmental regulation with the following sentiments, as expressed by Dieu-Donné: "Well then, the State needs to be strong; but not strong only in the sense of imposing things but also in the sense of educating, sensitizing, and teaching [people]. Because if State [officials] only prohibit things—'Don't do this, don't do that'—without giving any explanation, when [they] leave, people will simply go back to doing what they were doing because they won't have learned a thing." According to these project staff people, the State has to be involved, but it must be a State that is educative as well as punitive.

Education for Change:
The Direction of Flow

One final contrast between the narratives of project staff and those of local land-users is the emphasis the former give to this matter of educating local people, helping them to understand. We draw the focus group at Loko to a close by staff members stating the primary difficulty they face in their work, the number-one obstacle constraining their success. Five of the eight people present say that their primary difficulty is local people's lack of understanding, their failure to really connect with the reasons for and the value of the sustainable ecological practices Loko is trying to promote.

One person expresses it as local people "not giving much meaning to all that we have taught them." Another puts it more metaphorically: "For we are the ones pushing the people of the village. Because the people of the village are like children climbing a tree. They are not yet capable of climbing the tree by themselves from the ground. We have to give them a boost and keep pushing them until they are high enough to climb the tree. But while they are still young, if we stop pushing after they have only gone up a little ways, they will fall." Others speak of people's lack of conscience or their lack of motivation—whatever it is that is needed for people to make a change—to put things into practice in order that they might "move from this place that was bad so as to arrive at the one that is good."

Clearly evident here is the direction in which the majority of staff feel environmental education needs to travel: from them to the people. And there certainly is much important and useful knowledge among these highly trained staff from which local people have benefited and can continue to benefit. However, it also needs to be emphasized that education is a two-way street. The narratives of local people recorded in the preceding chapters reveal a great deal of environmental knowledge and wisdom from which project staff can also learn.

Furthermore, such environmental education efforts may be even more fruitful if staff also deeply examine the following questions: What might be the reasons for the failure of our work to really take root among local people? Is it people's lack of understanding? Or does the problem lie in the manner and substance of our teaching? Or are there extenuating circumstances that seriously impede people's ability to improve the environmental sustainability of their lives?

Perhaps one reason why many people in Congo have a hard time connecting with the goals of projects like Loko is because they have many pressing problems that for the moment take priority. Such extenuating circumstances certainly do play a role, according to Abbé Mosolo. When I ask him to evaluate the level of environmental awareness and concern among his students, he replies, "For the moment,… they are not really very aware of environmental problems…. That which is of greater interest now, the real issues, are social and political problems. These are what have really captured the attention of the students more than ecological problems. It has been this way ever since the structures of the society began to stop functioning. So then, people are much more aware of such problems than the problems concerning the environment. The environment is not yet really a problem for them."

I ask him when he thinks it will be and he replies, "When the problems become vital to people." Right now politics are vital, Mosolo contends, politics that struggle with the issue of neocolonialism. Mosolo feels that African countries have achieved independence only to find that many of their rulers are *nouveau colons* (new colonialists) who are leading their countries into worse places than those known before independence. Once some political stability is in place, it will be possible to give some reflection as well to nature, but for now politics remains in the fore.

"In the African situation," he continues, "students are also captivated by politics because the politician has everything to say. For example, say you are doing ecological research—okay, a good thing—but an African says, 'For what purpose? I will currently have little impact as a student of ecology.' I imagine that when there is more calm at the social level … a calm in which each one is more at ease in the society, then attention will be given to other aspects that are being neglected today like ecology. One neglects it because it is not the problem that troubles us the most today. But I am convinced that in the future, it will be a problem of interest to Africa."

Environmental problems, and hence environmental concern and awareness, are also not equally distributed in Africa, he points out. He gives the example of the Saharan countries where environmental problems have become vital. There, people's livelihood is endangered by the encroachment of the desert. But for him, here in Congo, environmental issues do not pose a vital problem. He does not feel endangered but at ease with regard to the environment. But he is not at ease politically and politics hold the strongest pull.

* * *

The voices of local academics and university-educated project staff recorded here contrast with those of local land-users and add new perspectives to the question of how efforts to cultivate environmental sustainability in Central Africa can be more firmly rooted in local cultural beliefs and practices. A primary difference between their views and those of local people is a deeper appreciation for just how difficult such rooting is—partially because the soil of culture has been sorely impoverished; partially because the severe economic drought facing Congo makes planting anything difficult; partially because economic and political winds originating from afar continue to disturb the fields of nature and culture from which local people draw their livelihoods; partially because the State proves to be a more fickle than faithful helping hand in the work of cultivation; and partially because there is little cross-fertilization of environmental education between staff and local people.

Even though such considerations make cultivating environmental sustainability in the soil of local culture very difficult, they do not make it futile. Dieu-Donné reflects a sentiment many project staff members and local academics share: Despite their skepticism, criticism, and pointing to the complexity of the issue, trying to apply the principle of *sankofa* (drawing on the past to prepare the future) to their work with the environment remains a valuable endeavor. "I think that if we really turn our vision there, we cannot reject everything of the past," he states. "We have tried to recover some things such as indigenous forest fruits in order that they become better known. Such things we see now. But I think if we look deeper we can find even more examples that are also good and that remain valuable today. Perhaps our path now needs to be one of searching for those ancestral values that can heal society. Perhaps they are there, not directly evident but they exist. They could have an impact and contribute in order to help us for the long haul. I don't have any examples of them right now, but perhaps they will come with time."

The following section of this chapter offers some of those examples as one means of showing how university-educated people and local people, despite their contrasting voices, also share some common ground.

Searching for Common Ground

The narratives above have already pointed to one piece of common ground beneath the feet of project staff and the local people with whom they work: the recognition that local environmental problems are linked to powerful external forces and that therefore solutions must be implemented at the problems' source points as well as at their end points. But other areas of common ground also exist. The work of *sankofa*, traveling the new path that Dieu-Donné refers to above, represents one means of finding those areas. Jacques Dongala, a CEFRECOF staff member, eloquently describes the nature of this task: "Now, it is necessary to go beyond to the real and total problem. It is necessary to go to the philosophical plane, to the plane of thought and mentality of the people who are here. What have they done to preserve this environment as it is for centuries and years? What can we draw from them to teach to others—their respect of the environment, their respect of their natural milieu? I believe that is the level at which we must situate ourselves."

Common Ground: The Need for Local Control

Not long after being at a loss for examples of ancestral practices having ecological benefits, Dieu-Donné comes up with one, an example that local land-users also raise. It is an example that meets the goal, as he puts it, of "looking for systems that stem from our [ancestral] values but are able to adapt themselves to the modern situation." He begins to describe the example of ancestral tenurial rights and how each clan (roughly equivalent to the modern political administrative category of the *groupement*) had their own domain, rights to which were respected by other clans. People knew the streams or hills that marked the boundaries of their lands and remained within their domains in order to avoid the risk of sparking conflict with neighbors. And within those domains they did not "hide the rules" but applied them to both those of the clan and those from without who might come asking to hunt or fish.

Dieu-Donné wonders if such tenure rights might provide a model to follow today, a model of keeping control over land and resources at the local level rather than losing that control to those higher up. He suggests reconstituting such ancestral tenurial systems, perhaps modifying them a bit, so as to enable people of modern day *groupements*, like Bogofo, to know again the extent of their domains. And it might even be possible, he suggests, that using such tenurial structures, local people could "set aside [some land] in which to do something like their own research in the sense of recognizing, 'Okay, this is our concession given us by our ancestors. Nevertheless, parts of it we want [to protect] by making them off-limits to people.'"

Dieu-Donné continues describing what such an arrangement would mean. "They would be the ones in control of such areas in order to prevent people, say, from Gbado [outside the *groupement*], from coming in and the same day beginning to cut a garden. It is they themselves who cut it. The same for hunting elephant or other animals—people from Gbado cannot hunt them here. It is necessary that they come and ask permission from the village before going into the forest. Then if someone comes to the village [leaders] they can orient him or her as to where hunting or fishing is permitted within the village domain.

"Such arrangements are something we really need, and they could be quite easy to set up to make them efficacious again in the present day. Because each *groupement* already has their designated area. And they themselves know that in order to assure their survival [for instance] here in the village of Bogofo, such and such an area is to be protected ... and such and such an area is for this purpose, and so on. And if people from outside the village come: 'Look, accompany them like this. Even if they are hunters, even if they have official papers from the State, no matter who they might be, entering that area over there is prohibited and it is only this area that they can use.'

"I think that such a system, coming in part from the ancestors, in part based on the State, will be quite easy to manage in order that ... land be designated to fulfill specific vocations. And it should be in the hands of the local collective, the local people, or the village headman, or the people of the village. Because that land is where their hope lies. It is only a matter of maintaining watch over it.

"For example, say the *commissaire* [*de zone*] feels the need for some hippo meat and he simply signs some papers [telling his hunters] 'Go kill a hippo in the Loko [River].' All right, the place from which he draws what he needs to live day by day is not here at Loko but there in Businga [the zone headquarters]. Nonetheless, he simply gives the order, 'Go kill a hippo.' However, those hippos have a direct relationship with the lives of people here in Bogofo. In one way or another they help them. But in any case, the person from Businga or from Gbado comes. And when he arrives, we the people tell him, 'Hunting such animals is forbidden there, or it is forbidden here.' In such a way the local collective has some control over that place.... It's like we hear about in other countries, that each canton has some degree of autonomy so that when something that threatens people's livelihoods gets introduced from outside, they stop it from continuing in order to first discuss things....

"So I think it is a matter of modifying these arrangements a bit ... such that all the people [of a *groupement*] know that this is their land. They all know it [and can say], 'All right, you in there, you are entering into the forest in a disorderly manner. However, we are trying to restore order by seriously exploring the question of how best to manage this land that belongs to our *groupement*.' I think that people in the village are doing this. It is somewhat of an ancestral system but one that has been revised a bit."

In response to Dieu-Donné's propelling argument, Ngalina maintains that trying to reconstitute the local tenure systems of ancestral times can actually block the work of sustainable development. Sometimes people with ancestral claims to land will just let that land lie and have no desire to "put it into value." Meanwhile, those next to them who really have such desire are blocked from extending their fields because the seemingly vacant land next door belongs to someone else.

Dieu-Donné is quick to raise a counter argument: "But that is indeed the point where we can try to find a solution by asking why those people want to extend their gardens in the first place. 'You were in this place. Why do you want to leave it?' Most likely it is because they have found that the soil is worn out. So then, in order to find a solution, [we need to ask them], 'If the soil there is worn out, is the only option for you to go and attack another piece of land? Or can you start to restore the soil here in the place where you are?' Such situations can help people see more clearly ... 'Okay, my area of land is limited. And I don't have another plot. Therefore, I must start to preserve the small amount of land that I have here.'... But in the case where no restrictions exist, [they might tend to feel that], 'All right, this land is worn out; I'll get some more over there. When that land is ruined, I'll get some over there. When that breaks down, I'll get more here,' and so on and so forth. But that only opens the way to waste."

Paluku expresses this need to focus things at the local level in another way. Like some of the farmers I speak with, he also feels it is best that actions for change—whether Loko's programs or the efforts to revitalize young people's initiation rites and teachings that Tambwe and Fulani spoke of (see Chapter 4)—remain centered at the level of the village or *groupement*. That is the level where one finds the elders who "truly know the meaning of their land," as Paluku puts it. "It is best to focus at the level of the *groupement*," he suggests, "because there one finds those who still hold to many of the ancestors' ways of thinking. So when you go out to the village for the first time, it is necessary to speak with the elders, the *chefs de groupement*. They are able to offer a good plan based on knowing how the village was in the past and how to improve it today. If we focus mostly on the *chefs de groupement*, then when they call together the people of the village, there is a greater chance that people will really hear what the elders have to say; it will really enter into their thoughts."

In contrast, he argues that if efforts are channeled through higher authorities such as the *commissaire de zone* or even the *chefs de collectivité*, they may be much less effective. This is due to the fact that such authorities "have come from afar. When they come and install themselves here they may sometimes get the idea ... 'Since I come from over there and not from here, this is not my place. That is good because now no matter how much I might mess things up, it will not bother my conscience.' And things will fall apart."

Paluku and Dieu-Donné's narratives indicate that among project staff one can find the same preference for local focus and local control as is voiced by local land-users. Both groups prefer to take a grassroots rather than a top-down approach to solving the environmental and social problems confronting their lives. Such common ground offers hope for the work of building sustainable livelihoods since that built from the bottom up usually outlasts what is built from the top down.

Common Ground: Ecological Benefits of Certain Ancestral Beliefs and Practices

Although Dieu-Donné feels that more examples of ancestral traditions on which to draw for the work of environmental sustainability will come to him with time, other project staff I speak with have them close at hand. Gilbert Mendo, the scientific warden at Epulu, like some local land-users, finds hope in the understanding and practice of totems and taboos that to some degree remain alive in the Ituri. He recounts how Lese farmers in the villages of Andikao and Andisoli explained to him that those place-names derive from local totems, the leopard and the sitatunga (*Tragelaphus spekii*) respectively (*andi*, against; *kao*, leopard; *soli*, sitatunga). "Even when such animals ravage our crops, we cannot kill them," he was told. From this he concludes, "So there are also these toponyms that reveal the cultural importance people have assigned to this conception [i.e., not killing certain animals]. And I think that in terms of consciousness-raising or revaluating culture, we could next propose how ... to give value to such cultural aspects as these that to me seem very important."

Malengo agrees with Gilbert and underlines how important it is for them to focus on cultivating the positive rather than the negative messages within various cultural traditions. He is optimistic that such positive meanings can help young people understand that indeed "such and such [i.e., a tradition] is something from which, if we observe it, we will also be able to benefit." He goes on to give the example of how among the Lese, Ndaka, and Bira the okapi is taboo—"truly king"—and deeply respected, not because it is dangerous but because its skin serves various ritualistic needs such as those associated with marriage and the installment of chiefs. Cases such as these, where cultural practices holding ecological benefits remain alive and well, present the most hope for efforts to ground conservation within local culture.

In the same vein as Gilbert and Malengo, Jacques suggests several other aspects of forest traditions that merit study and might very well contribute to CEFRECOF's conservation efforts. He refers to certain areas of forest where the Mbuti do not hunt because the places are believed to be inhabited by certain spirits. He suggests such places may serve as zones of "integral conservation" in which animals find refuge and a means to restore their numbers after periods of more intensive Mbuti hunting.

Another aspect he mentions—the periodicity of resource use—also figures highly in the narratives of local land-users (Chapters 3–5) as an ecologically beneficial practice. Jacques wonders if such periodicity among the Mbuti might not act as a "built-in" control on hunting. He asks whether during the honey season the Mbuti hunt in the same rhythm and timing as during other periods; or whether during the termite season they hunt as intensively as they do at other times of the year. When someone replies "no," he enthusiastically responds, "So, it is exactly these aspects that we need to make the most of!" CEFRECOF's research thus needs to ask such questions: "Why does hunting differ during different periods of the year? To what activities is it tied? Is it tied to the life cycle of honey or to other types of periodicity?"

In Jacques' opinion, the Mbuti understand very well these elements that influence the variation in the seasonal course of their lives. They know that the best times to go hunting are when the animals are fattest. And when the Mbuti are busy gathering honey, they may not hunt at all. And who knows, he asks, perhaps the honey season, when hunting may be down, correlates with the gestation period of certain animals? He seems genuinely excited about the prospect of doing research such as this that explores questions rooted within local cultural practices. Attitudes such as his offer some hope that project staff are taking the cultural traditions and practices of the people with whom they work seriously.

Common Ground: Ecological Benefits Within African Metaphysics

A bit farther away from forest and field, I hear other people with university training also voice positive assessment of the ecological benefits within certain elements of Central African thought and culture. Professor Kwela teaches at the Institut Supérieur de Pedagogie (ISP—somewhat equivalent to a postsecondary teacher's college) in the city of Bunia, the capital of the Ituri Subregion. I spend several days in the city on my way out of Congo and use the time to gain additional perspectives on the issues of culture and conservation among those teaching in local institutes of higher education. He warmly welcomes me into his office even though I have not been able to get a message to him of my coming. We are soon caught up in conversation about our shared interest in immigration into the forest, but he is also very interested in my research project and offers some views on the matter, which resonate with the narratives of the Ituri's land-users.[1]

"It is evident that cultural resources for conservation exist within African cultures and metaphysics," he begins. Continuing in his deeply resonant voice, he explains how African conceptions of the universe are very different from those found in Western countries. "Like the ancient Hebrews," he points out, "Africans tend to believe that everything comes from God's action. The existence of nature is therefore imbued with God, and correlatively, the body [i.e., the physical world] and spirit exist together. This is very different from the Western concept that requires one to mistrust the body in order to save the soul, a way of thinking that leads to a severe despoliation of nature.[2]

"Even though many Africans accepted the Western religion that destroyed some of these conceptions, they have also remained faithful to their metaphysic in many ways. For example, many still believe that God acts through the ancestors and that the ancestors remain alive in nature. Thus one must respect nature not in the sense of worshipping it but as a way of respecting the ancestors. If you transgress the ancestors who abide in nature, then you will be punished. Thus it remains forbidden to destroy certain trees and rivers that are a source of life."[3]

Professor Kwela, like the staff members at Loko and Epulu, conveys a healthy skepticism, but his is directed more against the values of modernity than against those of African traditions. Rhetorically, he asks, "Who really has the higher quality of life—modern New Yorkers living high up in their apartment buildings with their TVs, cars, and suits or the Lese living in the forest with no shoes on their feet?" According to him, the Lese may in fact be much more at ease, happier, suffering less anguish, and therefore have a higher quality of life than New Yorkers or Parisians who live in an artificial environment, engulfed in the worry and stress of pollution, nuclear weapons, and government taxes.

"Yes, we must change," Professor Kwela acknowledges, "but what type of change do we want?" The question for him is how Africans today can find some sort of equilibrium between the West's "brutal conversion to modernity" and a return to the African past. Africa needs development, he says, but it must be a development that keeps the environment intact and keeps from destroying the deep humanity that is a hallmark of African cultures. "Development is to enable the blossoming of man, but what man?" Kwela asks. "A Western-style man that has entered into a brutal change to modernity? Do we follow a German-Greek metaphysic or an African metaphysic?" He seeks some sort of balance between the two.

In terms of forest conservation, Kwela feels such a balanced perspective would draw on African metaphysics to seek a more appropriate vision of the forest as a place of spiritual as well as scientific value; a more appropriate vision of animals as "beings made alive by God"; and a more appropriate vision of people as respectful parts of nature rather than as separate from it. Foreign partners in the work of nature conservation must therefore base their efforts more on such African metaphysical elements rather than rely solely on Western concepts, he feels. They must be aware of the "real problems" on the ground and of the "human dimensions" of their work. In regard to foreign partners working in the RFO, Kwela recommends that they be "concerned with investing in Zaire [Congo] and not just with exporting okapi."

The ISP is also the home of Radio CANDIP, a local station that broadcasts a variety of news, educational, and cultural programs throughout the region. My time with Professor Kwela concludes with the opportunity for me, after all my interviewing, to see what it is like to be interviewed. Kwela expresses his wish to ask me a few questions for a program to be broadcast the following week. I agree, and after giving an overview of my research I use the occasion to say good-bye to all the people of Epulu and the Ituri who have helped make my stay so rich and enjoyable. Professor Kwela then proceeds, in one of those perfect radio voices, to summarize our conversation. He ends with a message to all the Ituri's gold diggers and wood sawyers that they must not destroy nature, and he poses a final question to the public: "What do you do for the protection of the environment?"

* * *

On the same side of town as the ISP is the Catholic mission where Abbé Mosolo resides. Priest and professor of ethics since 1981, Abbé Mosolo, like Professor Kwela, receives me with an openness that again is truly humbling. Also like Kwela, he speaks of ways in which African metaphysics can contribute to the task of cultivating environmental sustainability. In so doing, he also reveals some of the common ground shared by the university-educated and local land-users.

I ask him if he thinks it is true, as many African philosophers and theologians have argued, that Africans have not separated the sacred and the profane as we have done in the West and that in many African systems of thought all of nature is considered sacred. "If it is true," I ask, "could it possibly have some positive impact for an environmental ethic? Could it be a resource one could exploit for conservation purposes?"

"I think it is a resource that one might be able to exploit," he replies; "however, it would require that one speak in terms based on African logic. In Africa ... people and nature live within a certain degree of harmony, and often many of the things in nature that people don't understand, they simply attribute to God, or to the spirits. In the West, one proceeds to understand nature. One tries to explain each and every phenomenon. Here, for traditional Africans, that which they don't understand, they attribute directly to God or to the spirits. Thus there remains this bridge between nature and God. It is very spontaneous.

"So then, there could be a way to exploit that for a certain sort of ethic.... I think there are perhaps certain beliefs that could help to protect nature, but the question is one of entering deeply into African logic. Because if one arrives at things too much under the influence of Western categories, you risk getting quite a number of things wrong. However, there are some things that one can easily discover by beginning with precisely this approach; for example, by examining people's proverbs and sayings, most certainly important phenomena for discovering a certain logic existing within. Children have any number of things to talk with their parents about. Certain proverbs and such have an impact on a child's actions in regard to nature. Thus, in Africa, one must perhaps enter into its logic, into how logic functions within Africa, and see from there if there is not a way to protect nature itself."

I then ask Mosolo whether he agrees with the hypothesis that to Africans, or according to African logic, life is sacred. "Is this true or is it simply a fabrication of Western observers?"

"No, no ... life is sacred, that is true. In Africa, I believe, that is at least the hypothesis by which we live.... For example, here among us ... life is so sacred that if someone kills somebody else without anyone knowing, they would say that the blood [of the victim] spontaneously cries vengeance. Their blood cries vengeance. [Such murderers] will not be able to hide. They could perhaps hide their crime well, but even if others do not discover it, in the end they will turn themselves in for they have killed a human life. Moreover, it is not that someone witnessed it; nor even a matter of discerning it through investigation. But no, it is they themselves who end up turning themselves in. Because one cannot bury a crime against life. It is impossible to bury it. That is how they explain the remorse that works on someone. They say, 'No, when somebody messes with life to the point of killing someone, well, in the end they themselves will surrender.' Because life is too sacred for someone to make it disappear without others knowing."

"And according to African logic, is only human life considered sacred or does one consider all life sacred?" I ask.

"Yes, it is principally human life. But it is also the life, one could say, other lives, like animal life. Here, for example, there is a practice—when I was learning to drive a car, my father, who is still alive, saw me driving a car for the first time and said to me, 'My son, you must avoid running over a dog. Because if you run over a dog, if you run over it on purpose, then it is inevitable that the same day you will kill a person. Since you did not respect the life of the dog, of an animal, then the same day you will also be very likely to kill a person. Thus, avoid killing a dog.' Thus one could not run over a dog.

"There are some animals ... like ... snakes for example; those, one must kill.... But there are some animals that one must not touch. There are other animals that carry bad luck or good luck, like owls.... They let owls be. They do not kill them.... There are certain things that one respects, like the life of animals ... and the life of a person. There are some animals that one willingly considers as game, and one may kill them. But there are others that one does not kill. Thus that is what I ascertain. With a little reflection upon it, I don't know what one might be able to draw from it."

Common Ground: Nature Must Be Allowed to Serve Local People if Local People Are to Serve It

When I ask Mosolo his views on Congo's current conservation program, he does draw some conclusions similar to what I have heard before: People, whether those who work for parks and reserves or those who inhabit them, will not fully be able to conserve nature unless their own needs are first satisfied.

"Certainly, there are [conservation] structures that in essence are good," he explains, "but they are not efficacious in the sense that we put people in a park, a place especially for animals, and they eat them. Take the case of Virunga [National Park, along Congo's eastern border], for example; the people there ... are faced with vital problems just trying to live.... Those who work there, their own lives are not being taken care of. They do not earn a salary sufficient to live on. Thus, if they see a way of getting some extra money, they are ready to deliver even wild animals, even rare species to those willing to pay them money. Thus the government has set up a structure there but an ineffective one since those who work in those places do not receive the desired salary that would allow them to do their work at ease.... They will not protect nature because they are faced with vital problems in earning even their own subsistence."

I then raise the specific example of the RFO and ask him if he thinks a multiuse, human-inhabited wildlife reserve can succeed in the Ituri, in both a philosophical and practical sense, given the exigencies he has mentioned.

"Yes, I think it could succeed if the people are aware of the values. If one explains to them the various values that are present in a reserve, they would respect the reserve's policies. But they must be in a situation that permits them to forego the use of such values as resources. If they have their own resources, then I think they would be sensitive to those values.... But if there is something that pushes them to disturb wild animals, it is possible they would. But I think that if they were conscious of the values of nature, [such a reserve] is possible ... under the condition that they do not see these things [the natural resources of the park] as the sole means of making a little money. If they are at ease in their place, I am certain they would not disturb such resources but willingly protect them....

"Do the people of Epulu, [for example] know that [the forests of Congo] are the only place in all of Africa where one finds the okapi, and that it is very valuable, something to be proud of? I think they could [protect] them as long as one enables them to work in the service of that nature. The people in the area, if they could become tied into the same service, being paid for their service and tied in to the conservation of nature, then they would willingly carry it out. Because as they say in French, 'One does not cut the tree in which one is sitting,' or as we say, 'A wine drinker is not going to cut the palm that gives him wine.'

"Thus if they know that this nature that they protect also serves their interests, and that they have an interest in protecting this good, then they will not touch it.... And as it is that nature that allows them to live, they will not destroy it. Because they know it is of value, they have worked for it; therefore they must protect a part of it."

Similar sentiments about the need for nature to be able to serve people in order for them to serve it are voiced by staff at Epulu. Alphonse argues that the utilization of natural resources by the Ituri's indigenous peoples is really incompatible with the installation of a park. Therefore, he asks his colleagues: "Since we are in a reserve, the RFO, and we desire that indigenous people use the forest and remain there, that they not be evicted, is there not a way for us to place more emphasis in our work of coming up with a rubric that is more in line with these features; something other than a park?"

Alphonse's question sparks a heated discussion of the differences between a park and a reserve and to what degree each might be incompatible with people's use of the forest. At one point in the discussion, the RFO's chief warden turns and asks me if, in my research, I have considered the misery of the people—their poverty—since it is also a major factor that drives people to overexploit natural resources. "We cannot forget that," he reminds the group. He goes on to make the point that whether they are talking about a reserve or a park, they must still intervene in some way on people's behalf, both those living within the reserve as well as those living on its borders. He implies that if conservation projects do not become partners with local people in addressing local poverty, they risk losing much of what they are trying to conserve to the demands poor people are forced to make on the forest.

And finally, Étienne Songolo, an anthropologist working with CEFRECOF, relates how he has always been convinced that project personnel must get out and talk to local people if they want their conservation efforts to be successful. But recently he has also come to realize that beyond the utilitarian motives for such communication there are also moral reasons for doing so. From an ecojustice perspective, he feels it is only right that project personnel establish good relations with local people and try to involve them in the conservation of local natural resources. Étienne pauses, choosing his words carefully as he begins to tread some delicate turf. The sincerity of his quandary is written on every line of his face. He seriously struggles with the fact that for all his sensitive talking with and listening to local people he, and the project he works for, remain unable to offer them anything as just compensation for the ways in which the RFO has restricted their access to forest resources. "If we continue these contacts with local people but have nothing to provide in the way of compensation, we will truly face failure," he says. Therefore, Étienne feels strongly that CEFRECOF needs to change its strategy and begin some sort of pilot projects directed toward fulfilling at least some of the pressing needs local people continually tell him about. Without some such means of compensation, he feels the reserve will be hard-pressed to succeed.

* * *

Narratives such as these illustrate the common ground that can be found between project staff and local people, even in the midst of the very real conflicts they also face, such as those described in Chapter 5. Such narratives indicate that there *are* project staff members and local academics who share with local people a concern for the human dimensions of conservation, share the conviction that people will not be able to conserve unless, as Abbé Mosolo puts it, "their own lives are at ease" and their needs are being met. Perhaps such sensitivity on the part of local academics and project staff to the very real needs confronting local people can serve as a cornerstone for the careful and respectful negotiation of world views that I have argued needs to take place between Western conservationists and the people who live within the forests they seek to conserve.

Conclusion

This chapter has revealed differences and similarities between the views of project staff and local academics and those of local land-users in regard to the challenge of creating sustainable livelihoods from the lands and forests of northern Congo. As mentioned at the beginning of the chapter, environmental perceptions shared by the university-educated and local people provide some common ground on which to further the work of building environmental sustainability. Yet drawing out the points of contrast clarifies what each group can learn from the other as well as the lines along which further negotiation needs to take place.

Such analysis reveals that indeed each group has a great deal to teach the other. For beginners, project staff can gain additional insights into their ancestral heritage from local villagers who, in turn, can learn a great deal from those with a university education about the wider scope and complex dynamics of the external forces affecting their lives. But are project staff and local academics truly open to teaching and learning from local people, to learning more about the values within ancestral traditions? Are local people given the opportunity to educate those trained in Congo's universities? As mentioned above, although some university-educated staff and university educators speak favorably about the values of their heritage, in general they respond to ancestral practices and beliefs (such as those described by local people) with a great deal more criticism and skepticism than do the elders with whom I speak.

A common line of analysis as to why this might be may include the following reasons: Local elders are unable to be as self-critical about their ancestral traditions as are the university-educated, or they have less capacity for critical thinking in general; local elders want to present themselves and their traditions as favorably as possible to me, a white researcher perceived to have greater power; or perhaps it is that the formal education project staff and local academics have received teaches them to question, to be critical, and thus enables them to see things more uneducated people cannot.

Although such explanations may hold some of the truth, I would like to suggest another factor that may play a role in contributing to project staff members' skepticism of ancestral knowledge, beliefs, and practices. Just as the villager voices we have heard in this book do not present the whole picture and are blind to some of the evidence, so too are the voices of the university-educated. Although the Western education they have received opens new windows of knowledge, at the same time it closes others. One way in which it may close windows is to predispose project staff to believe in the superiority of scientific knowledge (both natural and social science) for solving the problems they confront. In turn they may presuppose that local or ancestral knowledge is inferior, outdated, and has very little to offer the altered conditions of the present. Under such an epistemological framework scientific and local knowledges are structured as two distinct bodies of knowledge arranged hierarchically. By preferencing the former, such a framework belittles the latter. More important, it seriously impedes the two-way flow of knowledge between local people and project staff that as this chapter intimates holds so much promise.

Perhaps a more fruitful strategy is the one suggested by Professor Kwela: to diligently seek an equilibrium between Western and African approaches, a position that draws on both Western scientific and local ancestral bodies of knowledge to further the work of environmental sustainability. Under such a framework, local staff and local people are both learners and teachers, and the gifts, skills, knowledges, and insights each group has to offer are respected as equal contributions toward achieving more sustainable livelihoods. Rather than a top-down transfer of knowledge from a superior episteme to an inferior one, such a framework fosters an atmosphere of mutual respect and appreciation in which knowledge travels horizontally and bidirectionally. To repeat the words of Jack Kloppenburg, quoted earlier, it is not a matter of "choosing between scientific knowledge or local knowledge, but of creating conditions in which these separate realities can inform each other" (1991, 540).

Part of creating those conditions, creating the mutual respect under which authentic dialogue can take place, is to develop greater appreciation for what, in the midst of their differences, local people and those with university education hold in common. As this chapter has shown, quite a stretch of common ground exists between the two groups. By focusing on the common ground between them, project staff and local people can gain a firmer foundation on which to build together the sustainable systems that will be so important to the future of Central Africa.

Another way in which to create the mutual respect necessary for dialogue is to develop a greater humility in regard to one's own episteme, to realize its limitations, and that it may not have all the answers. In the concluding chapter that follows, I point to specific ways in which the Western scientific episteme prevalent in much of the environmental work taking place in Central Africa indeed does not have all the answers and may in fact hinder the search for them.

Notes

1. Quotations from my conversation with Professor Kwela are reconstructed from my field notes, as I did not have a tape recorder with me.

2. Although Dr. Kwela describes this concept as "Western," it more accurately depicts the particular strain of Gnostic thinking that has had and continues to have a strong influence in Western metaphysics (see Jonas 1958). Gnosticism refers to a body of texts and variety of sectarian doctrines appearing with and around the time of Christianity, although there is some evidence that pre-Christian Jewish and Hellenistic pagan forms of Gnosticism also existed. Gnosticism emphasizes knowledge as the means of attaining an otherworldly salvation (Jonas 1958, 32; Merchant 1980, 17). One of Gnosticism's primary features is its radical dualism that permeates its conceptions of theology, cosmology, anthropology, eschatology, and morality. In terms of Dr. Kwela's reference to the "body" and "spirit," Gnosticism holds that the latter, or "pneuma," is "a portion of the divine substance from beyond which has fallen into the world.... In its unredeemed state the pneuma thus immersed in soul and flesh is unconscious of itself, benumbed, asleep, or intoxicated by the poison of the world" (Jonas 1958, 44). In terms of salvation, "the goal of Gnostic striving is the release of the 'inner man' [pneuma] from the bonds of the world and his return to his native realm of light" (Jonas 1958, 44).

Kwela's comment, like others within this book (including my own), reflects the difficulty in describing the "West" because it has come to mean so many different things.

3. This contrasts with what Dieu-Donné, a member of Loko's staff, says earlier about how Christianity succeeded in despiritualizing the natural world. The "truth" of the situation probably lies somewhere in-between Kwela's and Dieu-Donné's perspectives. I have encountered many Congolese who, despite exhibiting a strong belief in Christianity, also believe that aspects of nature possess spiritual qualities (e.g., nature spirits inhabit springs and waterfalls; some animals are really people who practice shape-shifting; certain species of trees hold ancestral or spiritual significance and must be left untouched; etc.) At the same time, I have also witnessed the demise of certain dietary taboos and the cutting of forest near water sources that Dieu-Donné describes. However, the extent to which these changes stem from Christian influences remains unclear.

7

Conclusion: Lessons for Environmental Practice, Theory, and Ethics

A people's vision is found in their thought and is mediated by their symbolizations. A work that is not grounded in people's thought and symbols risks remaining an abstract quest that people fail to find identity with.... An empowering framework moves from intellectual speculation to concrete situation by being grounded in people's thought and symbols.

—**Harvey Sindima**

Documenting the traditional philosophies and world views of African peoples is fruitful only when undertaken within the context of and out of an engagement with the concrete and actual problems facing the people of Africa.

—**Tsenay Serequeberhan**

The voices and stories heard in this book make very clear some of the concrete problems and challenges facing people and their surrounding environments in both the Ubangi and Ituri. The case study projects at Loko and Epulu are each making sincere efforts to address these problems. Yet both projects are experiencing difficulties, whether in getting people to "understand" and become engaged with their efforts, or by generating serious opposition among local people that could erupt into open conflict. Both sets of difficulties reflect the warning issued by Harvey Sindima that "work not grounded in people's thought and symbols risks remaining an abstract quest that people fail to find identity with" (1995, 228). By elucidating aspects of local people's thought and symbols pertaining to the natural world, this work aims to encourage such grounding in both the Loko and Epulu cases.

Although throughout the book I have alluded to ways in which such grounding may happen, in this final chapter I focus more directly on how local environmental perceptions and knowledge can be applied to what I perceive as the fundamental area of difficulty confronting each case study project. Focusing on fundamental problems rather than on the comprehensive set of issues raised in preceding chapters will allow for more in-depth discussion of the lessons to be learned by applying local environmental thought to the work of conserving rainforests and improving livelihoods in a sustainable and socially sound manner.

This research deals with two key aspects of Central African thought: First, from a Central African perspective, human beings relate to nature in a "both/and" holistic manner rather than in a manner characterized by "either/or" dualism; second, such "both/and" holistic perspectives also characterize Central African social thought, particularly in regard to another ancient philosophical question—the relationship between the individual and society.[1] Furthermore, as argued earlier, these two sets of relationships are inextricably linked; that is, the relationship between the individual and the community holds very real implications for the human-environment relationship and for the environment itself. It is also within these two key relational dilemmas that the fundamental difficulties facing each case study project lie.

In this chapter I first focus on Loko and the CEUM and the lessons to be learned by applying Central African "both/and" understandings of the individual-community relationship to their work of promoting environmentally sustainable development in the Ubangi. Next, I discuss the implications Central African holistic understandings of the human-environment relationship hold for conservation practice within the Ituri's Réserve de Faune à Okapis. I then conclude the chapter with some thoughts on what Central African ideas and experiences of nature can contribute to Western environmental thought and ethics. I begin with a narrative account of my visit to one of Loko's community development initiatives, an experience that illustrates some of the difficulties Loko and the CEUM face in their work.

Loko, the CEUM, and Individual-Community Relations

It would have been a tight squeeze to fit all three of us in the cab of the Toyota pickup, I am thinking, as my elbows flail around the steering wheel, spinning it this way and that to avoid the road's precipitous potholes and gaping gullies. Mamuya, one of Loko's two community development organizers, had offered to sit up in back, leaving the cab seat to Mama Tonga, who organizes the CEUM's development work with the women of the region. Having left Loko just as the sun was rising, the three of us are on our way to the village of Bonudana for a meeting of one of the regional development committees (RDCs) recently created by the CEUM's Department of Development.

As described in Chapter 3, the CEUM has set up an extensive hierarchy of regional and parish development committees (PDCs) through which it hopes to realize various communally owned projects such as communal gardens, cattle herds, fish ponds, and reforestation plots. Such actions are part of the major shift the CEUM's

Department of Development has recently made from an individual- to a community-based development approach. As explained above, until recently the CEUM's strategy was to seek out individuals with exceptional drive, interest, and ability with whom to work or whom it could train at Loko and Karawa. It was hoped that such a tactic of "betting on the strong" would have the greatest success in enabling CEUM development efforts to take root in the village and grow into programs benefiting the entire community. Due to a variety of difficulties with this approach, CEUM leaders and missionary staff are now trying a different tactic focused on creating group rather than individually owned projects, the benefits and labor to be shared by all. The purpose of the Bonudana meeting today is for RDC members to take stock of what they have accomplished and discuss their successes, problems, and needs with Mamuya and Mama Tonga.

Some 20 kilometers down the road from Loko, Mama Tonga and I are absorbed in a conversation about how our respective societies have defined the relationship between the individual and the community in very different ways. I remark how in the West we have tended toward one extreme or the other. We *either* put all the emphasis on the individual to the detriment of the community, as in our capitalist liberal democracies, *or* we do things the other way around—place extreme emphasis on the collective and thereby eclipse the rights and welfare of the individual, as exemplified in the former Eastern bloc countries (which one might argue are simply another expression of the West). I then broach the observation (certainly not original) that many African societies instead conceive and practice a much more balanced relationship between the individual and the community. I ask Mama Tonga what she thinks, and she tends to agree with me. Rather than envisioning and enacting the relationship as an "either/or" choice between the two, many African cultures have developed complex means of preserving the good of *both* the individual *and* the community simultaneously.

Other Congolese express a similar assessment. Mutuza Kabe, professor in the faculty of law at the University of Kinshasa, writes:

> The Western conception, which we would qualify as individualistic, privileges the individual. Meaning that within that conception, things are good or bad according to whether they favor or counteract the interests of the individual.
>
> The socialist conception, as the term indicates, gives primacy to the society over the individual. Indeed, just the opposite of the individualistic conception, here things are good or bad according to whether or not they serve the interests of society.
>
> Finally, the African conception, which in order that it make sense to us we would call "vitalistic," but to use a better known term, we would qualify as communalistic, privileges life, considered in terms of the individual and of the society. This means that things are good or bad according to whether they contribute to developing or to deadening the life of the society and that of the individual....
>
> We can say that the liberal and socialist visions of the world are dichotomous and unilateral in the sense that they divide life by privileging one of these aspects. We have seen how both of them are founded upon the idea of opposition and contradiction....

Conversely, the communalistic vision of the world which accords primacy to life, to us appears more whole and more entire; more universal and more fraternal (1987, 121–124).²

In Chapter 4 I outlined additional perspectives on African understandings of the relationship between the individual and the community drawn from the writings of various African and Africanist philosophers, theologians, and historians. Like Kabe, these writers argue that in many African societies individualism and communalism stand not in "either/or" opposition but rather in a "both/and" mutuality whereby each reinforces the other. I also gave some concrete examples from the realms of land tenure and labor organization of how individual-community relations are enacted on the ground. By holding land in communal trust, many African societies assure that no one in the community who needs land goes without; at the same time, they allow each individual within the community to have their own piece of land that truly belongs to them. Similarly, African communal labor groups such as *likilemba* or *mutualités* simultaneously provide space for both individual projects and the fostering of community. Finally, in Chapter 4 I alluded to the fact that many of the development difficulties Loko, the CEUM, and other NGOs working in Central Africa experience seem to be rooted in fundamentally different metaphysical understandings, on the part of Westerners (or Western-educated development workers) and African villagers, regarding the individual's relation to the community and the relation of both to the environment on which they depend.

By the end of the meeting at Bonudana it has become clear that these committee members, working through the hierarchical system of RDCs and PDCs, are having a hard time generating the energy, motivation, and inspiration to initiate and maintain the communal projects envisioned for the region and for individual parishes. The same is true for many other regions, I later learn from Gary Sutter, adviser to Loko's agroforestry program. Of course, regardless of approach, many of these efforts, like all projects in Congo, are seriously constrained by the country's economic debility. But beyond that, I would argue that part of the problem also stems from the particular "way of seeing" being employed by the CEUM, especially in regard to the relationship between the individual and the community.

The Problem: Individuals or Community Rather Than Individuals and Community

It is a long walk out to Pindwa's garden. I accompany him to plant his rice and harvest his peanut crop. Walking the winding trail that alternately skirts forest and savanna, we have plenty of time to talk. We enter into a fascinating discussion of individual-based versus community-based development approaches. I ask him his thoughts on the recent shift the CEUM development department has made from the former to the latter. Both approaches seem to hold disadvantages not easy to remedy, as Pindwa proceeds to explain.³

"The problem with the individual-based approach is that people forget to help those around them, and development too easily becomes a private individualistic enterprise based on *le moi* (the ego, self). If individuals would then turn and help their fellow villagers, even just to show them that they, too, could begin a similar project through contacting staff at Loko, then the individual-based approach might really work to help the whole community." Pindwa pauses a moment to flick a millipede off the trail with his shoe. "Unfortunately, this rarely happens," he continues, shaking his head. "Instead, often individual efforts are complicated by feelings of jealousy within the community, sometimes leading to an individual's project being sabotaged."

I recall what happened to Fulani during the building of his microhydro but am quickly drawn back to our conversation as I hear Pindwa describing the problems with the opposite approach. "On the other hand," he explains, "the question the CEUM faces with the communal approach is, 'Who is really going to take responsibility for taking care of the project?' In belonging to everyone, in effect it belongs to no one. And with the intense workloads people already face with their own gardens, they are not quick to sacrifice precious time and resources to build something that may not offer them direct returns."

We soon reach Pindwa's garden, but it takes a lot longer for the full import of his thoughts to filter down into any sort of coherent analysis of the problem on my part. That evening on the computer I jot some notes, which I later refine before entering into any discussion of the topic with members of Loko's staff. A few days later, I meet with several staff members to exchange ideas and talk about some of the things I am learning. I mention how although the CEUM's new emphasis on a communal approach holds attractive and commendable features, the goals it aspires too are pretty hard to achieve in any society, much more so in a society and economy such as rural Congo's where there is so little to work with and no supporting infrastructure. But, more important, I suggest that the way in which the CEUM is trying to achieve these goals seems quite out of step with certain characteristics of African thought. Its strategic shift from the individual to the communal approach ignores the African facility with "both/and" rather than "either/or" thinking. An approach more appropriate to the African context might be to combine both the individual and the communal rather than attempt to shift from one to the exclusion of the other.

Lessons from Central African Individual-Community Relations for Actual Development Initiatives

How might these ideas translate into the very real development dilemmas facing the CEUM and other organizations working toward environmentally sound development in Central Africa? Rather than completely reject an individual-based strategy and switch to a totally communal strategy, the CEUM and other organizations need to find a way to combine both in a single system, a middle way that allows the individual and the communal to co-exist. The African land tenure systems and cooperative labor groups mentioned above may in fact provide good models for achieving such a middle path.

For example, rather than trying to build communal projects owned by everyone yet owned by no one, it might work better to encourage and support individuals in their personal projects (fish ponds, vegetable gardens, fruit tree orchards, reforestation plots, agroforestry gardens, animal husbandry, etc.), yet also encourage a communal system of labor to build such projects. Modeled after the *likilemba*, such a system could help provide the people power often needed to get individual projects off the ground. Under such a "both/and" system, each person would also have the certainty that they would benefit directly from development, thus providing the necessary motivation and responsibility to make projects successful. At the same time, communal or cooperative labor that rotated between individual initiatives could keep development from becoming a completely individualistic, moneymaking enterprise based on *le moi* and enhance the communal spirit and association for which many grassroots development organizations strive.

Efforts to improve livelihoods are difficult if not impossible to instill from the top down through a series of committees and extension agents organized hierarchically. Rather such efforts achieve greater success if they begin with real-live individuals who truly desire to undertake certain development initiatives. Perhaps grassroots development projects such as those of the CEUM need to start with such individuals— encourage them, teach them, learn from them, and provide the seeds for them to realize their own individual projects. But as with the cooperative labor group Ntembe na Mbeli, whose story I told in Chapter 4, they also need to promote projects that can be achieved only through individuals coming together to help each other by working cooperatively. Out of that cooperative labor, people might then begin to learn from each other and to meet together, doing so not because it is required by the committee or center above them in the hierarchy but because they really have a reason to meet, they really have a desire to share ideas generated by their individual projects, or really want to find out whether others might know how to solve this problem or that. Meeting and sharing stories thus develop out of the actual work people are engaged in rather than work developing out of top-down promotional programs.

Under such a "both/and" system modeled upon indigenous systems of natural resource management and ownership, Loko's programs may well develop sooner and better because individual projects, just like individual gardens, instill and require motivation and responsibility. Under such a system, the work of creating sustainable livelihoods grows because this person sees what that person has done and, curiosity sparked, goes and asks her or him about it and then tries to imitate it. At the same time, such a "both/and" system can foster the growth of community because people recall things like the *likilemba* within their own culture and realize that it takes a whole village to raise the quality of livelihood. Under such a system, community develops because people learn how much easier it is to achieve their livelihood goals if they help each other out, if they create cooperative labor groups, and, out of those and the time captured, thereby create projects that benefit and belong to the whole community.

Epulu, the RFO, and Human-Environment Relations

The same sort of "both/and" holistic relationship between the individual and the community can be observed in Central African understandings of the relationship between people and the natural world itself. A major theme throughout Chapters 3–5 has been to show how Central African peoples, rather than separate people and nature into dualistic opposition, are more inclined to view themselves and nature as intimately connected parts of a greater whole. Similar to how Central African peoples see both the individual and the community as vital parts of the larger society, so too do they see themselves and the rest of creation as vital parts of the larger whole that is life. In the words of Malawian theologian Harvey Sindima,

> The African world is concerned with the fullness of life; it is in its fullness that life's meaning is realized.... However, the question of human life cannot be understood apart from nature, to which it is bonded. Nature plays an important role in the process of human growth by providing all that is necessary, food, air, sunlight, and other things. This means that nature and person are one, woven by creation into one texture or fabric of life (1995, 126).

Life, or *moyo*, in Sindima's native Chichewa, is an all-encompassing symbol by which his own and many other African peoples interpret the cosmos, nature, and themselves.

> *Moyo* is the *all* in *all* value and meaning of creation ... the overarching symbol of all creation.... The African attitude toward nature is based on the understanding that all creation is knit in one fabric or texture, namely, *moyo*.... Since human life depends on the generosity of nature, the Achewa consider nature as *moyo*. This connectedness means that people and nature share the same destiny, just as they live and belong to the same origin, that is, the Divinity, the giver of *moyo*, who is *Moyo*. From this follows the understanding that both people and nature are sacred since they have a divine origin. Respect and care of nature is grounded on this sense of oneness (Sindima 1995, 207, 209).

The narratives of local land-users in both the Ubangi and Ituri support such ideas and, as discussed earlier, confirm that in Central African contexts people consider themselves very much part of nature rather than separate from it. Such connectedness to nature, along with other factors, has contributed to maintaining a dynamic balance between their needs and the needs of nonhuman nature throughout a great deal of their history. However, as explained in Chapter 3, the metaphysical position of being part of, rather than apart from, nature certainly does not mean that Central Africans have played no role in changing nature. Evidence of such roles and of Central African peoples making conscious choices in regard to nature throughout their history is inarguable (Vansina 1990, Harms 1987, 1981).

Thus it is important to emphasize that this book has steered away from works lying within the functionalist approach to cultural ecology that view historical equilibrium between humans and nature as being "worked out in unconscious ways, lying above human cognition" (Fairhead and Leach 1996, 9). It has argued instead that Central Africans have acted to maintain some degree of harmony between people and the natural world in a conscious manner. Land use knowledge and practices such as those described in Chapter 4 have stemmed from people's conscious recognition of consequences and benefits, from their continual process of learning, changing, and adapting to find what works and what doesn't work to guarantee the continued existence of both the human community they belong to and the nonhuman community they depend on. This book has also shown how such knowledge and practices have been backed up by certain values, symbols, and beliefs that are also conscious aspects of society. For example, people's conscious belief that God acts through the ancestors and that the ancestors remain alive in nature (expressed by Professor Kwela in Chapter 6) necessitates a respect for nature, not in the sense of worshipping it but as a way of respecting the ancestors.

In short, although I have argued that Central African peoples have created more harmony than disharmony in their relationship to the land, I remain in agreement with pluralistic ecological perspectives on human-nature equilibrium such as those of researchers James Fairhead and Melissa Leach. In their recent rereading of local peoples' role in deforestation in Guinea, they point out that "in reducing 'human organisation and consciousness to a regulative mechanism for preserving an equilibrium,' functional equilibrium models rob people of their action and consciousness" (Fairhead and Leach 1996, 9 citing Amanor 1994, 19). Pluralistic perspectives such as theirs, coupled with understandings of nature's disequilibrium and discontinuity found within the "new ecology" (see discussion in Chapter 2), "enable better comprehension of local ecological reasoning, and reappraisal of people's roles in achieving environmental stability or directing change" (Fairhead and Leach 1996, 10).

Fairhead and Leach go on to elegantly reveal how local people have consciously increased the extent of forest cover in Guinea's Kissidougou Prefecture, contrary to the opposite conclusion that has been entrenched in the minds of both ecologists and policymakers for more than a century. Functionalist equilibrium models of social science, they argue, have complemented the conventional perspective of the natural sciences that "more people equals less forest" so as to maintain erroneous perspectives and policies regarding land degradation. They make very clear how "local peoples' own theories concerning social and ecological issues carry very different implications for issues of agency, cause and responsibility in explaining vegetation change" (Fairhead and Leach 1996, 9).

As Fairhead and Leach have done in a different African context, this book has attempted to bring forward such local "theories" in order to help increase confidence on the part of researchers and policymakers in African peoples' ability to manage their own environments.

The Problem: Two Models of the Human-Environment Relationship Meeting Head-to-Head

With regard to the various peoples inhabiting the RFO, this book has revealed many varied factors, in addition to the metaphysical perception of being part of nature, that have enabled people to coexist more or less harmoniously with the environment for the better part of their history. Key have been the limited nature of human needs; the absence of commercial markets; the diversification and periodicity of resource use (farming, fishing, hunting) such that no one activity has been entered into full-time in such a way as to undermine the environmental foundation on which it stands; conscious controls on population growth; and the strong social ethic that has governed the individual's interactions with the natural world. For instance, for many years land in Central Africa has been held in communal trust, and natural wealth, although not divided equally, has been distributed throughout the community through the highly developed and complex practice of sharing. As a result, less has needed to be taken from the environment, and every member of the community has had hope of getting something. As noted above, this communal ethic extends both into the past and into the future—one's way of interacting with the environment cannot undermine one's obligations to both the ancestors and future generations. As one Ituri farmer expresses it to me, "If we kill all the animals, *kesho watoto wetu watakula nini* [tomorrow what will our children eat]?"

Thus in these indigenous systems of resource management, society and nature coexist in a holistic relationship in which people consider themselves more as part of, rather than apart from, nature. People manage their environment as if they and their human community are part and parcel of the whole picture. Thus, under such a model, humans living in these forests for the past hundreds of years are most accurately conceived of as necessary parts of the total functioning of the natural system, rather than as intruders into a complex of parts that works best without them. Humans' hunting, foraging, and farming, along with the activities of other animals, have contributed to a relative balance being maintained throughout the whole system such that one group of animals does not override another. An example of this (detailed in Chapter 4) is how over hundreds of years of experiment and experience, forest farmers have developed a complex and intricate system of correlating shifting cultivation with hunting and trapping as a means to control animal damage to their crops, obtain close sources of protein, and indirectly leave large areas of forest farther afield unexploited.

Under such a theory and practice of resource management, the idea of a nature reserve that severely limits local people's roles in managing their environments by replacing them with a management program based largely on Western conceptions of conservation and administered by the State in partnership with international conservation organizations seems incongruous if not ethically controvertible. Indeed, it is the discrepancies between such contrasting natural resource management models that form the basis of the fundamental problem confronting the RFO: The failure to cultivate authentic cooperation and partnership with a broad spectrum of local people is now resulting in enhanced conflict between the reserve and those who live in and around it.

As I've noted, my return to Epulu in 1995 came after a six-year absence. Chapter 5 has made clear how far community-reserve relations, already strained in 1989, have now deteriorated. Unless some changes are soon made, local people's current resistance to conservation initiatives may erupt into open conflict. In order to avert such conflict, I have argued, those in power in the RFO need to make a more concerted effort to sit down and negotiate their differences with local people. First, however, it may be helpful to take a deeper look at the metaphysical roots of such differences.

Western Conservation Management as Misplaced Pristineness

It is important to stress how these growing conflicts and forest management problems are rooted in the metaphysical understanding of the human-environment relationship that governs the research and policymaking of those in power in the RFO, an understanding fundamentally different from that of local people. Western dualism meets African holism head-to-head in the search for management options. Although as Chapter 6 illustrates, local Congolese staff and academics hold some serious doubts about the West, they have also been influenced by it. And much more significantly, the international organizations who really hold power over the course of policies in the RFO remain entrenched within a Western dualistic metaphysic that separates people and nature into distinct camps.

It should come as no surprise that such a metaphysic continues to operate in the Ituri and in numerous other African contexts where Western agencies tend to be in control of natural resource management decisions. The roots of such a metaphysic run deep within our Western heritage and stem from ancient influences including Platonic idealism, Gnosticism, and Enlightenment thought (Glacken, 1967).[4] Today such dualistic understandings of human's separation from nature continue to pervade scientific research and policy on land management and nature conservation in non-Western cultures, as Fairhead and Leach explain:

> This problem stems, at a fundamental level, from the framing of scientists' and inhabitants' explanations within very different root assumptions concerning the relationship between social and ecological processes.
>
> In short, since the Enlightenment, western science has conceptualised natural and social phenomena as being of a different order; as *a priori* separate. It is assumed that "natural" phenomena can be investigated as separate from human society, except in as much as people and their social world are subject to "nature" and act on "it." Boundary problems are interesting, but do not undermine the conceptual scheme. It is partly this perspective which circumscribes the concerns of ecological science ... , encouraging ecologists to break down their consideration of vegetation into "natural" forms and

processes (involving plants, animals, soils, water and so on) as if uninfluenced by people and society—so-called anthropic factors. Ecologists who seek out untouched nature—pristine forest—against which to assess human impact are drawing on and reaffirming this divide; they may be disappointed, but not conceptually challenged, when they find pottery sherds in their soil pits beneath "natural" vegetation.... This ideal of nature has, of course, been central to conservation policies which have commonly deemed the exclusion of people as necessary for the preservation, or reestablishment, of nature (1996, 5,6).

Conservation policies built on such dualistic understandings of humans and nature fall prone to what we might call the "Fallacy of Misplaced Pristineness," to paraphrase Alfred North Whitehead.[5] They stem from a management model that takes human beings out of the nature picture and manages for an abstract "pristine nature" sans *Homo sapiens*. The danger and inadequacies of such a model built on abstraction have been well-stated by Whitehead: "The disadvantage of exclusive attention to a group of abstractions, however well-founded, is that, by the nature of the case, you have abstracted from the remainder of things. Insofar as the excluded things are important in your experience, your modes of thought are not fitted to deal with them" (1925, 200). Furthermore, such a model often leads conservationists to go beyond simply excluding humans from the rest of creation. It can also contribute to their weighing the artificial nature-culture dualism normatively such that the human cultural component always comes out in the pejorative. This stems in part from viewing humans as ipso facto destroyers of nature's pristineness, but perhaps it also comes from the common tendency within abstractive science to view that which doesn't fit one's model unfavorably; in extreme cases, as something that must be "tweaked" to fit the model.

Conservationists within certain Western countries (especially the United States) have socially constructed an idea of wilderness that leaves little room for human beings. This is now being exported to and forced upon cultures and peoples in other parts of the globe deemed desirable for conservation—cultures and peoples whose perceptions of the human-nature relationship are quite dramatically different. Environmental historian William Cronon has painstakingly documented the historical intricacies by which "wilderness" in America has come to be constructed as a pristine place, as virgin uninhabited land free from human influence of any kind. Ironically, it is just the opposite, he argues: "Far from being the one place on earth that stands apart from humanity, it is quite profoundly a human creation" (1995, 69), one that has had and continues to have extremely destructive consequences for many people. Cronon's work is so important for the points being discussed here that it merits elaboration.

Cronon's "troubles with wilderness" are numerous, but most pertinent to this discussion are the ways in which such American conceptions and valuing of wilderness continue to play themselves out in contemporary concerns to preserve biological diversity and protect endangered species. According to Cronon, such a convergence "has helped produce a deep fascination for remote ecosystems, where it is easier to imagine that nature might somehow be 'left alone' to flourish by its own pristine devices. The classic example is the tropical rain forest, which since the 1970s has become the most powerful modern icon of unfallen, sacred land—a veritable Garden of Eden—for many Americans and Europeans" (1995, 82).

Iconic visions of tropical rainforests indeed come into play in the Ituri and hold consequences there not unlike those Cronon delineates in a more general sense. He writes:

> Protecting the rain forest in the eyes of First World environmentalists all too often means protecting it from the people who live there. Those who seek to preserve such "wilderness" from the activities of native peoples run the risk of reproducing the same tragedy... that befell American Indians. Third World countries face massive environmental problems and deep social conflicts, but these are not likely to be solved by a cultural myth that encourages us to "preserve" peopleless landscapes that have not existed in such places for millennia. At its worst, as environmentalists are beginning to realize, exporting American notions of wilderness in this way can become an unthinking and self-defeating form of cultural imperialism (1995, 82).

Chapter 5 documents prime examples of managing for "misplaced pristineness" within the RFO. To name only a few, these include reserve policies that separate the Mbuti from their village farmer partners by prohibiting farmers from following the Mbuti into the forest; policies that limit garden clearing only to areas of secondary forest; and de facto policies that restrict subsistence hunting and traditional means of controlling animal damage to gardens. I realize that these policies operate within a context that has been influenced by external forces such as modern capitalist markets and the recent human immigration from socially and environmentally stressed areas to the east of the Ituri. Yet rather than helping to restore some degree of balance into natural resource use systems, such policies tend to promote a further distancing from the relatively balanced human-environment relationship that existed prior to such influences.

In addition, such policies hold real-life costs for people living within the RFO. For example, as Chapter 5 records, many village farmers recount how the loss of crops to garden-ravaging mangabeys, pigs, and elephants is causing serious economic strife for their households. By removing the means to control animal damage, such policies may in fact force farmers to overexploit other areas of forest in order to regain the household production they've lost from the destruction of their crops. Over the long term, the effects of such policies represent an "unnatural" upset of the previous degree of relative balance that had been maintained by the complex resource management systems documented above.

Finally, especially disturbing is the fact that local and Western natural resource management models clash within an unequal power balance under which Western conservation organizations, backed by the State, hold the upper hand. In the ensuing difficulties, many local people stand to be seriously marginalized and further oppressed by reserve policies unless respectful and careful negotiation of such policies with local people takes place immediately.

In the meantime, Central African forest peoples and the ancestral heritage they hold have much to teach us in the West if we but take the time and develop the patience to listen. On both a practical and philosophical level, they can help us learn a great deal about how to live within the natural systems of which we are part, rather than separate ourselves from a nature we either iconicize or plunder.

Lessons from Central African Human-Environment Relations for Actual Conservation Initiatives

Central African "both/and" understandings of the human-environment relationship hold lessons for those working on conservation projects in the region; in fact, they hold lessons for anyone interested or engaged in actions to halt the depletion of the world's biodiversity. First, they provide a management principle that some systems ecologists in the West are also coming to realize, namely, we are misguided to manage for a "pristine" nature because nature does not exist "pristinely" (i.e., without human influence or presence). We only place our desire for pristineness upon it. Central African holistic thought reminds us that we are nature, we cannot get ourselves out of it.

Furthermore, such a fact also makes us aware that neither can we view nature outside of ourselves. We will always be looking at it through some degree of subjectivity. Therefore, we would do well to examine and know what our subjectivities are and how they influence what we see. When we try to manage according to the subjectivity of "natural pristineness," we often end up moving more against nature's grain than with it, not to mention the damaging effects such management models hold for local people living in management areas.

Thus a Central African model of environmental management can correct our Western dualistic style of management by emphasizing that we must manage for a whole system, humans included. It can remind us that nature and humans are interactive parts of one whole and that any environmental management program must seriously consider the ramifications of such interactions. It behooves Western environmental managers therefore to take seriously local people's points of view; points of view such as those of the Mbo farmers quoted in Chapter 5 who speak of a principle found in the thought of many other indigenous peoples as well. They remind us that God made people and animals together and that if people leave the forest the animals will also disappear. People and animals, one very significant component of nonhuman nature, need each other.

One might even propose that the dynamic ecological balance that ensued from these complex interdependencies between forest animal and human populations actually, to some degree, depended on human uses of the environment. As mentioned earlier, the patchwork of primary and secondary forest resulting from human perturbations such as garden clearings provides habitat for a greater number of both plant and animal species than does primary forest (Bailey and Peacock 1988, 92, 113). Although much more research is required, it is not out of the question to hypothesize that human perturbations in the rainforest, such as the complex farming-trapping systems I described, are analogous to a variety of ecosystemic perturbations (both natural and human-induced), which ecologists have come to recognize as vital in maintaining the integrity of such systems. These perturbations (e.g., natural and human-controlled burns) are part of the system and not a destructive glitch. Forest-dwelling farmers and their correlated hunting-trapping-farming systems are part of the system, a "natural" perturbation if you will, and not a destructive glitch. Therefore, management models that subtract out the beneficial roles people play in the working of nature can detract from, rather than enhance, the possibilities for fostering the sustainability of the entire ecological system.

Another way of thinking about these issues is to look more closely at the word "conservation." Cal DeWitt has pointed out that in addition to the standard meaning of the word—"preservation from loss, injury, decay, or waste" (Webster 1989)—we can also think of conservation as "con-servancy" or "con-service"—"serving together" (personal communication). This idea of "reciprocal service" (*kosalisana* in Lingala) comes closer to a Central African understanding of the term. In return for the forest's serving up food and construction materials to Central Africa's rainforest peoples, they in turn serve the forest in ways that foster its continuance as a server of food and materials.

Such an understanding is poignantly illustrated by the tree-planting Eucharistic liturgy of the Association of African Earthkeeping Churches mentioned in Chapter 4. The liturgy is a meaningful expression of mutual service between African peoples and nature. Moke's narrative in Chapter 5 on "enticing the forest" also intimates a sense of mutual service between the forest and people. His words express how the Mbuti have been taught by their ancestors to "coax, soothe, and entice the forest's favor as a lover, to treat it a bit kindly; to coax it, along with caring for it." The purpose of such actions is "to allow [the forest] to give birth again."

Similar ideas of reciprocal service are expressed in the Ngbaka proverb *Mɔ dala zu'bu, ɛ zu'bu dala mɔ*, which means, "If you feed your garbage pit, then your garbage pit will feed you." Figuratively the proverb refers to the counsel that if you care for your children, then in your old age they will care for you. However, in thinking about the literal translation of the proverb, I was reminded of the role garbage pits have played in Central African household economies and ecologies as described by Vansina (1990). They were actually a form of composting, taken care of in order that in turn they might serve as fertile soil on which to grow food and medicines for the household, even to serve as "experimental plots" on which to test new plants. According to Vansina,

> Every woman ... had a kitchen garden ... next to her house. This was fertilized with detritus of all kinds, and crop rotation was practiced here. One or a few specimens of useful plants such as peppers, greens, and medicinal plants were grown there. Historically these kitchen gardens are quite important because they served as experimental stations to test new plants. Here new crops and new varieties of old crops were tested for natural and labor-related requirements as well as for yields (1990, 85).

Thus, if you served your garbage pit by tending it literally as a garden, a part of nature in a way, then it also served you as a producer of food and medicines and as a ground for improving one's knowledge of plants and farming.

Central African "both/and" understandings of the human-environment relationship can also do much to clarify that the real problem, the real destructive glitch, is *not* human beings per se but distinct, human-created socioeconomic institutions that foster unsustainable uses of the environment. Modern capitalist markets, one example of such institutions, interacting with a complex of other forces, including technology and human (African as well as Western) greed, have been and continue to be key factors in destroying the dynamic balance that has existed between humans and the natural forest environment that has supported them.

Tata Fulani, whose story began this book, shares with me a poignant example of how these "bad ways" penetrated and changed the relatively balanced systems of land use that had existed in the time of his father: "This problem of poison in the waters," he explains, "it came really only with this civilization of the Europeans. They have this poison to put in the soil next to the crops in order to kill pests, but crafty people have taken it and put it in the rivers and streams to kill fish. People took it for a good thing, but it is only ruining our waters, some is even killing people. These ways, they began to change, ... well some of it is due to the whites, those who came to us. It was their knowledge that began to change our knowledge. We saw how much easier it was to get things with these bad ways. We see the ease and we jump into it and even though the rivers may be ruined, I get my fish and I sell it and I get wealthy."

In destroying traditional resource use patterns, this commercialization of nature also succeeded and continues to succeed in destroying the natural ecosystems on which all of life, human and nonhuman, depends. In short, commercialized use, more than indigenous peoples' use of the forest, lies at the root of Africa's, including the Ituri's, environmental problems. Central African models of environmental management suggest that we would do better to try to control the market forces that lead to overexploitation of the environment rather than unjustly restrict the subsistence practices of people who have lived in these forests much longer than ourselves.

In the case of the Ituri, the RFO, like other Western conservation initiatives, has opted for a management model that rather than directly restraining this commercialization of nature establishes State (and, in the minds of many of the villagers I talked with, "American") control over vast areas of forest seen by local people as God's gift to them to help sustain their lives. Not only does such a model, in Cronon's words, "privilege some parts of nature at the expense of others" (1995, 86), it also opens the door for State exploitation of the local population that top-down control has always facilitated.

Furthermore, regardless of official RFO policy, State control and exploitation have resulted in local people's loss of access to certain provisions the forest gave them. In response, the conservation projects at Epulu have failed to provide local people with significant compensation in the way of alternative means to replace the resources lost, fearing that any such "development" efforts will draw more people to the Ituri and put more pressure on the forest. Instead the projects have sought to gain local people's participation in, and support of, their conservation initiatives by forming a network of *comités permanents de consultation locale* (CPCL—standing committees of local consultation), village-based groups created as mediators between local people and the RFO. In theory, CPCLs provide a means for local people to voice their complaints to project staff while also providing the projects with representatives at the village level who can promote conservation and report any actions that are deemed contrary to it.

In reality, participation in the CPCLs has been very difficult to maintain. Perhaps part of the reason for this is that the projects at Epulu have again remained hesitant, or feel they do not have the means, to provide people with incentives to participate, incentives such as those mentioned by Nkomo in Chapter 5—schools, dispensaries, and houses for chiefs made from durable materials. Thus, although the CPCLs appear to be a means of carrying out participatory conservation, they seem to more closely resemble the State systems of control to which local people in the Ituri have been subjected since well before Congo's independence. Again, local people have been engaged to act on behalf of the State within and sometimes against their own communities—writing reports and policing their neighbors and the immigrants who come to settle among them for any actions they have been told are antithetical to conservation (such as those described in Box 5.1).

Such considerations re-emphasize the point made in Chapter 5 that the key social issue on which the success of a multiuse reserve such as the RFO ultimately depends is whether such a model of conservation is something local people really want, something of which they are interested in taking ownership. If it is not, then perhaps it is more honest for the conservation projects at Epulu to admit that their programs, linked to larger, more powerful Western organizations, are taking part in the sort of imperialism Cronon mentions above, a cultural imperialism that results not in economic exploitation but in nature's appropriation for purposes reflecting their own interests more than those of the people who have lived in the Ituri the longest. In such a case, is it better to make the RFO a national park outright and remove local people but offer them fair and just compensation for their land and assure them an equal life elsewhere? Although such an option also involves numerous complexities, it may produce fewer conflicts and less guilt about controlling the land people live from for conservationists' own purposes. But, of course, such a more honest policy the State and Western conservation organizations also cannot afford; and they are not willing to take it upon themselves. And so, as alluded to in Chapter 5, the RFO is in the middle of a true ecojustice quandary—that is, unless ethics and justice really do not matter.

If such is the case, then the quandary will be easily solved by local people, in response to the growing difficulty of their lives in and around the reserve, getting up and moving out of their own volition, leaving the RFO to its animals, plants, and a few Western-trained scientists who like to study them. But, of course, ethics and justice do matter. And so the conservation projects at Epulu need to find a viable solution to the ethical quandary they face, a way to put authentic ecojustice into practice on the ground. I tend toward the following attempt at a solution.[6]

Toward the end of my research experience with these two case study projects, I found myself often thinking, *The RFO needs a little bit of Loko, and Loko needs a little bit of the RFO*. I think the underlying meaning was this: In exchange for some of the professional knowledge of tropical ecology held by its staff, the RFO could use some of Loko's expertise in providing local people with alternative resource options. Although Loko's experience comes from working in ecologically impoverished environments such as those in the Ubangi, its expertise could also apply to the RFO, where natural resources have been taken out of reach of local people for conservation purposes. In exchange for the removal of such resources, the conservation projects at Epulu must start building with local communities some alternative means to provide for their needs. Loko's experience and expertise with options such as rotational agroforestry systems, animal husbandry, aquaculture, fruit tree propagation, and other means of ecologically sound agricultural production could have a lot to offer the RFO.

Funding such additional activities is difficult, conservation organizations are quick to point out. But perhaps, as Nkomo suggests, the unique natural resource

An experimental agroforestry garden at Epulu begun by staff of the ICCN and GIC. Such pilot projects hold promise and need to be greatly encouraged as a means of promoting ecojustice within the RFO.

found in the Ituri—the okapi—could provide adequate funds for such projects on a sustainable basis. It is important to remember how valuable these animals are to Western zoos and private animal collectors.[7] Should not some of the money gained through the captive breeding and export of okapi that has been going on at Epulu since the early 1950s come back to benefit the local people who bear the brunt of the burden resulting from the RFO's restrictions?

Finally, other possible ways of building ecojustice in the RFO include:

- Creating with local communities authentic institutionalized means for comanagement of the forest by its indigenous inhabitants and conservationists.
- Establishing specified zones for human activity and for conservation as a means of limiting animal damage to gardens while protecting animal corridors.
- Beginning on a pilot basis, designing and implementing microprojects that fulfill some of the needs expressed by local villagers, for example, employing mobile health teams to provide local people with the basic health care they state as being their number-one need.
- Actively seeking out and engaging the expertise of other conservation or environmental projects that have greater experience with ways of integrating conservation and meeting local human needs. In addition to Loko, prime candidates for such consultancy include ADMADE in Zambia (Lewis and Carter 1993), CAMPFIRE in Zimbabwe (Dix 1996), and the exciting work being done by the Biodiversity Conservation Network (Stevens 1998).
- Making more diligent and explicit efforts to control the market forces that commercialize nature and are a primary cause of the Ituri's environmental problems.
- Of most importance: creating a viable mechanism by which local people and the Epulu-based conservation projects can undertake immediate and sustained negotiation of the conflicts and differences that exist between them such as those highlighted in this book.

In sum, the weight of the RFO's restrictions really does fall most heavily on the "small guys," as Père Toma says; those without a great deal of power or wealth—the Mbuti and indigenous farmers. Those of us who desire to conserve the Ituri's natural landscapes must realize that such desires come with responsibilities to these people who have lived in the Ituri much longer than ourselves.

Lessons from Central African Human-Environment Relations for Western Environmental Theory and Ethics

Central African "both/and" understandings of the human-environment relationship also hold implications for Western environmental thought and ethics, particularly in regard to the quest for a "world ethic for living sustainably" referred to in Chapter 2 (Engel 1994a, 5). Such "both/and" understandings can help correct certain dualistic debates about humans' relationship to nature that are part of this quest yet hinder its progress.

One of these dualistic debates has been outlined by social ecologist Murray Bookchin (1990, 19–30). Bookchin remarks on the tendency of Western environmental thinking to fall prone to either of two extreme and fallacious views on the relationship between human society and nature: One extreme is the view that society and nature are totally separate realms (the hallmark within capitalist and materialist schools); at the opposite extreme is the view that dissolves all differences between nature and society such that nature absorbs society (prevalent among sociobiologists and extreme biocentrists).

As this book has shown, such dualistic and reductionist views would be quite foreign to Central Africans whose understanding of the relationship between society and the natural world is more complex and holistic. Drawing certain lessons from the nature of interaction between Central African peoples and the forests they inhabit, Jan Vansina points to the more satisfying understanding of the human-environment relationship such interactions both exemplify and induce:

> A landscape is not a variable independent from the people who inhabit it. Both people and their habitats are part of a single reciprocal system. People and their natural environments constantly influence each other. Thus a natural disaster ... could force populations to abandon a locality or district. But people could also change landscapes....
> Any one-way model of this constant, dynamic, and complex ecological interrelationship oversimplifies the picture (1990, 255).

Such a mutual and complex interaction between forest dwellers and their surroundings has allowed (perhaps even forced) them to develop a rich knowledge of the environment that goes beyond the purely utilitarian. Widespread knowledge provides them with the room and directions in which to innovate in the face of change. If something no longer works, if a natural disaster wipes out a certain resource, they know what else to try. If one year the forest gives only a little of the preferred *asali* honey, they know where to look for the less sweet but also good *apiso*. If hunting proves poor in one locale, switching camps to a new area is not difficult. In other words, nature and humans interrelate with some degree of flexibility and slack.

Unlike Western biocentrists who tend to view humans as victims under the heavy hand of nature as taskmaster, Central Africans see nature as offering them some freedom of choice rather than forcing their fate upon them. The experience of Central Africans again provides us with a lesson. It affirms that although we cannot do with nature whatever we please, neither does nature leave us freedomless. Instead there exists the opportunity and (dare we say) the responsibility for us to play a creative role in shaping the future of the natural and social evolutionary process. We are cocreators, not simply victims of natural deterministic forces. Although the position of cocreator holds great obligations, it also provides greater hope than does the role of victim, for actually crafting a more ethical relationship to the earth, both locally and globally.

Central African "both/and" understandings of the human-environment relationship hold ramifications for another of the dualistic debates prevalent in Western discourse on global environmental ethics: the relationship between anthropocentrism and biocentrism. Indian ecologist Ramachandra Guha has referred to the anthropocentric-biocentric distinction, so preciously guarded by Western deep ecologists, as a largely spurious means by which they may have appropriated the moral high ground yet, in the process, done American and global environmentalism a serious disservice (1989, 83). From a Central African perspective the distinction also seems a bit misguided.

Malawian theologian Harvey Sindima, in describing the African concept of creation, writes, "The African understanding of the world is life-centered. For the African, life is the primary category for self-understanding and provides the basic framework for any interpretation of the world, persons, nature, and divinity" (1990, 142). Elsewhere he speaks of African cosmology as stressing the "bondedness, the interconnectedness, of all living beings" (1990, 137).

Although their denotative meanings are the same, the African idea of life-centeredness contrasts and corrects the Western meaning that has been given to biocentrism. The latter has not escaped the trap of dualism such that it has often come to imply a certain misanthropic and oppositional understanding of the relationship between humans and nature or, conversely, a relationship offering no distinctions at all between the two. Environmental ethicists such as Paul Taylor uphold a biocentrism in which the human species has no special status vis-à-vis other species, and *Homo sapiens*, like all other species, must be judged only on a morally individualistic basis (1986).

Instead of focusing on the "either/or" debate between anthropocentrism and biocentrism, life-centeredness focuses on the bondedness of all of life. Rather than analyzing the place and standing of different human and nonhuman life forms on the basis of their comparative rights, African life-centeredness focuses on life itself, in a holistic rather than analytic fashion. Under such a holistic world view, concern for life does not exist within a zero-sum game in which regard for one type of life means neglect of another. Philosopher Kwame Gyekye writes that in African thought "reverence for nonhuman entities is not necessarily considered to lessen one's devotion to the welfare and interests of human beings in this life" (1987, 208). Thus it is not a matter of seeing what is most important, or of deciding if one thing is more important than another, but of believing and acting on the basis that all of life is important; even more, that all of life is sacred.

Further, life-centeredness is oriented less toward individual entities (rocks versus trees versus animals versus people) and more toward the relations between them. More attention is paid to processes and the flow of forces between entities than to the entities themselves. Emphasis falls on relating rather than existing since it is the nature and quality of relationship that determines whether the whole will sink or swim. The relationship between any two living entities affects all the rest of life since

Conclusion: Lessons for Environmental Practice, Theory, and Ethics   273

all of life is bonded. Therefore, deciding whether humans or life is central is, according to Central African understanding, a nonquestion. Similar to Guha's assessment, a Central African evaluation would conclude that such a debate does the quest for ethical ways of relating to the earth more harm than good.

Finally, such holistic interactions between Central Africans and nature also hold lessons for Western environmental epistemology, specifically in regard to how lived experience in nature relates to the development of knowledge about the natural world. Vansina's critique of one-way models of the human-environment relationship (cited above) also implies a critique of one-way models of the relationship between the physical and the cognitive, between experience in nature and knowledge about nature and, in a broader sense, between practice and theory. The bias of such models toward lineal causality distorts and reduces the full complexity to be found in all of these relationships.

In place of a unilineal understanding of such relationships, Central African thought views the components of each pair as standing in dynamic interrelationship. Regarding the relationship between lived experience in nature and the development of ecological knowledge, Central African thought is less concerned about which holds the greater influence and more concerned with how both are constantly and simultaneously influencing each other. People's cognition, perceptions, attitudes, beliefs, and cosmologies—all epistemological media—combine to influence their actual experience in nature, while in turn people's experience in nature—and the workings of nature itself—combine to influence their environmental knowledge, attitudes, and perceptions. In short, Central African peoples' theories about nature and their practices within nature are constantly interacting, each reshaping and reinfluencing the other. Rather than people's theories about nature playing an inordinate role in determining what they do in and with nature, in Central African thinking, nature is simultaneously lived in and thought about.

Furthermore, part of this process of simultaneously living in and thinking about nature involves an epistemological give-and-take with nature itself. Cal DeWitt has described this interactive epistemology as a "con-learning," a mutual process by which people and other parts of nature "learn together" (personal communication). For example, Central African farmers learn that if they place their gardens right in the middle of established elephant corridors they will likely be raided, whereas elephants learn that if they continue wreaking havoc in farmer's gardens they may likely be killed. As mentioned in Chapter 3, such interactive learning has contributed to humans and animals coexisting within a dynamically balanced system.

Central African understandings of the relationship between natural experience and ecological knowledge holds important correctives for the ways in which Western theoretical preferences have historically thwarted and continue to thwart various human experiences in sustainable living. Africanist scholar Robin Horton makes the observation that in the West experience is significantly molded by a priori theoretical positions (1967, 173). When applied to experiencing and theorizing about nature, such an assessment gains support from Cronon's historical study of the Western concept of wilderness referred to above. Cronon's work reveals how theoretical ideas about nature developed without much actual experience in nature[8] have had and continue to have serious detrimental effects on the real-life experience of millions of people. Imagining and socially constructing nature as a pristine place free of all human influence contributed to the forced removal of thousands of Native Americans from land they had always considered their home so that "tourists could safely enjoy the illusion that they were seeing their nation in its pristine, original state, in the new morning of God's own creation" (Cronon 1995, 79). And, as mentioned above, such a priori theories about nature continue to threaten the life experience of people in places far away from where such theories were developed—places such as Central Africa's tropical rainforests.

But Cronon also makes the important point that such theories of nature also hold serious consequences for Westerners—indeed for all of humanity—given the global scale of and linkages among the environmental challenges we face. He argues that by glorifying and always keeping wilderness available to us as the place where we "really belong," "we give ourselves permission to evade responsibility for the lives we actually lead" in "the homes we actually inhabit" (1995, 81). Beyond the flagrant hypocrisy and disconnections we thus evidence is the disturbing fact that such a stance in the end gives us no lessons, no wisdom, no understanding for how to actually live in and with nature in a sustainable manner. Cronon sums it up in these words:

> To the extent that we celebrate wilderness as the measure with which we judge civilization, we reproduce the dualism that sets humanity and nature at opposite poles. We thereby leave ourselves little hope of discovering what an ethical, sustainable, *honorable* human place in nature might actually look like....
>
> Idealizing a distant wilderness too often means not idealizing the environment in which we actually live, the landscape that for better or worse we call home. Most of our most serious environmental problems start right here, at home, and if we are to solve those problems, we need an environmental ethic that will tell us as much about *using* nature as about *not* using it. The wilderness dualism tends to cast any use as *ab*-use, and thereby denies us a middle ground in which responsible use and non-use might attain some kind of balanced, sustainable relationship.... Only by exploring this middle ground will we learn ways of imagining a better world for all of us: humans and nonhumans, rich people and poor, women and men, First Worlders and Third Worlders, white folks and people of color, consumers and producers—a world better for humanity in all of its diversity and for all the rest of nature too. The middle ground is where we actually live. It is where we—all of us, in our different places and ways—make our homes (1995, 80, 85).

In Central Africa, people's knowledge and theories about nature have been developed less from the position of an uninhabited "natural" wilderness and more from the "middle ground" of a nature they have inhabited, used, experienced, changed, respected, and honored for millennia. Such a heritage has much to offer Western environmental thought and practice in terms of learning to live sustainably within the

Primary forest and plantain gardens intermingled in the eastern Ituri. Much of the Ituri's landscape represents the "middle ground" Cronon refers to—nature that local people have inhabited, used, experienced, changed, and respected for millennia so as to maintain a dynamic yet sustainable balance between their own needs and those of the rest of nature.

immediate nature that is our home, learning to control our use of nature's resources in order to avoid bringing ruin to ourselves and the natural fabric that sustains us.

Conclusion

I began this book with the story of the microhydro and the Mami wata to illustrate how important it is for those working in Africa to take spiritual and symbolic realms seriously when dealing with issues of land. I then related the stories local Congolese land-users in the Ubangi and the Ituri shared with me about their own perceptions of, and practices on, the land. From such narratives, I drew certain principles, ideas, and ways of relating to the land that could help local environmental projects cultivate environmental sustainability more firmly in the soil of local culture. I also recounted the perspectives of local Congolese staff and academics, highlighting ways in which their views are both similar to and different from those of local land-users. I then went on to discuss the lessons such local narratives hold for environmental practice by reviewing the fundamental difficulties facing each of the two case studies and suggesting ways local cultural perspectives can help to remedy them. Finally, I considered the various contributions Central African cultures have to offer Western environmental theory and ethics.

This book has begun an ongoing discovery of Central African perceptions and ways of using the forest and what they have to teach us about living sustainably on the earth. Other avenues of investigation remain, such as those proposed in Appendix B, through which to increase our understanding of the contributions Central African environmental thought, values, and practice can make to the quest for sustainable livelihood at both local and global scales. Let me conclude this initial exploration with a summary of what we have learned thus far.

First, over a long history of habitation, Central African forest dwellers have developed complex systems of land use to sustain themselves without jeopardizing the land's ability to provide for future generations. Such knowledge remains alive even though it is being continuously eroded by various market and demographic forces. It behooves us to pay it attention and respect in our efforts to build sustainability in Central Africa.

Second, in Central Africa nature is certainly not considered dead matter or even simply a living material system but, in many ways, a socially and spiritually charged entity. Therefore, how one relates to it holds repercussions on numerous fronts. Unlike our own Enlightenment traditions that removed much of this deeper meaning from the natural world, reducing it to inert matter, in Central African thought and practice, the natural world is alive in many different ways and one must take caution in relating to it or bear certain consequences, both social and cosmological.

Third, one's relation to nature cannot be separated from one's relation to the human community to which one belongs. That is to say, a separate and discrete environmental ethic may be hard to find in Central Africa, but not because environmental wisdom does not exist. Rather wisdom in relating to the environment springs from the wisdom and ethics that Central African cultures have developed to govern the social realm. A deep valuing of social harmony and communalism has real implications for the human-environment relationship. Individual property, desires, needs, and uses of the environment were and are allowed, but seldom if ever are they free from the restraints of communal obligations, obligations that extend to both the past and the future. In one's relation to the environment, one must think of and act with regard to both ancestors and offspring, both one's family and the common good; one can extract from the natural world only in certain places and in quantities not to surpass the needs of one's family and one's social obligations. Limits definitely were, and in some cases still are, very real, and what is taken from nature must be shared. Thus

Central African traditions remind us that the search for environmental solutions requires an elaboration of problems in social as well as ecological terms.

Finally, the environmental wisdom of Central African forest peoples stems from the knowledge and belief that nature and humans are never separate entities but parts of one system. We are part of nature not set apart from it. Nature and culture, humans and environment, social ethics and environmental ethics, ecology and justice—all go hand in hand. Rather than humans or nature being central, it is *life* that is primary, and that includes the entire community of life, for all of life is important, all of life is bonded, all of life is sacred.

Notes

1. See Chapter 2, note 6.
2. Translated from French by author.
3. The following quotations are reconstructed from my field notes.
4. Dualism weaves a wide course through all of these schools of thought and, in each of them, contributes to a separation of humans from the world, or nature. In Platonic idealism, humans are called to perfect themselves by paying greater attention to the superior abstract world of ideas over the lesser material world of concrete objects, bodily appetites and passions, and, by inference, the earth itself. What is "truest" in humans exists captive in an earthly prison, yearning to soar back again to the divine life of heaven (Loomis 1942).

Gnosticism distorted Platonic idealism to produce a much more radical separation between humans and the world. As Hans Jonas explains, "The gnostic God is not merely extra-mundane and supra-mundane, but in his ultimate meaning contra-mundane. The sublime unity of cosmos and God is broken up, the two are torn apart, and a gulf never completely to be closed again is opened: God and world, God and nature, spirit and nature, become divorced, alien to each other, even contraries. But if these two are alien to each other, then also man and world are alien to each other, and this in terms of feeling is very likely even the primary fact. There is a basic experience of an absolute rift between man and that in which he finds himself lodged, the world" (1958, 251).

Finally, the Enlightenment theories of mechanism and reductionism also strengthened the separation of humans from nature. Nature was largely reduced to a machine governed by universal laws discernible through scientific analysis and "objective" measurement done by human observers removed from the nature they observed, all for the purpose of allowing humans to control and master nature so that it might serve them (Merchant 1980).

5. What Alfred North Whitehead called the "Fallacy of Misplaced Concreteness" refers to the error of misplacing the concrete for the abstract. It is often committed by deriving conclusions through a process of successive abstraction and then applying those conclusions to concrete situations with little awareness of the danger involved, i.e., one's abstractions may be so far removed from conditions in the "real world" that the course of action they recommend is totally at odds with what the concrete demands (Whitehead 1925; also see Daly and Cobb 1989).

In much the same way, conservationists have applied policies to the concrete situations of Central African rainforest peoples, policies they have derived from abstract ideas of a "pristine nature" (in their extreme, these ideas imply that nature is *either* pristine *or* it's not really nature, i.e., it has been tainted by human influence), often with little awareness of the degree of misfit nor of the very real dangers incurred to others.

6. My choice of tentative language in this sentence is deliberate since I am as unsure as anyone else who is involved in the Ituri of how to really go about solving the complex issues facing the Epulu-based conservation projects and local people. Consequently, I offer the following ideas as suggestions only, not ultimatums bespeaking a certainty insensitive to the excruciating challenges cases such as the RFO present.

7. I was not able to locate recent figures on how much an okapi costs, but during my previous research in the Ituri (1989), the sale of one okapi could earn as much as $250,000 (Peterson 1991, 112).

8. Cronon makes this point very clear. He writes, "The dream of an unworked natural landscape is very much the fantasy of people who have never themselves had to work the land to make a living—urban folk for whom food comes from a supermarket or a restaurant instead of a field, and for whom the wooden houses in which they live and work apparently have no meaningful connection to the forests in which trees grow and die. Only people whose relation to the land was already alienated could hold up wilderness as a model for human life in nature, for the romantic ideology of wilderness leaves precisely nowhere for human beings actually to make their living from the land" (1995, 80).

Appendix A
Glossary of Local Terms

Akongo   one of the Ngombe names for God (Ngombe)
andi   against (KiLese)
apiso   one type of wild honey found in the Ituri (KiBira)
asali   another type of wild honey found in the Ituri (Swahili)
balabala   road (Lingala)
bankoko   ancestors (Lingala)
baraza   meeting or reception shelter; place of palaver and giving children instruction (Swahili)
-bondela nzoto   lit. to implore or petition the body; to discipline oneself (Lingala)
boro   *Colobus guereza* or Guereza; black and white colobus monkey (KiBira)
bouloneurs   gold diggers (local French slang)
Eh bandeko   lit. brothers, sisters, relatives; idiom expressing surprise, astonishment (Lingala)
elima   girls' initiation rite (KiBira)
esobe   savanna, plain covered with grass; in area around Loko, the most common species of savanna grass is *Imperata cylindricum* (Lingala)
esumba   Mbuti ritual of celebrating the forest, restoring peace and order, or calling on the ancestors for help in times of stress (KiBira)
etaba   various species of wild yam of the genus *Dioscorea* (KiBira)
-funga   to close, to shut, to tie (Swahili)
-funga pori   to close or tie the forest; to make it impossible to get an animal on the hunt (KiNgwana—the Ituri dialect of Swahili)
gaza   initiation rites, rites of passage (Ngbaka and many other languages in the Ubangi)
gbagulu Ngbaka   important Ngbaka sayings, proverbs, stories (Ngbaka)
gbodo   *Triplochiton scleroxylon*, a species of large rainforest tree common in forest-savanna ecotones (Ngbaka)
goigoi   lazy; a *goigoi* chair refers to a foldable wooden sling chair (Lingala)
hôpital   hospital or dispensary (local French)
kao   leopard (KiLese)

kapita   village or neighborhood headman or leader (Lingala)
karibuni   welcome (pl.; Swahili)
kasadui   *Marantaceae* sp. (?), a broad-leaved understory plant (KiBira)
kimia   peace, calm (Lingala and Swahili)
kizungu   of or pertaining to white people, European (Swahili)
-kokota mukila   lit. to pull the [hunting] nets; to call the hunt (KiNgwana)
-komea   to mature, ripen, become fully grown (Swahili)
kpa   large tree of the *Afzelia* spp.; common in forest-savanna ecotones (Ngbaka)
kuku   chicken (Swahili)
kusa   *Manniophyton fulvum*, a forest vine the bark of which is used to make netting and trapping twine (KiBira and many other languages spoken in Central Africa's forest region)
lendu   *Cephalophus dorsalis*, or bay duiker; a small forest antelope (KiBira)
likambo   problem, affair (Lingala)
likembe   thumb piano (Lingala and Swahili)
likilemba   cooperative labor group (Lingala and Swahili)
mafuku   *Artocarpus incisa*, or breadfruit tree (Lingala and Ngbaka)
makayabo   dried salt fish (Lingala and Swahili)
makemba   plantain, tree and fruit (Lingala)
malako   fishing shelter (Lingala)
Mambela   boy's initiation rite (KiBali)
Mami wata   water spirits (many African languages)
mangongo   *Marantaceae*, a broad-leaved understory plant (Lingala and Swahili)
mayele   knowledge, wisdom, intelligence (Lingala)
mbangi   *Chlorophora excelsa*, a species of large rainforest tree common in forest-savanna ecotones
mboka   village, home (Lingala)
mboloko   *Cephalophus monticola*, or blue duiker, a small forest antelope (Lingala and Swahili)
mindele   white people, Europeans (Lingala)
mohai   cooperative labor group (Mbuja)
moimbo   *Cephalophus sylvicultor*, or yellow-backed duiker, the largest of the Ituri's duikers (KiBira and other languages of the Central African forest region)
moyo   life (Chichewa and other Bantu languages)
mpini   handle, usually of a metal tool such as a knife, hoe, or axe (Swahili)
mpondu   cooked manioc greens (Lingala)
Mputu   Europe or America (Lingala)
mtu or mutu   person (Swahili and Lingala)
mukpala, bakpala   Bantu or Sudanic farmers who share an exchange relationship with Mbuti foragers in the Ituri (KiNgwana and other languages of the Ituri region)

Appendix A

mungele   *Cephalophus callipygus weynsi*, or Weyn's duiker, a small forest antelope (KiNgwana)
mupe   Catholic priest (Lingala and KiNgwana)
mzee, wazee   elder, elders (Swahili)
ndibili   *Canarium schweinfurthii*, a large rainforest tree with edible olivelike fruit and flammable sap used for torches (Ngbaka)
ndo!   exclamation of surprise (many languages of the Ituri region)
nguluma, mbalu, mbau   *Gilbertiodendron dewevrei*, a large rainforest tree often found in monodominant stands (Ngbaka, Lingala, and Swahili respectively)
njelu   cooked ripe plantain (KiNgwana)
nkoko   grandparent, grandchild; also brother or sister of grandparent (Lingala)
Nkumbi   boys' initiation rite (KiBira and other languages of the Ituri region)
Nzambe   name used for God in many languages of Central Africa; use greatly extended through missionary activities (Lingala and other languages)
pamba pamba   for no good reason, without purpose; in a disorderly fashion (Lingala)
pusiya   *Treculia africana*, a species of rainforest tree with large fruit, the seeds of which can be roasted and eaten (KiBira)
safu   *Pachylobis edulis*, a species of rainforest tree with edible fruit resembling a wild plum (Lingala and other languages of the Central African forest region)
sakasa   traditional traps that use chord made from the raffia palm (KiNgwana and other languages of the Ituri region)
-salisa   to serve; to do or make for someone or something (Lingala)
-salisana   to serve one another; reciprocal service (Lingala)
sankofa   to draw on the past to prepare people for the future (Akan)
Satongé   name for God among the Fulu people (Fulu)
Seto   one of the Ngbaka names for God but also commonly the name of the trickster in Ngbaka oral traditions (Ngbaka)
sibili   *Atherurus africanus*, or African brush-tailed porcupine (KiBira and other languages of the Ituri region)
siri   secret, hidden thing, mystery, secrecy (Swahili)
soli   *Tragelaphus spekii*, or sitatunga, a type of antelope (KiLese)
sombe   cooked manioc greens (KiNgwana)
sonlu   scientific identification n.a.; type of reed grown in rivers to attract fish (Ngombe)
soukous   type of popular music; also called Kinshasa jazz (Lingala slang)
Tata   father or paternal uncle (Lingala)
-teka   to sell (Lingala and other languages of the Central African forest region)
ti-i-i   for a long ways, for a long time (Lingala)
umunthu   personhood (Chichewa and other Bantu languages)
wageni   guests (Swahili)
wala   long-handled planting stick with a small shovellike blade (Ngbaka)
wali   cooked rice (Swahili)
wazungu   white people, Europeans (Swahili)

Appendix B
Areas Awaiting Further Investigation

As often happens in a writing project, the completed work differs from what one had originally proposed. Once into the research and writing, one often finds that the fields of inquiry prove to be even richer than originally imagined. Such has been the case here. The voices of local people, project staff, and local academics were to have been included in one chapter and half of another; only an entire book adequately addresses their richness, intricacy, and insights. Thus some of the other research areas I originally proposed await further investigation. Indeed this book, focusing on local perspectives on Central African land ethics, is but the beginning of what I envision as a lifelong work. This appendix sketches several areas of further investigation, including some of the methodological considerations they provoke.

Voices Within African Oral Traditions

In each language there is also a public corpus of narrative material which is, especially in nonliterate societies, the common heritage of all the members. These materials, "narratives" as we shall call them, myths and tales, told and retold generation after generation, may thus serve as the primary resources for world view analyses.
—**T. Overholt and J. Baird Callicott**

The words of the land are things that tie together, things that transform [like the clouds], things that are high [like the sky].
—**Lega proverb, eastern Congo**

In addition to local peoples' contemporary narratives, Africa's diverse and vibrant oral traditions offer a means of articulating local knowledge and values about nature. In such capacity they may also serve as a window through which to view local land ethics. The idea of myth serving as a window on cultural cognition and logic has been a frequent theme of structuralist anthropology. French anthropologist Claude Lévi-Strauss thought myth could also reveal a culture's ethical systems in such a way as to "give us a lesson in humility" (1979, 507). Ritual, often an accompaniment to oral traditions, can also reveal ethics, particularly the ethics of humans' relation to the natural world. The basis of this claim is well described by naturalist Barry Lopez, writing about the Navajo:

> Among the Navajo and, as far as I know, many other native peoples, the land is thought to exhibit a sacred order. That order is the basis of ritual. The rituals themselves reveal the power in that order.... An indigenous philosophy—metaphysics, ethics, epistemology, aesthetics, and logic—may also be derived from people's continuous attentiveness to both the obvious (scientific) and ineffable (artistic) orders of the local landscape....
>
> Among the various sung ceremonies of this people ... is one called Beautyway, ... in part, a spiritual invocation of the order of the exterior universe, that irreducible, holy complexity that manifests itself as all things changing through time.... The purpose of this invocation is to recreate in the individual who is the subject of the Beautyway ceremony that same order, to make the individual again a reflection of the myriad enduring relationships of the landscape (1988, 67).

Lopez captures very well the idea mentioned at the beginning of Chapter 2—that oral tradition encapsulated within myth, story, ritual, and symbol can serve as a means of both revealing and transforming a culture's and an individual's relationships to the land. Furthermore, oral traditions and accompanying rituals emanate quintessentially from the subjective. Therefore, they serve as another way to hear the voices of African subjects speaking about land and human beings' relation to land. However, collections of African oral traditions and the literature about them are now plethoric and thus demand that one be selective. This fact calls for an exploration of how one might approach the study of Central African oral traditions to see what they reveal about peoples' relationship to and ethics toward the land.

Africanist historian Jan Vansina is recognized universally as an interpreter of African oral traditions, most notably for his work on oral tradition as history. He has done more than any other Africanist to lay the foundations—the ideas, methods, and schools of thought—for what is now called the field of African history. Indeed Vansina was among the first Western researchers seeking to change non-African perceptions of Africa and to debunk Western myths about African societies, a prime one being that Africa had no history, as it had no writing. On the contrary, Vansina considers African oral traditions to be key revealers of this history, and much of his work has been on the methodology by which to construct history from oral traditions. Through years of fieldwork carefully listening to, collecting, and cross-checking oral traditions, Vansina has successfully recounted dimensions of the very real history that Africa holds (Vansina 1961, 1985).

In Chapter 2, I discussed Congolese philosopher V.Y. Mudimbe's critique of existing "inventions of Africa" and his call for an indigenously African epistemological method. Vansina's work serves as one of several predecessors to Mudimbe's project in that it has significantly contributed to dethroning the Western episteme as the only true way to knowledge, or the value standard against which to judge other knowledges. Nevertheless, Mudimbe, along with other African scholars such as Kenyan philosopher D.A. Masolo, find Vansina going not quite as far as they would like when it comes to the legitimacy of African subjects as producers of knowledge. Masolo assesses Vansina somewhat critically, as the following quote reveals:

> In his treatment of oral tradition as a source of history, Vansina regards narrative as a body of discourse in which an objective body of knowledge, especially "historical knowledge," can be discerned by the scholar. He therefore treats the storyteller as a mere resource from whom the scholar extracts and constructs his mute knowledge....

Appendix B

> In ... Vansina ... knowledge is defined as an objective product of the expert. This position runs clearly counter to the current debate which defines knowledge as a common product of the dialogue between the scholar, the cultural practitioners or experts, and the social actors of everyday life (1994, 186).

Mudimbe offers similar critiques of oral tradition methodologies in which "narratives were submitted to a theoretical order, and rather than accounting for their own being and their own meaning, they were mainly used as tools to illustrate grand theories concerning the evolution and transformation of literary genres" (1988, 182). Instead he encourages researchers to use methods whereby "narratives presented in the truth of their language and authenticity become texts of real peoples and not merely the results of theoretical manipulations" (1988, 182). Such methods require the researcher to let narratives stand on subjects' own terms and fall along the lines of demarcation that subjects themselves choose. Mudimbe quotes Barnes to show how such an "enterprise from within" is far more fruitful for explanation, as "such a demarcation is part of the actors' perception of the situation; and action is intelligible only as a response to that perception. Making a demarcation by external standards, on the other hand, is useless for explanatory purposes" (Barnes 1974, 100).

In short, Mudimbe and Masolo, like Jack Kloppenburg (1991) in a very different setting, are seeking ways to understand "what's going on" by listening to subjects' narratives (including both current narratives of ordinary people caught up in the concrete dynamics of every day life, as well as narratives passed on from long ago) as they relate them, rather than subjecting them to external epistemological grids under the mistaken notion that only then will such narratives yield knowledge. Their work provides critical methodological guidance to any investigation of Central African oral traditions for their insights into an ethical relationship to the earth.

Guidance can also be gleaned from non-Africanist scholars working with the oral traditions of other indigenous peoples to discover components of culturally grounded land ethics. Thomas Overholt and J. Baird Callicott have discerned aspects of an Ojibwa world view through analysis of various Ojibwa tales (1982). They define their approach as follows:

> A less rigorous (and less forbidding) means ... might utilize an approach which is both usual in philosophical studies and engaging for the general reader. This approach assumes that the narrative legacy of a culture embodies in an especially charming way its most fundamental ideas of how the world is to be conceptually organized and integrated at the most general level, and that part of the special function of narratives within a culture is to school the young, remind the old, and reiterate to all members how things at large come together and what is the meaning of it all. The narratives of a culture, in other words, provide not only a most enjoyable diversion, they serve pedagogical and archival purposes as well (1982, 20).

However, Overholt and Callicott also make it clear that any automatic presumptions that native peoples hold an "environmental ethic" must be modified by defining and dealing with ecoethical issues in culturally specific terms from the beginning. The culturally specific terms they employ are the narratives themselves. Rather than present only their analyses or bits and pieces of narratives, the majority of their book is taken up with unaltered reprints of Ojibwa tales as they were originally collected around the turn of the century. In a similar vein, offering complete accounts of Central African narratives can help to let the narratives speak for themselves, "so that some of the flavor and feeling ... may be intuitively garnered" (1982, 21).

Going beyond presenting narratives to culling out from them ecoethical principles and attitudes becomes more problematic. It requires some guidelines, yet they must be flexible enough to avoid the error of using narratives only to illustrate preconceived theoretical positions. One possible guideline, which Overholt and Callicott use, is to compare the environmental attitudes displayed in narratives to the first deliberate and self-conscious articulation of an environmental ethic undertaken in the West, that of Aldo Leopold (1949, 237–264). From Leopold, they draw three main features of a "land ethic": (1) an extension of community beyond fellow human beings to include nonhumans and the land; (2) entering into social relations with other members of this extended community that are governed by rules of conduct or "a limitation on freedom of action" rather than by expediency; and (3) going beyond self-interest to respect, love, and admire nonhuman creatures and the land itself (1982, 153–155).

These criteria seem quite worthy, and as the voluminous works about Leopold's ideas attest, the ecoethical standards he sets are rich for interpretation and comparison. However, using such criteria does raise the problem mentioned in my previous discussions of Callicott, Mudimbe, and Michael Jackson—that of privileging Western science or, as Mudimbe says, "using categories and conceptual systems which depend on a Western epistemological order" (1988, x). Thus, although Leopold's criteria can inform explorations of Central African oral traditions, it is problematic to impose them directly onto those narratives. Instead one must interpret Central African oral traditions within the context of evidence gleaned from research carried out at the local level, within the context of one's own intuitions and experiences of African culture, and by drawing on the works of experienced African and Africanist folklorists such as Scheub (1985, 1990) and Mbele (1986). Throughout such interpretation, one must be on constant vigil in regard to Mudimbe's concerns above as well as Jackson's reminder—that seeing patterns and order can often be more a matter of wishful thinking than portraying what is really there.

As I have said, the way around this is, first, to present the narratives as they are and let them speak for themselves; and second, to qualify any interpretation of them as being only one interpretation, not *the* interpretation. Overholt and Callicott share this concern for humility and uncertainty when it comes to recording and interpreting the oral traditions of another culture: "The contemporary investigator of an alien world view will consistently try to compensate for the conceptual dispositions imposed by his or her own primary cultural outlook. But in the last analysis there is no purely neutral or objective place to stand, so to speak, no acultural vantage point, and thus a measure of uncertainty must in principle be admitted" (1982, 22).

As to where one might find Central African oral traditions to examine, the extensive collections of the University of Wisconsin–Madison libraries provide a wonderful place to begin (see Appendix C). These include collections of tales, myths, proverbs, lullabies, songs, fables, prayers, and two major epic traditions from Congo's forest region, the Mwindo Epic of the Lega people, and *Nsong'a Lianja*, the epic of the Mongo. Field research can, of course, yield more obscure written collections of oral traditions belonging to local ethnic communities, in addition to narratives recorded directly with local people. Missionary linguists working in the area can often be of great help in these field research tasks. With regard to Congo, the rich resources of Centre Aequatoria, a small but very impressive library and research center located at Bamanya Catholic mission just outside of Mbandaka, holds especially unique source material often unobtainable anywhere else.[1]

In sum, if it is true, as I have claimed above, that premodern traditions were born out of humans' intimate contact with the natural world—and that human "oughts" were closely tied to natural "is's"—then it is highly probable that within the vast number of oral traditions produced by Central African cultures there will be some that convey, albeit subtly, environmental attitudes as well as principles for guiding individual and community ways of relating to the earth. The territory encompassed by those traditions therefore proves to be an important area meriting further investigation.

Voices Within the Writings of Various African Thinkers

Another area that demands more investigation is that within the writings of various African philosophers, theologians, writers, and conservationists. Works by Africans that deliberately address the question of environmental ethics are growing, but African writings in philosophy, theology, literature, and conservation that bear on environmental ethics already abound.

Sources that address environmental ethics directly include Antonio (1994), Asante (1985), Berhane-Selassie (1994), Daneel (1994, 1991), Mpanya (1987), Mutombo (1985), Ngwey (1985), Nkombe (1980), Odera Oruka (1994), Omari (1990), Omo-Fadaka (1990), Sindima (1990), and Zimba (1994). Some of the works that indirectly bear on environmental ethics I have already referred to quite extensively (Mudimbe 1988, Masolo 1994). Other works of African Philosophy that hold promise for this area of investigation are several of Mudimbe's other works (1994, 1991b, 1982), Appiah (1992), Gyekye (1987), Hountondji (1994), Odera Oruka (1990), Serequeberhan (1994, 1991), and Wiredu (1980). Key works in African religion and theology that include relevant material are Mbiti (1988), MacGaffey (1981), Paris (1995), Ray (1976), Sindima (1995, 1994), Somé (1993), and Zuesse (1979). The realm of African literature is vast, and therefore one might limit exploration to those works in which natural themes dominate. These include Labou Tansi (1979), Lopes (1990, 1987), several poems from Moore and Beier (1984), Mudimbe (1993,1991a), Nagenda (1986), Ngoie-Ngalla (1988), and Pepetela (1983). Finally, African conservationists are starting to do important work that both critiques and corrects conventional Western conservation practices in Africa by grounding conservation more appropriately in indigenous conceptions of land and people's relations to land. Many of these works can be found in Lewis and Carter (1993) and in Western and Wright (1993). Bonner (1993), although not writing as an African, also presents a helpful critique and some constructive alternatives for African conservation.[2]

Investigation of these works would basically follow that used for doing any research on secondary materials—a combination of analysis, synthesis, critique, and interpretation. Throughout, one must seek to cull from the wide scope of ideas covered by these authors only those principles and understandings that pertain to peoples' relationship to land. Most of these authors have received a Western-style education, currently work, teach, or write from a position of being very familiar with, if not entrenched within, a modern Western world view, and do not directly depend on the land. It would be interesting to compare their views concerning the natural world to those expressed within the oral tradition and to those of people still living on the land whose actions as well as words speak to the issue of Central African land ethics.

Local and Regional Voices: Contributions to and Confrontations with Global and Historical Agendas

One final area awaiting further investigation is outlined by the process of bringing local and regional Central African voices on land ethics into dialogue with the various discussions of global ecological ethics taking place in international fora. Although I begin to enter that area in this book, there is much more ground to be covered. As mentioned earlier, bringing the rich variety of ecoethical principles that stem from Central African thought and practice into the debate on world ethics for sustainable living can help to remedy the current insufficiency of African contributions to global ecological ethics.

At this widest juncture, a fruitful approach would be to also probe into *why* African voices have been more silent than others in the debate. As I have hinted, part of the reason for this may be the quick assumption on the part of many policymakers and theoreticians that Africa is more a source of environmental problems than a source of solutions. However, such an assumption fails to examine the historical legacy and international economic structures that help to explain why the African environment is in the condition it is. The exclusion of African voices represents the tendency to jump to global ecological ethics without incorporating global ecological justice.

One might begin to uncover some of the reasons for this paucity of African voices by examining the degree to which international documents and writings concerning global ecological ethics (Callewaert 1994, Engel and Denny-Hughes 1994, Engel 1994a, IUCN 1994, IUCN General Assembly 1994, Trzyna 1994, UNCED 1993) reflect the ecojustice themes emanating from Central Africa's colonial history and current configuration within the international political economy. In order to get at these themes, one would need to review some of the key dynamics within colonial history and the global economy affecting Central African environments. This would entail drawing on some historical sources that analyze the colonial legacy's impact on the region's forests and lands (Harms 1982; Jewsiewicki 1983; Rodney 1982; Vansina 1990; and Young 1994), as well as drawing on some works that situate the conditions of Central Africa's environments within a postcolonial political economic context (Bianga 1982; Bulu-Bobina 1984; Lumpungu 1977; MacGaffey 1988, 1987; Newbury and Schoepf 1989; Nzongola-Ntalaja 1986; Schoepf 1984; Schoepf and Schoepf 1987; Newbury 1986, 1984; and Vwakyanakazi 1982).

However, rather than provide exhaustive historical and political economic contextualizations, one could focus analysis on the implications such contextualizations hold for the concept of ecojustice as it is being articulated within global ecoethical debates. For example, one might ask what particular insights Central African stories and Central Africa's story bring to the increasingly heated North-South debate over historically conditioned disparities in levels of consumption (Athanasiou 1992, Engel 1994b). What do Central African stories and Central Africa's story have to say to those who, within the international arena of ecoethical policymaking, often choose to ignore what M.A. Partha Sarathy, chair of the IUCN Commission on Education, calls the "ethics of consumption," an ethics demanded by the reality that one-quarter of the earth's peoples are consuming three-quarters of its resources (1994, 32)? Indeed the Central African ethic of sharing what comes from the land and the forests (discussed in detail in Chapter 4) holds serious implications for such debates. Sources for this area of investigation, in addition to those in the literature, include conferences on global environmental ethics being held quite regularly.[3]

Finally, this area of investigation might also include further documentation of the signs of hope and resistance that are rising up from African soils. By researching the hope-filled stories of communities in other areas of Africa that are drawing on both the past and the present to build ecoethical alternatives to environmental destruction, applications for practice can be garnered from beyond the scope of this book. Such signs of hope also hold applications for

Appendix B

theory, for it is not just doing that needs changing but thinking as well, especially thinking premised on distorted images and stereotypes about Africa. What better way to engage concrete problems, whether practical or theoretical, than to be reminded and encouraged by how others are successfully engaging them? Documenting more of such stories and learning more from the models of resistance they offer are vital for the work of cultivating social and environmental sustainability in Central Africa, for the hope they bring to the struggle is the very fuel by which the struggle continues.

As a summary of the material provided in this appendix, Figure B.1 provides a model of how research on Central African land ethics can proceed from the inside out, moving from local-level to regional and global-level investigations using a variety of methods appropriate to each scale.

Notes

1. The center provides excellent logistical as well as library resources for anyone interested in doing research that pertains to the Central African region (see Vinck 1987).

2. See Appendix C for additional references.

3. One such conference, the Institute for Ecology, Justice, and Faith, took place in March of 1995 in Chicago. The Institute consisted of a 4-day intensive cross-disciplinary seminar during which much discussion centered around the problem of global ecological ethics and ecojustice. The journal *Earth Ethics*, published by the Center for Respect of Life and Environment, is a good source for learning about similar conferences. The address is 2100 L Street, N.W., Washington, D.C. 20037; (202) 778-6133.

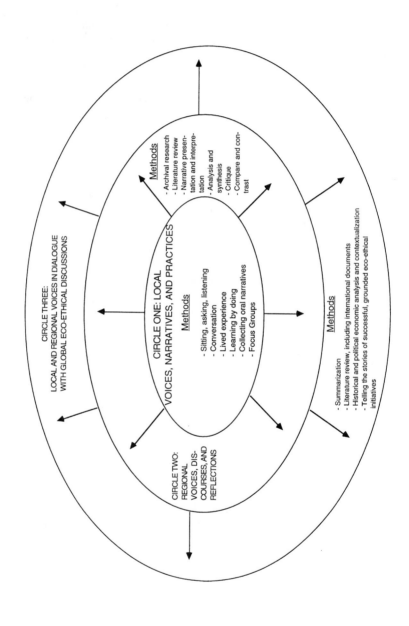

FIGURE B.1 Circles of Investigation.

Appendix C
References[1] for Areas of Further Investigation

Central African Oral Tradition

Bascom, W., ed. 1975. *African dilemma tales*. Den Haag: Mouton.
Bayomba, M. 1987. Acculteration et transformation des traditions. *Scientia* 3:29.
Benveniste, A. 1981. Changements et contrastes: Nouvelles traditions et liens traditionnels. *Recherche Pedagogie et Culture* 53/54: 77–78.
Biebuyck, Daniel, and K. Mateene. 1969. *The Mwindo epic from the Banyanga*. Berkeley: University of California Press.
Biebuyck, Daniel, and Brunhilde Biebuyck. 1987. *We test those whom we marry: An analysis of thirty-six Nyanga tales*. Budapest: African Research Program, Loránd Eötvös University.
Colldén, Lisa. 1979. *Trésors de la tradition orale Sakata: Proverbes, mythes, légendes, fables, chansons, et divinettes des Sakata*. Uppsala: Almquist and Wiksell International.
Courlander, Harold. 1975. *A treasury of African folklore: The oral literature, traditions, myths, legends, epics, tales, recollections, wisdom, sayings, and humor of Africa*. New York: Crown Publishers.
De Rop, A. 1978. *Versions et fragments de l'épopée Mongo: Partie I, versions 1–6*. Bruxelles: Academie Royale des Sciences d'Outre-mer.
De Rop, A., and E. Boelaert. 1983. Versions et fragments de l'épopée Mongo: Nsong'a lianja (partie 2). *Études Aequatoria* 1: na.
Dzokanga, A. 1978. *Chansons et proverbes Lingala*. Paris: Conseil International de la Langue Française.
Ekombe, E. 1987. Tension entre le traditionnel et le moderne dans la littérature orale traditionnelle. *Etudes Aequatoria* 7: 243–250.
Eno Belinga, S.M. 1986. Civilisation du fer et tradition orale bantu. *Muntu* 4: 11–46.
Frobenius, Leo, and Douglas C. Fox. 1983. *African genesis*. Berkeley: Turtle Island Foundation.
Gisangi, Sona. 1974. *Dieu nous a tout confie excepte cette forêt!: Mythes Mbala*. Bandundu: Ceeba Publications.
Heim, P. 1968. La tradition en Afrique noir: Déclin et survie des mythes ancestraux. *Euraf* 7: 9–10.
Hulstaert, G. 1942. Littérature indigène. *Aequatoria* 5: 39–40.
_____. 1949. Musique indigène et musique sacré. *Aequatoria* 12: 86–88.
_____. 1956. Chants de Portage. *Aequatoria* 19: 56–64.

_____. 1958. Proverbes Mongo. *Annales de Musée Royal de l'Afrique Centrale: Series Sciences de l'Homme* 15.

_____. 1965. *Contes Mongo*. Bruxelles: Academie Royale des Sciences d'Outre-mer (ARSOM), n.s. 30 (2).

_____. 1970. *Fables Mongo*. Bruxelles: Academie Royale des Sciences d'Outre-mer (ARSOM), n.s. 37 (1).

_____. 1971. *Contes d'ogres Mongo*. Bruxelles: Academie Royale des Sciences d'Outre-mer (ARSOM), n.s. 39 (2).

_____. 1972. *Poèmes Mongo modernes: Recueillis, traduits, et annotés*. Bruxelles: Academie Royale des Sciences d'Outre-mer (ARSOM), n.s. 39 (3).

_____. 1977. *Berceuses Mongo*. Bandundu: Ceeba Publications, Series II 36.

_____. 1978. Poèmes Mongo anciens. *Annales de Musée Royal de l'Afrique Centrale: Series Sciences Humaines* 93.

_____. 1979. *Traditions orales Mongo*. Bandundu: Ceeba Publications, Series II 50.

_____. 1982a. *Chants Mongo*. Bandundu: Ceeba Publications, Series II 76.

_____. 1982b. *Chansons de danse Mongo*. Bandundu: Ceeba Publications, Series II 78.

_____. 1989. Chants Funières Mongo. *Annales Aequatoria* 10: 224–240.

Igot, Yves, and Marie-Jeanne Igot. 1970. *Contes et fables du centre de l'Afrique*. Paris: Didier.

Kakule, M.K. 1976. Analyse d'un conte sexuel Nande. *Africanistique* 4: 1–17.

Kibal, N.N. 1982. Proverbes Dinga. *L'Africain* 20 (4): 19–23.

Kibanda, M. 1988. Problematique du myth en littérature orale africaine. *Scientia* 3 (2): 111.

Kibulya, H.M. 1976. *Folk tales of Bwamba*. Nairobi: East African Publishing House.

Kizobo, O.O. 1981. Pour une analyse plus rentable des traditions orales des sociétés africaines actuelles. *A.D. Cerdac* 9 (1/2): 25–37.

Krzywicki, J. 1984. Contes didactiques Bira. *Africa (Rome)* 3: 416–434.

Laye, Camara. 1980. *The guardian of the word (Kouma Lafôlô Kouma)*. Translated by James Kirkup. New York: Vintage Books.

Lomami, W. 1974. Essai de description de trois genres de contes d'animaux dans la littérature swahili (Kindu). *Culture au Zaire et en Afrique* 4: 47–87.

Maes, V. 1994. Les souvenirs des refuges sur les grands arbres chez les peuples de l'Ubangi. *Annales Aequatoria* 15: 33–49.

Makuta, N. 1984. Le sort de la tradition orale africaine face au développement technique. *Education, Science, et Culture* 9: 129–140.

Mbaka, M. 1989. Trésors de la littérature orale: Cas des proverbes Leele. *Piste et Recherches* 4 (3): 287–315.

Mudekereza, C., and B. Koba. 1984. Tradition africaine et science: L'example de Bushi. *Africanistique* 14:121–145.

Ndiaye, A.R. 1980. Les traditions orales et la quête de l'identité culturelle. *Présence Africaine* 114: 3–17.

Niang, M. 1979. Note sur le symposium Leo Frobenius: Role des traditions dans le développement de l'Afrique. *Présence Africaine* 111: 113–115.

Opaa, Kiladi. 1976. *Les arbres se mirent a danser: Mythes Mputu*. Bandundu: Ceeba Publications.

Palangi, M.E. 1979. Essai d'analyse d'un conte Luba: "Cilembi wa mayi." *Africanistique* 7: 73–95.

Roulon, P., and R. Doko. 1987. Entre la vie et la mort: Le parole des oisseaux. *Journal des Africanistes* 1/2: 175–206.

Ryckmans, A. 1968. Contribution à la littérature orale Kongo. *Cultures et Développement* 1: 83–118.
Scheub, Harold. 1971. *Bibliography of African oral narratives*. Madison: African Studies Center, University of Wisconsin–Madison.
Senghor, Leopold, et al. 1979. Tradition et développement en Afrique. *Présence Africaine* 111: 141–166.
Theuws, Jacques A. 1983. *Word and world: Luba thought and literature*. Bonn: Anthropos-Institut.
Thomas, Jacqueline M.C. 1970. *Contes: Proverbes, devinettes ou énigmes, chants, et prières Ngbaka-Ma'bo (République Centrafricaine)*. Paris: Éditions Klincksieck.
Vinck, Honoré. 1988. Essai de bibliographie sur la littérature orale Mongo. *Annales Aequatoria* 9: 257–268.

African Philosophy

Bodunrin, P.O., ed. 1985. *Philosophy in Africa: Trends and perspectives*. Ibadan: University of Ife Press.
Chukwudum, B.O. 1983. African philosophy: A process interpretation. *Africana Marburgensia* 16 (2): 3–14.
Crahay, Franz. 1965. Conceptual take-off conditions for a bantu philosophy. *Diogenes* (52): 55–78.
Gyekye, Kwame. 1975. Philosophical relevance of Akan proverbs. *Second Order* 4:2 (July): 45–53.
Hountondji, Paulin, and Kwasi Wiredu, eds. 1984. *Teaching and research in philosophy: Africa*. Vol. 1, Studies in Teaching and Research in Philosophy Throughout the World Series. Paris: UNESCO.
Kagame, Alexis. 1976. *La philosophie bantu comparée*. Paris: Présence Africaine.
Kambembo, D. 1969. Le MuKongo et le monde qui l'entourait: Cosmogonie Kongo. *Cahiers des Religions Africaines* 3 (6): 315–318.
Laleye, I.P. 1976. Le myth: Création ou recréation du monde?: Contribution à l'elucidation de la problematique de la philosophie en Afrique. *Présence Africaine* 99/100: 41–59.
Mazrui, Ali A. 1986. *The Africans: A triple heritage*. London: BBC Publications.
Mate, K. 1987. Le beau dans la culture Nande: Essai d'une esthétique. *Zaire-Afrique* 214: 239–250.
Momoh, C.S., ed. 1989. *The substance of African philosophy*. Auchi, Nigeria: African Philosophy Project's Publications.
Nduka, Otonti. 1974. African traditional systems of thought and their implications for Nigeria's education. *Second Order* 3: 1 (January): 95–110.
Ntumba, K.C. 1983. Dialectique de la spontanéite et de la contrainte: Le cas de la morale traditionnelle africaine. *Revue Philosophique de Kinshasa* 1 (2): 71–84.
Okere, Theophilus. 1983. *African philosophy: A historico-hermeneutical investigation of the conditions of its possibility*. Lanham, MD: University Press of America.
Okolo, Okonda. 1980. Tradition et destin: Horizons d'une herméneutique philosophique Africaine. *Présence Africaine* 114: 18–26.
Okolo, Chukwudum B. 1987. *What is African philosophy?* Enugu, Nigeria: Freemans Press.
Ruch, E.A., Omi Anyanwu, and K.C. Anyanwu. 1981. *African philosophy: An introduction to the main philosophical trends in contemporary Africa*. Rome: Catholic Book Agency.

Tempels, Placide. 1959. *Bantu philosophy*. Translated by Colin King. Paris: Présence Africaine.
Tshiamalenga, N. 1983. L'espace théorique du système des valeurs. *Revue Philophique de Kinshasa* 1 (2): 99–110.
Vincke, Jacques L. 1973. *Le prix de péché: Essai de pschoanalyse existentielle des traditions Européenne et Africaine*. Vol. 6, Essais Series, ed. V.Y. Mudimbe. Kinshasa: Edition du Mont Noir.
Wright, Richard A., ed. 1977. *African philosophy: An introduction*. Washington, DC: University Press of America.

African Religion and Theology

Appiah-Kubi, Kofi, and Sergio Torres, eds. 1979. *African theology en route: Papers from the pan-African conference of Third World theologians, December 17–23, 1977, Accra, Ghana*. Maryknoll, NY: Orbis Books.
Blakely, Thomas D., Walter E.A. van Beek, and Dennis L. Thomson, eds. 1994. *Religion in Africa: Experience and expression*. London: James Curry.
Bujo, B. 1993. L'Afrique en passe d'élaborer une théologie originale. *Select* 27: 157–171.
Dickson, Kwesi A. 1984. *Theology in Africa*. Maryknoll, NY: Orbis Books.
Kahang'a, R. 1986/87. Le destin de la verité de l'homme et le pathos negro-africaine de la totalité comme source du sacré. *Cahiers des Religions Africaines* 20/21 (39–42): 251–272.
King, Noel Q. 1986. *African cosmos: An introduction to religion in Africa*. Belmont, CA: Wadsworth Publishing.
MacGaffey, Wyatt. 1972. Comparative analysis of Central African religions. *Africa (London)* 42: 21–31.
_____. 1986. *Religion and society in Central Africa: The Bakongo of Lower Zaire*. Chicago: University of Chicago Press.
Mudimbe, V. Y. 1983. An African criticism of Christianity. *Geneve-Afrique* 21 (2): 89–100.
Mufuta, Kabemba. 1981. Langage littéraire et pedagogique de l'éthique chez les Bantu. *Cahiers des Religions Africaines* 15 (30): 255–286.
Mulago, Cikala M. gwa. 1980. *La religion traditionnelle des Bantu et leur vision du monde*. Kinshasa: Faculté de Théologie Catholique.
Muyengo, M. 1988. La bioéthique: Quelques perspectives africaines. *Revue Africaine de Théologie* 23/24: 183.
Muzorewa, Gwinyai H. 1985. *The Origin and development of African theology*. Maryknoll, NY: Orbis Books.
Oduyoye, Mercy Amba. 1986. *Hearing and knowing: Theological reflection on Christianity in Africa*. Maryknoll, NY: Orbis Books.
Societé Africain de Culture, n.d. *Les religions Africaines comme source de valeurs de civilisation: Colloque de Cotonou*. Paris: Présence Africaine.
Thomas, Louis-Vincent. 1967. L'Africain et le sacré. *Bulletin de IFAN (Institut Fondamentale Afrique Noir)* 3/4: 619–677.
Thomas, Louis-Vincent and René Luneau. 1969. *Les religions d'Afrique noire: Textes et traditions sacré*. Paris: Fayard/Denoël.

Van Binsbergen, Wim, and Matthew Schoffeleers, eds. 1985. *Theoretical explorations in African religions*. London: KPI.

Wan-Tatah, Victor. 1989. *Emancipation in African theology: An inquiry on the relevance of Latin American liberation theology to Africa*. New York: Peter Lang.

Zahan, Dominique. 1979. *The Religion, spirituality, and thought of traditional Africa*. Translated by Kate Ezra Martin and Lawrence M. Martin. Chicago: University of Chicago Press.

Works by Africanist Anthropologists, Historians, and Political Scientists

Bahuchet, Serge. 1992. *Dans la forêt d'Afrique Centrale: les pygmées Aka et Baka*. Paris: Peeters-Selaf.

Beidelman, T.O. 1986. *Moral imagination in Kaguru modes of thought*. Bloomington: Indiana University Press.

Bosc, P.M. 1992. Culture attelée et environnement: Reflexions à partir d'experiences ouest-africaines. *Afrique Contemporaine* 161: 197–209.

Davidson, Basil. 1992. *The blackman's burden: Africa and the curse of the nation-state*. New York: Times Books.

Ellis, S. 1992. Defense d'y voir: La politisation de la protection de la nature. *Politique Africaine* 48: 7–21.

Forde, Daryll, ed. 1954. *African worlds: Studies in the cosmological ideas and social values of African peoples*. London: Oxford University Press.

Gillon, Y. 1992. Empreinte humaine de facteurs du milieu dans l'histoire écologique de l'Afrique tropicale. *Afrique Contemporaine* 161: 30–41.

Hladik, C.M., S. Bahuchet, and I. de Garine, eds. 1989. *Se nourir en forêt equatoriale: Anthropologie alimentaire des populations des regions forestières humide d'Afrique*. Paris: UNESCO.

Horton, Robin. 1993. *Patterns of thought in Africa and the West: Essays on magic, religion, and science*. Cambridge: Cambridge University Press.

Jackson, Michael. 1982. *Allegories of the wilderness: Ethics and ambiguity in Kuranko narratives*. Bloomington: Indiana University Press.

Karp, Ivan, and Charles S. Bird, eds. 1980. *Explorations in African systems of thought*. Bloomington: Indiana University Press.

Miller, Joseph C., ed. 1980. *The African past speaks: Essays on oral tradition and history*. Folkestone, England: Wm. Dawson and Sons.

Preiswerk, Yvonne, and Jacques Vallet, eds. 1990. *La pensée métisse: Croyances Africaines et rationalité occidentale en questions*. Paris: Presses Universitaires de France.

Rivière, C. 1992. Attitudes africaines face à l'environnement. *Anthropos* 87: 365–378.

Vansina, Jan. 1972. Once upon a time: Oral traditions as history in Africa. In *Historical studies today*, eds. Felix Gilbert and Stephen R. Graubard, 413–439. New York: W.W. Norton.

Vansina, Jan. 1983. Is elegance proof? Structuralism and African history. *History in Africa* 10: 307–348.

Willis, Roy. 1974. *Man and beast*. London: Hart-Davis, MacGibbon.

Wolfe, Alvin W. 1961. *In the Ngombe tradition: Continuity and change in the Congo*. Evanston: Northwestern University Press.

Yoder, John C. 1992. *The Kanyok of Zaire: An institutional and ideological history to 1895*. Cambridge: Cambridge University Press.

Notes

1. These sources are not cited in Chapters 1–7 or Appendix B of this book and are additional to those in the reference list that follows this Appendix C.

References Cited

Abruzzi, W.S. 1979. Population pressure and subsistence strategies among the Mbuti pygmies. *Human Ecology* 7 (2): 183–189.

Achebe, Chinua. 1958. *Things fall apart.* London: Heinemann.

Agar, Michael, and James MacDonald. 1995. Focus groups and ethnography. *Human Organization* 54 (1): 78–86.

Almquist, Alden, Ian Deshmukh, Paula Donnelly-Roark, George Frame, Barbara Pitkin, and Fred Swartzendruber. 1993. *African biodiversity: Foundation for the future: A framework for integrating biodiversity conservation and sustainable development.* Washington, DC: Biodiversity Support Program.

Almquist, L. Arden. 1993. *Debtor unashamed: The road to mission is a two-way street.* Chicago: Covenant Publications.

Amanor, K.S. 1994. *The new frontier: Farmer responses to land degradation.* London: UNRISD and Zed Books. Quoted in James Fairhead and Melissa Leach. *Misreading the African landscape: Society and ecology in a forest-savanna mosaic*, 9. Cambridge: Cambridge University Press, 1996.

Anderson, Bernhard. 1984. Creation and the Noachic covenant. In *Cry of the environment: Rebuilding the Christian creation tradition*, eds. Philip N. Joranson and Ken Butigan. Santa Fe: Bear.

Antonio, Edward P. 1994. Letting people decide: Towards an ethic of ecological survival in Africa. In *Ecotheology: Voices from south and north*, ed. David G. Hallman, 227–234. Maryknoll, NY: Orbis Books.

Appiah, Kwame Anthony. 1992. *In my father's house: Africa in the philosophy of culture.* New York: Oxford University Press.

Asante, Emmanuel. 1985. Ecology: Untapped resource of pan-vitalism in Africa. *African Ecclesial Review (AFER)* 27:5 (October): 289–293.

Athanasiou, T. 1992. After the summit. *Socialist Review* 22:4 (October-December): 56–92.

Bailey, R., and N.R. Peacock. 1988. Efe pygmies of Northeast Zaire: Subsistence strategies in the Ituri Forest. In *Coping with uncertainty in the food supply*, eds. I. de Garine and G.A. Harrison, 88–117. Oxford: Oxford University Press.

Bailey, R., S. Bahuchet, and B. Hewlett. 1990. Development in Central African rain forest: Concerns for tribal forest peoples. Unpublished report prepared for the World Bank. August 8, 31 pp.

Barnes, B. 1974. *Scientific knowledge and sociological theory.* London: Routledge and Kegan Paul. Quoted in V.Y. Mudimbe. *The invention of Africa: Gnosis, philosophy, and the order of knowledge*, 199. Bloomington: Indiana University Press, 1988.

Berhane-Selassie, T. 1994. Ecology and Ethiopian orthodox theology. In *Ecotheology: Voices from south and north*, ed. David G. Hallman, 155–172. Maryknoll, NY: Orbis Books.

Berry, Wendell. 1972. *A continuous harmony: Essays cultural and agricultural.* New York: Harcourt, Brace, Jovanovich.
Bianga, W. 1982. Peasant, state, and rural development in post-independent Zaire: A case-study of *Reforme Rurale*, 1970–1980. Ph.D. dissertation. University of Wisconsin–Madison.
Bonner, Raymond. 1993. *At the hand of man: Peril and hope for Africa's wildlife.* New York: Alfred A. Knopf.
Bookchin, Murry. 1990. *Remaking society: Pathways to a green future.* Boston: South End Press.
Boulaga, Fabian Eboussi. 1984. *Christianity without fetishes: An African critique and recapture of Christianity.* Translated by Robert Barr. Maryknoll, NY: Orbis Books. Quoted in Harvey J. Sindima. *Africa's agenda: The legacy of liberalism and colonialism in the crisis of African values,* 197. Westport, CT: Greenwood Press.
Brecher, Jeremy, John Brown Childs, and Jill Cutler, eds. 1993. *Global visions: Beyond the new world order.* Boston: South End Press.
Brokenshaw, D., D. Warren, and O. Werner, eds. 1980. *Indigenous knowledge systems and development.* Lanham, MD: University Press of America.
Bruce, J. 1988. A perspective on indigenous land tenure systems and land concentration. In *Land and society in contemporary Africa,* eds. R.E. Downs and S.P. Reyna, 40–62. New Hampshire: University Press of New England.
Brueggemann, Walter. 1977. *The land: Place as gift, promise, and challenge in biblical faith.* Philadelphia: Fortress Press.
———. 1989. *Finally comes the poet: Daring speech for proclamation.* Philadelphia: Fortress Press.
Bulu-Bobina, B. 1984. Les problèmes des terres cultivables en milieu rurale: Cas de la zone de Kasangulu (Bas Zaire). *Zaire-Afrique* 189: 539–552.
Burnett, G.W., and Kamuyu wa Kang'ethe. 1994. Wilderness and the Bantu mind. *Environmental Ethics* 16 (Summer): 145–160.
Callewaert, John H. 1994. International documents and the movement toward a global environment ethic (preliminary draft). Chicago: IUCN Ethics Working Group. Photocopied.
Callicott, J. Baird. 1994. *Earth's insights: A survey of ecological ethics from the Mediterranean basin to the Australian outback.* Berkeley: University of California Press.
Carneiro, R.L. 1960. Slash and burn agriculture: A closer look at its implications for settlement patterns. In *Men and Cultures,* ed. A.F.C. Wallace, 229–234. Philadelphia: University of Pennsylvania Press.
Chambers, R. 1983. *Rural development: Putting the last first.* London: Longman.
Chief Seattle. 1854. Oration. Originally appeared in the *Seattle Sunday Star,* October 29, 1882, under a heading by Dr. Henry A. Smith.
Coles, Robert. 1971. *Migrants, sharecroppers, mountaineers.* Vol. 2, *Children of crisis.* Boston: Little, Brown, and Company.
Communauté Pour le Développement Intégral de Bodangabo (CODIBO). 1995. Statut. Boyasegeze: CODIBO. Typewritten.
———. n.d. Plan d'action to milongo na misala na kolanda. Unpublished document. Photocopied.
Cornell, George. 1992. American Indian philosophy and perceptions of the environment. In *Human values and the environment: Conference proceedings,* by the Institute for Environmental Studies, University of Wisconsin–Madison. Madison: Wisconsin Academy of Sciences, Arts, and Letters, 41–46.

Cox, Harvey. 1994. Healers and ecologists: Pentecostalism in Africa. *Christian Century* (November 9): 1042–46.

———. 1995. *Fire from heaven: The rise of pentecostal spirituality and the reshaping of religion in the twenty-first century.* Reading, MA: Addison-Wesley.

Cronon, William. 1995. The trouble with wilderness; or getting back to the wrong nature. In *Uncommon ground: Toward reinventing nature*, ed. William Cronon, 69–90. New York: W.W. Norton.

Daly, Herman E., and John B. Cobb Jr. 1989. *For the common good: Redirecting the economy toward community, the environment, and a sustainable future.* Boston: Beacon Press.

Daneel, M.L. 1991. The liberation of creation: African traditional religious and independent church perspectives. *Missionalia* 19:2 (August): 99–121.

———. 1994. African independent churches face the challenge of environmental ethics. In *Ecotheology: Voices from South and North*, ed. David G. Hallman, 248–263. Maryknoll, NY: Orbis Books.

Dix, Anne. 1996. *CAMPFIRE: Communal areas management programme for indigenous resources: An annotated bibliography (1985–1996).* Harare: Centre for Applied Social Sciences, University of Zimbabwe.

Douglas, Mary. 1954. The Lele of Kasai. In *African worlds: Studies in the cosmological ideas and social values of African peoples*, ed. Daryll Forde, 1–26. London: Oxford University Press. Quoted in J. Baird Callicott. *Earth's insights: A survey of ecological ethics from the Mediterranean basin to the Australian outback*, 158. Berkeley: University of California Press, 1994.

Drennon, C. 1990. Agricultural encroachment in two of Uganda's forest and game reserves. M.S. thesis. University of Wisconsin–Madison.

Drewal, Henry John. 1988. Performing the other: Mami wata worship in Africa. *Drama Review* 32 (Summer): 160–185.

Engel, J. Ron. 1990. The ethics of sustainable development. In *Ethics of environment and development: Global challenge, international response*, eds. J. Ronald Engel and Joan Gibb Engel, 1–23. London: Belhaven Press.

———. 1993. The role of ethics, culture, and religion in conserving biodiversity: A blueprint for research and action. In *Ethics, religion, and biodiversity: Relations between conservation and cultural values*, ed. Lawrence S. Hamilton, 183–214. Cambridge: The White Horse Press.

———. 1994a. Our mandate for advancing a world ethic for living sustainably. Paper presented at *Ethics and covenant workshop: Session one*, IUCN General Assembly, Buenos Aires, January, 6 pp.

———. 1994b. Sustainable development. A manuscript prepared for the *Encyclopedia of bioethics: New edition*, 17 pp.

———. 1994c. Class lecture. Divinity RE 638: Ecology, Justice, and Faith, October 28. Meadville-Lombard School of Theology, Chicago.

Engel, J. Ron, and Julie Denny-Hughes, eds. 1994. *Advancing ethics for living sustainably: Report of the IUCN ethics workshop.* Sacramento: International Center for the Environment and Public Policy.

Escobar, Arturo. 1995. *Encountering development: The making and unmaking of the Third World.* Princeton: Princeton University Press.

Fairhead, James, and Melissa Leach. 1996. *Misreading the African landscape: Society and ecology in a forest-savanna mosaic.* Cambridge: Cambridge University Press.

Flax, Jane. 1990. *Thinking fragments: Psychoanalysis, feminism, and postmodernism*. Berkeley: University of California Press.
Foucault, Michel. 1977. *Language, counter-memory, practice*. Ithaca: Cornell University Press. Quoted in D.A. Masolo. *African philosophy in search of identity*, 183. Bloomington: Indiana University Press, 1994.
Galeano, Eduardo. 1991. *The book of embraces*. Translated by Cedric Belfrage with Mark Schafer. New York: W.W. Norton.
Geertz, Clifford. 1973. *The interpretation of cultures: Selected essays*. New York: Basic Books.
George, Susan, and Fabrizio Sabelli. 1994. *Faith and credit: The World Bank's secular empire*. Boulder: Westview Press.
George, Susan. 1977. *How the other half dies: The real reasons for world hunger*. Montclair, NJ: Allanheld.
Girard, Denis. 1962. *The new Cassell's French dictionary*. New York: Funk and Wagnalls.
Glacken, Clarence J. 1967. *Traces on the Rhodian shore: Nature and culture in Western thought from ancient times to the end of the eighteenth century*. Berkeley: University of California Press.
Grinker, Richard R. 1994. *Houses in the rainforest: Ethnicity and inequality among farmers and foragers in Central Africa*. Berkeley: University of California Press.
Guha, Ramachandra. 1989. Radical American environmentalism and wilderness preservation: A Third World critique. *Environmental Ethics* 11 (Spring): 71–83.
Gyekye, Kwame. 1987. *An essay on African philosophical thought: The Akan conceptual scheme*. Cambridge: Cambridge University Press.
Hamilton, Lawrence S., ed. 1993. *Ethics, religion, and biodiversity: Relations between conservation and cultural values*. Cambridge: The White Horse Press.
Haraway, Donna. 1988. Situated knowledges: The science question in feminism and the privilege of partial perspective. *Feminist Studies* 14 (3): 575–600.
Harding, Sandra. 1986. *The science question in feminism*. Ithaca: Cornell University Press.
Harms, R. 1981. *River of wealth, river of sorrow: The Central Zaire Basin in the era of the slave and ivory trade, 1500–1891*. New Haven: Yale University Press.
———. 1987. *Games against nature: An eco-cultural history of the Nunu of equatorial Africa*. New York: Cambridge University Press.
Hart, John A. 1978. From subsistence to market: A case study of the Mbuti net hunters. *Human Ecology* 6 (3): 325–353.
———. 1979. Nomadic hunters and village cultivators: A study of subsistence interdependence in the Ituri Forest of Zaire. M.A. thesis. Michigan State University.
Hart, T.B., and J.A. Hart. 1984. Political change and the opening of the Ituri Forest. *Cultural Survival Quarterly* 8 (3):18–20.
———. 1986. The ecological basis of hunter-gatherer subsistence in African rainforests: The Mbuti of eastern Zaire. *Human Ecology* 14 (1): 29–55.
Hart, Terese B. 1990. Monospecific dominance in tropical rain forests. *Trends in Ecology and Evolution* 5 (1): 6–11.
Hoffer, Eric. 1969 *Working and thinking on the waterfront: A journal, June 1958-May 1959*. New York: Harper and Row.
Horton, Robin. 1967. African traditional thought and Western science. *Africa* 37: 50–71 (Part 1), 155–187 (Part 2).
Hountondji, Paulin J., ed. 1994. *Les savoirs endogènes: Pistes pour une recherche*. Paris: Karthala.
Hughes, Diane, and Kimberly DuMont. 1993. Using focus groups to facilitate culturally anchored research. *American Journal of Community Psychology* 21 (6): 775–806.

Hunt, Nancy Rose. 1997. "Le bébé en brousse": European women, African birth spacing, and colonial intervention in breast feeding in the Belgian Congo. In *Tensions of empire: Colonial cultures in a bourgeois world*, eds. Frederick Cooper and Ann Laura Stoler, 287–321. Berkeley: University of California Press.

Institute of Kiswahili Research, University of Dar es Salaam. 1981. *Kamusi ya Kiswahili sanifu*. Dar es Salaam: Oxford University Press.

Inter-Territorial Language Committee for the East African Dependencies. 1939. *A standard Swahili-English dictionary*. Nairobi: Oxford University Press.

IUCN. 1994. *Draft international covenant on environment and development*.

IUCN General Assembly. 1994. *Papers from the ethics and covenant workshop*. Buenos Aires, January 1994.

Jackson, Michael. 1989. *Paths toward a clearing: Radical empiricism and ethnographic inquiry*. Bloomington: Indiana University Press.

———. 1995. *At home in the world*. Durham: Duke University Press.

Jackson, Wes. 1996. *Becoming native to this place*. Washington, DC: Counterpoint.

Jackson, Wes, and Marty Bender. 1984. Investigations into perennial polyculture. In *Meeting the expectations of the land: Essays in sustainable agriculture and stewardship*, eds. Wes Jackson, Wendell Berry, and Bruce Colman, 183–194. San Francisco: North Point Press.

Jackson, Wes, and William Vitek, eds. 1996. *Rooted in the land: Essays on community and place*. New Haven: Yale University Press.

Jewsiewicki, B. 1983. Rural society and the Belgian colonial economy. In *History of Central Africa*, Vol. 2, eds. D. Birmingham and P. Martin, 95–125. London: Longman.

Johnson, Alan. 1996. 'It's good to talk': The focus group and the sociological imagination. *Sociological Review* 44 (August): 517–538

Jonas, Hans. 1958. *The Gnostic religion: The message of the alien God and the beginnings of Christianity*. Boston: Beacon Press.

Jurion, F., and J. Henry. 1969. *Can primitive farming be modernized?*. Brussels: INEAC.

Kabe, Mutuza. 1987. *Éthique et développement: Le cas du Zaire*. Kinshasa: Editions Zenda.

Keats, John. 1958. *The letters of John Keats, 1814–1821* (edited by H.E. Rollins), vol. 1. Cambridge: Cambridge University Press. Quoted in Michael Jackson. *Paths toward a clearing: Radical empiricism and ethnographic inquiry*, 16. Bloomington: Indiana University Press, 1989.

Kloppenburg, J. 1991. Social theory and de/reconstruction of agricultural science: Local knowledge for an alternative agriculture. *Rural Sociology* 56 (4): 519–548.

Koch, Henri. 1968. *Magie et chasse dans la forêt camerounaise*. Paris: Berger-Levrault.

Krueger, Richard A. 1988. *Focus groups: A practical guide for applied research*. Newbury Park, CA: Sage Publications.

Labou Tansi, Sony. 1979. *La vie et demie*. Paris: Éditions du Seuil.

Leopold, Aldo. 1949. The land ethic. In Part 4 of *A Sand County almanac*. New York: Ballantine Books.

Lévi-Strauss, Claude. 1979. *The origin of table manners*. New York: Harper and Row. Quoted in V.Y. Mudimbe. *The invention of Africa: Gnosis, philosophy, and the order of knowledge*, 31. Bloomington: Indiana University Press, 1988.

Lewis, Dale, and Nick Carter, eds. 1993. *Voices from Africa: Local perspectives on conservation*. Washington, DC: World Wildlife Fund.

Loomis, Louise Ropes. 1942. Introduction to *Plato: Five great dialogues (Apology, Crito, Phaedo, Symposium, and Republic)*, edited by Louise Ropes Loomis and translated by B. Jowett. Classics Club Series. New York: Walter J. Black.

Lopes, Henri. 1987. *Tribaliks (Tribaliques)*. Translated by Andrea Leskes. London: Heinemann Educational Books.
———. 1990. *Le chercheur d'Afriques*. Paris: Éditions du Seuil.
Lopez, Barry. 1988. *Crossing open ground*. New York: Vintage Books.
Lumpungu, G.K. 1977. Land tenure system and the agricultural crisis in Zaire. *African Environment* 2 (4) and 3 (1): 56–71.
MacGaffey, J. 1987. *Entrepreneurs and parasites: The struggle for indigenous capitalism in Zaire*. New York: Cambridge University Press.
———. 1988. Zaire. In *Coping with Africa's food crisis*, eds. N. Chazan and T.M. Shaw, 169–174. Boulder: Lynne Reinner.
MacGaffey, Wyatt. 1981. African ideology and belief: A survey. *African Studies Review* 24 (2/3): 227–274.
Mark, Joan. 1995. *The king of the world in the land of the pygmies*. Lincoln: University of Nebraska Press.
Masolo, D.A. 1994. *African philosophy in search of identity*. Bloomington: Indiana University Press.
Matthews, Freya. 1994. Cultural relativism and environmental ethics. *IUCN Ethics Working Group Circular Letter* Nº. 5 (August): 5–7.
Mbele, Joseph. 1986. The hero in the African epic. Ph.D. dissertation, University of Wisconsin–Madison.
Mbiti, John S. 1975. *Introduction to African religion*. London: Heinemann. Quoted in J. Baird Callicott. *Earth's insights: A survey of ecological ethics from the Mediterranean basin to the Australian outback*, 157. Berkeley: University of California Press, 1994.
Mbiti, John S. 1988. *African religions and philosophy*. London: Heinemann.
McCoy, Charles S. 1991. Creation and covenant: A comprehensive vision for environmental ethics. In *Covenant for a new creation: Ethics, religion, and public policy*, eds. Carol S. Robb and Carl J. Casebolt, 212–225. Maryknoll, NY: Orbis Books.
McFague, Sallie. 1993. *The body of God: An ecological theology*. Minneapolis: Fortress Press.
Merchant, Carolyn. 1980. *The death of nature: Women, ecology, and the scientific revolution*. San Francisco: Harper.
Merleau-Ponty, M. 1962 *Phenomenology of perception*. Translated by Colin Smith. London: Routledge. Quoted in Michael Jackson. *At home in the world*, 172. Durham: Duke University Press, 1995.
Merleau-Ponty, M. 1964. *Signs*. Translated by R.C. McLeary. Studies in Phenomenology and Existential Philosophy Series. Evanston: Northwestern University Press. Quoted in Michael Jackson. *Paths toward a clearing: Radical empiricism and ethnographic inquiry*, 14. Bloomington: Indiana University Press, 1989.
Midgley, Mary. 1995. Idealism in practice: Comprehending the incomprehensible. In *The right to hope: Global problems, global visions: Creative responses to our world in need*, ed. The Right to Hope Trust, 39–44. London: Earthscan Publications.
Miracle, M. 1967. *Agriculture in the Congo basin*. Madison: University of Wisconsin Press.
Misambo, Claude B. 1994. Bilenge ya Mwinda: Signe des temps, signe de notre temps. *Renaître* 19 (October): 13–14.
Momaday, N. Scott. 1994. Native American attitudes to the environment. In *The environmental ethics and policy book*, eds. Donald VanDeVeer and Christine Pierce, 102–105. Belmont, CA: Wadsworth Publishing Company.
Moore, Gerald, and Ulli Beier, ed. 1984. *Modern African poetry*. 3rd ed. London: Penguin Books.

Mpanya, Mutombo. 1987. African land ethics: Contributions to a Christian perspective on land. Paper presented at Au Sable forum: A Christian land ethic, Au Sable Institute, Mancelona, Michigan, August 5–8, 20 pp.

Mudimbe, V.Y. 1982. *L'odeur du père: Essaie sur des limites de la science et de la vie en Afrique noire*. Paris: Présence Africaine.

———. 1988. *The invention of Africa: Gnosis, philosophy, and the order of knowledge*. Bloomington: Indiana University Press.

———. 1991a. *Between tides (Entre les eaux)*. Translated by Stephen Becker. New York: Simon and Schuster.

———. 1991b. *Parables and fables: Exegesis, textuality, and politics in Central Africa*. Madison: University of Wisconsin Press.

———. 1993. *The rift (l'Écart)*. Translated by Marjolijn de Jager. Minneapolis: University of Minnesota Press.

———. 1994. *The idea of Africa*. Bloomington: Indiana University Press.

Muir, John. 1944. *My first summer in the Sierra*. n.p., 1911; reprint, Boston: n.p. Quoted in Donald Worster. *Nature's economy: A history of ecological ideas*, 429. 2nd ed. Cambridge: Cambridge University Press, 1994.

Mutombo, Huta. 1985. La faune et le rôle de l'homme dans l'organisation sociale. In *Philosophie Africaine et l'ordre social: Actes de la 9ème semaine philisophique de Kinshasa, 1–7 décembre, 1985*, by the Faculté de Théologie Catholique. Kinshasa: Faculté de Théologie Catholique, 247 ff.

Nagenda, John. 1986. *The seasons of Thomas Tebo*. London: Heinemann Educational Books.

Nelson, Richard K. 1983. *Make prayers to the raven: A Koyukon view of the northern forest*. Chicago: Chicago University Press.

Newbury, C. 1984. Dead and buried or just underground? The privatization of the state in Zaire. *Canadian Journal of African Studies* 18 (1): 35–54.

———. 1986. Survival strategies in rural Zaire: Realities of coping with crisis. In *The crisis in Zaire: Myths and realities*, ed. Nzongola-Ntalaja, 99–112. Trenton, NJ: Africa World Press.

Newbury, C., and B.G. Schoepf. 1989. State, peasantry, and agrarian crisis in Zaire: Does gender make a difference? In *Women and the state in Africa*, eds. J.L. Parpart and K.A. Staudt, 91–110. Boulder: Lynne Rienner.

Ngoie-Ngalla, Dominique. 1988. *Lettre d'un pygmée à un Bantou*. Brazzaville: C.R.P.

Ngwey, Ngond'a Ndenge. 1985. La tâche écologique de la philosophie en terre africaine. In *Philosophie Africaine et l'ordre social: Actes de la 9ème semaine philisophique de Kinshasa, 1–7 décembre, 1985*, by the Faculté de Théologie Catholique. Kinshasa: Faculté de Théologie Catholique, 51–56.

Nkombe, Oleko. 1980. La nature de l'éthique et l'éthique de la nature. In *Éthique et société: Actes de la 3ème semaine philisophique de Kinshasa, 3–7 avril, 1978*, by the Faculté de Théologie Catholique. Kinshasa: Faculté de Théologie Catholique.

Nzongola-Ntalaja, ed. 1986. *The crisis in Zaire: Myths and realities*. Trenton, NJ: Africa World Press.

Odera Oruka, H., ed. 1990. *Sage philosophy: Indigenous thinkers and modern debate on African philosophy*. Leiden: E.J. Brill.

———. 1994. *Philosophy, humanity, and ecology: Philosophy of nature and environmental ethics*. Nairobi: ACTS Press.

Omari, C.K. 1990. Traditional African land ethics. In *Ethics of environment and development: Global challenge, international response*, eds. J. Ronald Engel and Joan Gibb Engel, 167–175. London: Belhaven Press.

Omo-Fadaka, Jimoh. 1990. Communalism: The moral factor in African development. In *Ethics of environment and development: Global challenge, international response*, eds. J. Ronald Engel and Joan Gibb Engel, 176–182. London: Belhaven Press.

Onyewuenyi, Innocent. 1991. Is there an African philosophy? In *African philosophy: The essential readings*, ed. Tsenay Serequeberhan, 29–46. New York: Paragon House.

Overholt, Thomas W., and J. Baird Callicott. 1982. *Clothed-in-fur and other tales: An introduction to an Ojibwa world view*. Lanham, MD: University Press of America.

Pappu, S.S. Rama. 1994. Humans, animals, and the environment: Indian perspectives. In *Philosophy, humanity, and ecology: Philosophy of nature and environmental ethics*, ed. H. Odera Oruka, 295–300. Nairobi: ACTS Press.

Paris, Peter J. 1995. *The spirituality of African peoples: The search for a common moral discourse*. Minneapolis: Fortress Press.

Pashi, Lumana, and Alan Turnbull. 1994. *Lingala-English dictionary*. Kensington, MD: Dunwoody Press.

Pepetela. 1983. *Mayombe*. Translated by Michael Wolfers. Harare: Zimbabwe Publishing House.

Peterson, R.B. 1991. To search for life: A study of spontaneous immigration, settlement, and land use on Zaire's Ituri Forest frontier. M.S. thesis. University of Wisconsin–Madison.

Putnam, Patrick. 1948. The pygmies of the Ituri Forest. In *A reader in general anthropology*, ed. E. Coon, 322–342. New Haven: Yale University Press.

Ray, Benjamin C. 1976. *African religions: Symbol, ritual, and community*. Englewood Cliffs, NJ: Prentice-Hall.

Rockefeller, Steven C., and John C. Elder, eds. 1992. *Spirit and nature: Why the environment is a religious issue*. Boston: Beacon Press.

Rodney, Walter. 1982. *How Europe underdeveloped Africa*. Washington DC: Howard University Press.

Ruether, Rosemary R. 1975. *New woman new earth: Sexist ideologies and human liberation*. New York: Seabury Press.

———. 1978. The biblical vision of the ecological crisis. *Christian Century* (November 22): 1129–1132.

———. 1981. *To change the world: Christology and cultural criticism*. New York: Crossroad.

———. 1992. *Gaia and God: An ecofeminist theology of earth healing*. San Francisco: Harper.

———. 1995. Ecology, justice, and religion in light of women's experience. Paper presented at Institute for Ecology, Justice, and Faith, Chicago, March 15–18.

Sale, Kirkpatrick. 1991. *Dwellers in the land: The bioregional vision*. Philadelphia: New Society Publishers.

Salmon, P. 1969–70. Les carnets de campagne de Louis Leclercq. *Revue de l'université de Bruxelles* 22: 233–302. Quoted in Jan Vansina, *Paths in the rainforests: Toward a history of political tradition in Equatorial Africa*, 244. Madison: University of Wisconsin Press, 1990.

Sarathy, Partha M.A. 1994. The contribution of yesterday to the ethics of tomorrow. In *Advancing ethics for living sustainably: Report of the IUCN ethics workshop*, eds. J. Ron Engel and Julie Denny-Hughes, 31–32. Sacremento: International Center for the Environment and Public Policy.

Sayer, Jeffrey A., Caroline S. Harcourt, and N. Mark Collins, eds. 1992. *The conservation atlas of tropical forests: Africa*. New York: Simon and Schuster.

Schebesta, Paul. 1933. *Among Congo pygmies*. London: Hutchinson.

_____. 1936a. *My pygmy and negro hosts*. London: Hutchinson.
_____. 1936b. *Revisiting my pygmy hosts*. London: Hutchinson.
Scheub, Harold. 1985. A review of African oral traditions and literature. *African Studies Review* 28 (2/3): 1–72.
_____. 1990. *The African storyteller: Stories from African oral traditions*. Dubuque, IA: Kendall/Hunt.
Schoepf, B.G. 1984. Man and biosphere in Zaire. In *The Politics of agriculture in tropical Africa*, ed. J. Barker, 269–290. Beverly Hills, CA: Sage Publications.
Schoepf, B.G., and C. Schoepf. 1987. Food crisis and agrarian change in the eastern highlands of Zaire. *Urban Anthropology* 16 (1): 5–37.
Schoff, Gretchen Holstein. 1988. Place: A condition of the spirit. In *Two essays on a sense of place*, 15–25. Madison: The Wisconsin Humanities Committee.
Scott, James. 1976. *The moral economy of the peasant: Rebellion and subsistence in Southeast Asia*. New Haven: Yale University Press.
Serequeberhan, Tsenay. 1991. African philosophy: The point in question. In *African philosophy: the essential readings*, ed. Tsenay Serequeberhan, 3–28. New York: Paragon House.
_____. 1994. *The hermeneutics of African philosophy: Horizon and discourse*. New York: Routledge.
Shipton, Parker. 1994. Land and culture in tropical Africa: Soils, symbols, and the metaphysics of the mundane. *Annual Review of Anthropology* 23: 347–377.
Simbotwe, M.P. 1993. African realities and Western expectations. In *Voices from Africa: Local perspectives on conservation*, eds. Dale Lewis and Nick Carter, 15–21. Washington, DC: World Wildlife Fund.
Sindima, Harvey J. 1990. Community of life: Ecological theology in African perspective. In *Liberating life: contemporary approaches to ecological theology*, eds. Charles Birch, William Eakin, and Jay B. McDaniel, 137–147. Maryknoll, NY: Orbis.
_____. 1994. *Drums of redemption*. Westport, CT: Greenwood Press.
_____. 1995. *Africa's agenda: The legacy of liberalism and colonialism in the crisis of African values*. Westport, CT: Greenwood Press.
Skolimowski, Henryk. 1981. *Eco-philosophy: Designing new tactics for living*. Boston: Marion Boyars.
Somé, Malidoma Patrice. 1993. *Ritual: Power, healing, and community*. Portland, OR: Swan/Raven.
Spear, Thomas. 1997. *Mountain farmers: Moral economies of land and development in Arusha and Meru*. Berkeley: University of California Press.
Stegner, Wallace. 1986. *The sense of place*. Madison: The Wisconsin Humanities Committee.
Stevens, C.J. 1998. Section news: Anthropology and environment section. *Anthropology Newsletter* (February): 29–30.
Tabaziba, Clever. 1994. A world ethical statement. Paper presented at *Ethics and covenant workshop: Session two*, IUCN General Assembly, Buenos Aires, January, 7 pp.
Taylor, Paul. 1986. *Respect for nature*. Princeton: Princeton University Press.
Thiesenhusen, W.C. 1991. Implications of the agricultural land tenure system for the environmental debate: Three scenarios. *Journal of Developing Areas* 26 (1): 1–23).
Trzyna, Ted. 1994. A plan to promote sustainability by building ethics into decision-making. *Environmental Strategy (Newsletter of the IUCN Commission on Environmental Strategy and Planning)* 9 (December): 1, 4–14.

Tuan, Yi Fu. 1968. Discrepancies between environmental attitude and behavior: Examples from Europe and China. *Canadian Geographer* 12 (3): 176–191.
_____. 1989. *Morality and imagination: Paradoxes of progress*. Madison: University of Wisconsin Press.
_____. 1994a. Class lecture. Geography 519: Environment, environmentalism, and the quality of life, November 7. University of Wisconsin, Madison.
_____. 1994b. Class lecture. Geography 519: Environment, environmentalism, and the quality of life, November 9. University of Wisconsin, Madison.
Turnbull, Colin M. 1961. *The forest people: A study of the pygmies of the Congo*. New York: Simon and Schuster.
_____. 1965. *Wayward servants: The two worlds of the African pygmies*. Garden City, NY: Natural History Press.
_____. 1983. *The Mbuti pygmies: Change and adaptation*. New York: Holt, Rinehart, and Winston.
Urwin, Kenneth. 1968. *Langenscheidt's compact French dictionary*. New York: Langenscheidt.
Van Gelder, Sarah. 1993. Remembering our purpose: An interview with Malidoma Somé. *In Context* (34): 30–34.
Vansina, Jan. 1961. *De la tradition orale: Essai de méthode historique*. Tervuren: Musée Royale de l'Afrique Centrale.
_____. 1984. Western Bantu expansion. *Journal of African History* 25: 129–145.
_____. 1985. *Oral tradition as history*. Madison: University of Wisconsin Press.
_____. 1986. Do pygmies have a history? *Sprache und Geschicte in Afrika* 7 (1): 431–445.
_____. 1990. *Paths in the rainforests: Toward a history of political tradition in Equatorial Africa*. Madison: University of Wisconsin Press.
_____. 1992. Habitat, economy, and society in the Central African rain forest. *Berg Occasional Papers in Anthropology* 1: na.
Vinck, Honoré. 1987. Le Centre Aequatoria de Bamanya: 50 ans de recherches africanistes. *Zaire-Afrique* 212: 79–102.
Vwakyanakazi, M. 1982. African traders in Butembo, eastern Zaire (1960–1980): A case study of informal entrepreneurship in a cultural context of Central Africa. Ph.D. dissertation. University of Wisconsin–Madison.
Webster's encyclopedic unabridged dictionary of the English language. 1989 ed. New York: Portland House.
Western, David, and R. Michael Wright, eds. 1993. *Natural connections: Perspectives in community-based conservation*. Washington, DC: Island Press.
Whitehead, Alfred North. 1925. *Science and the modern world*. New York: Macmillan.
Wilkie, David. 1987. Cultural and ecological survival in the Ituri Forest. *Cultural Survival Quarterly* 11 (2): 72–74.
_____. 1988. Hunters and farmers of the African forest. In *People of the tropical rainforest*, eds. J.S. Denslow and C. Padoch, 111–126. Berkeley: University of California Press.
Wilkie, David, and J.T. Finn. 1988. A spatial model of land use and forest regeneration in the Ituri Forest of northeastern Zaire. *Ecological Modeling* 41: 307–323.
Wilson, E.O., ed. 1988. *Biodiversity*. Washington, DC: National Academy Press.
Wiredu, Kwasi. 1980. *Philosophy and an African culture*. Cambridge: Cambridge University Press.
_____. 1994. Philosophy, humankind, and the environment. In *Philosophy, humanity, and ecology*. Vol. 1, *Philosophy of nature and environmental ethics*, ed. H. Odera Oruka, 30–48. Nairobi: Acts Press.

Worster, Donald. 1994. *Nature's economy: A history of ecological ideas*. 2nd ed. Cambridge: Cambridge University Press.

Young, Crawford. 1994. *The African colonial state in comparative perspective*. New Haven: Yale University Press.

Zimba, Roderick F. 1994. Environmental ethics from an African perspective. Paper presented at Eco-Ethical Symposium, Department of Psychology, University of Saarland, Kirkel, Saarbrücken, Germany, July 29–August 2, 21 pp.

Zuesse, Evan M. 1979. *Ritual cosmos: The sanctification of life in African religions*. Athens: Ohio University Press.

Index

ADFL. *See* Alliance of Democratic Forces for the Liberation of Congo-Zaire
ADMADE (Zambia), 270
African conservation/philosophy/religion
    key works in, 287
African metaphysics
    ecological benefits within, 245–248
Afrocentricity, 8, 9–10
Aggrandizement
    within Central African societies, 131, 132
    environmental consequences of, 136–137. *See also* Money; Wealth
Agriculture, 93, 130, 197, 198. *See also* Gardens and gardening
AIDS, 65
Aldrin, 138–139, 185
Alliance of Democratic Forces for the Liberation of Congo-Zaire, xiii
Amanzala, Mama (Bapukele farmer)
    on hunger, 213
    on relations between *bakpala* and Mbuti, 199–200
Ancestors, 63, 68
    in dreams, 204
    forests and fishing protected by, 10, 69, 71, 139
    garden-ravaging animals handled by, 211
    lack of hunting restrictions during time of, 160
    lack of selling of forest by, 162
    laws and advice of, 73–74, 75, 83, 123–124
    nature and respect for, 245, 260
    prohibitions against killing animals by, 166, 173, 181–182
    and wars of the past, 105
    youth and abandonment of ways of, 129
Ancestral traditions/values
    Christianity's role in destruction of, 231–232
    colonialism and impact on, 129
    and disciplining of children, 80, 81, 82
    ecological benefits of, 244–245
    initiations at heart of, 132
    and marriage, 101, 102
    project staff's perspectives on, 227–229
    reinvigorating, 142–143, 240
Animals, 61, 112
    ancestors ways of caring for, 139
    and anthropocentrism, 126–127
    conservation of, 212–213
    within cosmological realm, 124–125
    gardens ravaged by, 152, 154, 178, 180, 209–211, 216, 218, 264
    and hunting periodicity, 91–92, 180, 244, 261
    hunting restrictions on, 72–73, 74, 160, 173, 192
    impact of guns on, 97
    killing of, only for eating, 212–214
    Mbo elders on balance between humans and, 169–170
    overhunting of, 126, 135, 136, 180–181
    sacredness of, 247–248
    totemic, 244
Anthropocentrism, 126, 127, 272
Association of African Earthkeeping Churches, 119, 266

Baboons, 180, 181, 209, 216
Badengaido
    history of, 178–179
    living within the RFO, 158–159
Bakpala (Ituri farming peoples)
    animals killed by, 195
    effects of RFO policy on Mbuti relations with, 207, 216–217
    relations between Mbuti and, 196
Beautyway, 284
Biases, 3, 42
Bilenge ya Mwinda (Young People of Light), 141, 142
Biocentrism, 271, 272
Biocommunitarianism, 78
Biodiversity, 3, 4, 7, 36
    and cultural diversity, 28
    within Ituri Forest, 148–149
    local people and conserving of, 63
    and socioeconomic poverty, 153

309

Biodiversity Conservation Network, 270
Birth-spacing customs, 103, 105
Bodangabo
    Tata Fulani's story about changes in, 130–131
    tree plantings in, 137
Bogofo (near Loko)
    discussions in, 67, 68–69, 71
    focus group with land-users in, 56
    Ubangian farmers in, 16
"Both/and" holistic perspective
    for actual conservation initiatives, 265, 267
    and development initiatives, 258
    "either/or" dualism contrasted with, 254, 256, 257
    on relations between nature and people, 259
    and Western environmental thought/ethics, 270–271, 272
Bricolage, 42
Bronx Zoo
    informational placard about RFO at, 220(photo), 221(photo)
Buffalo, 69, 71, 124, 164, 192

Callicott, J. Baird, 24, 36
    on African biocommunitarianism, 78
    on narratives, 283
    on Ojibwa world view, 285
    on one/many problem, 29
    on postmodern ecological science, 30
    on transformation toward ecological sustainability, 38
CAMPFIRE (Zimbabwe), 270
Capitalism
    and creation of consumption, 235–237
    disparities in wealth under, 134
    individualism emphasized by, 106, 107
Caterpillar gathering/harvesting, 88, 89(photo), 90
Catholic Church
    and "Africanization" of Christianity, 141
    ancestral traditions diminished by, 231, 232
CEFRECOF. *See* Centre de Formation et Recherche en Conservation Forestière
Center for Indigenous Knowledge for Agriculture and Rural Development, 19(n2)
Central Africa
    communal labor groups in, 114
    environmental ethics in, 24–26
    land ethic rooted in forest culture of, 5
    land tenure in, 109–110, 115
    oral traditions of, 284–287
    repair and reconstruction: signs of hope in, 47–48
    sharing ethic in, 288
    summary of environmental thought in, 276–277
    trade in, 98–99
Centre Aequatoria (Bamanya Catholic mission)
    unique Congo source materials in, 286
Centre de Formation de Recherche en Conservation Forestière
(Center for Training and Research in Forest Conservation), xvi, 16, 151, 157
    garden damage by animals recorded by, 211
    need for communication between local people and, 249–250
    notice to employees of, 197–199
CEUM. *See* Communauté Evangelique de l'Ubangi-Mongala
Children
    and ancestral tradition, 73–74
    discipline of and respect for nature, 80, 81, 82
    forest preservation for, 128–129
    influence of Western culture on, 134
    and inherited land, 115
    preserving animals for sake of, 129
    spacing of, 105
Choral Mutchatcha, 183
Christian evangelism
    role of, in destroying ancestral traditions, 231–232
Christianity
    "Africanization" of, 141
CIKARD. *See* Center for Indigenous Knowledge for Agriculture and Rural Development
Circumcision, 132–133
"Closing the forest," 193–194, 207, 216, 224(n18)
CODIBO. *See* Communauté pour le Développement Intégral de Bodangabo
Coles, Robert
    direct transmission of who people are, 60, 62
    Migrants, Sharecroppers, Mountaineers, 62
    on narrative approach, 62
Colonialism
    birthrate policies under, 105–106
    and Central Africa eco-ethical ideals, 58
    deforestation and agricultural markets under, 131
    destruction of initiations under, 132
    hope rising out of destruction from, 47
    impact of, 8, 44–45, 99, 129
Comités permanents de consultation locale (standing committees of local consultation), 268
Commercial markets
    ethics surrounding killing animals for, 213–214, 215
    and fishing/hunting, 89, 155
    and resource degradation, 98

Index

Common ground
  ecological benefits of ancestral beliefs/practices, 244–245
  ecological benefits within African metaphysics, 245–248
  and need for local control, 241–243
  between project staff and local people, 241–250
  reciprocity between nature and local people, 248–250
Communauté Evangelique de l'Ubangi-Mongala, xvi, 16, 66, 109, 140, 145
  Loko, community relations, and, 254–256
Communauté pour le Développement Intégral de Bodangabo, 142
Communication
  necessity of, between local people and project personnel, 249–250, 251
Community/community relations
  African idea of, 107–109
  CEUM, Loko, and, 254–256
  and individual relationships, 106–107
Congo, Democratic Republic of
  economic/political conditions in, 61
  official renaming of, xiii
  political administrative divisions/research bases, 17(map)
Conservation, 87
  and culture, 14, 15, 244, 245–248
  local people's view of Ituri reserve policy on, 216–217
  and making a living, 87–88
  participatory, 268
  for whom?, 2–3. *See also* Environment; Restraints and limits
Conservation initiatives
  "both/and" holistic perspective for, 265, 267
  lessons from Central African human-environment relations for, 265–270
Consumption and consumerism
  "ethics of," 288
  impact of, on Central African youth, 46
  role of capitalist market and creation of, 235–237. *See also* Money
Conversation
  lived experience, radical empiricism and, 48–52
  origins of word for, 49, 59(n10)
Cooperatives
  gardening, 114
  labor, 258
Cosmology
  interconnectedness of nature, society and, 16, 86–87, 125–126, 136, 144
  land and society within realm of, 116–135

Covenant, meaning and metaphor of, 32–34
Cox, Harvey, 29
  on African religious ecological ethic, 120
CPCLs. *See* Comités permanents de consultation locale
Creation
  African concept of, 272
  purpose of, 126–127, 128, 160, 162, 169–170, 215
Cronon, William, 273
  on idealizing distant wildernesses, 274, 278(n8)
  on "troubles with wilderness," 263–264, 267
Culture
  and conservation, 14, 15, 244, 245–248
  and environmental sustainability, 4, 240
  myths within, 283
  revaluating, 244

DANZER (German logging firm), 234
Death of Nature, The (Merchant), 31
Debo (Mbo elder), 183
  on celebrating the forest, 168
  on early white settlers, 166
  on negative impact of RFO hunting restrictions on local people, 171, 175–176
Deforestation, 77
  in Bodangabo, 130–131
  and desire for money among youth, 134–135
  in Guinea, 260
  and logging, 234
  and population growth, 233
De Medina, Jean, 211, 216
Democracy
  impact of, on forests, 113
Democratic Republic of Congo. *See* Congo, Democratic Republic of
"Destructive" warfare, 105
Development initiatives
  lessons from Central African individual-community relations for, 257–258
Dewey, John, 49, 51
Diet, 215
  impact of RFO hunting restrictions on local, 176–177, 178
Dieu-Donné
  ancestral patterns of mobility confirmed by, 230
  on government environmental regulation, 238
  on local control, 241–242, 243
  on recovering ancestral values, 240
  on role of Christian evangelism in destroying ancestral //traditions, 231
  on sedentariness, 231
"Disturbance-impressed" ecology, 35

Dongala, Jacques, 157, 158
Douglas, Mary, 77
Dungu (Mbandja elder)
    on land tenure, 110–111

Ecojustice, 33, 153, 158, 249, 288
Ecology, 34
    and communal land tenure, 114–115
    intersection of culture, conservation, and, 14, 15
    juncture of cosmology and, 118, 119
    values, value transformation and, 22–23
Ecology Group (University of Chicago), 35
Education
    and economic hardships, 64–65
    effect of depopulation on, 152
    Hoffer on, 42
    impact of Westernization on, 226
Elders
    respect for, 132
    social/environmental advice by, 139–140. *See also* Ancestors
Elephants, 78, 124
    gardens ravaged by, 209, 264
    hunting of, 71
    in Ituri Forest, 148
    prohibitions against killing of, 173, 174, 192
Engel, J. Ron, 24
    on dialectic between universalism and pluralism, 27
    on value transformation, 22, 23
Enlightenment, the, 262
    and isolation from/control of nature, 31, 76
Environment
    and Afrocentricity, 9–10
    costs of aggrandizement on, 136–137
    and individual-community relations, 109
Environmental breakdown/degradation, 48
    linking of, to breakdown of ancestral world, 80–82, 83
    population growth and role in, 100, 233
    possible solutions to, 137–138
Environmental education
    need for two-way flow between locals and project staff, 238–239
Environmental ethics, 6, 21, 144
    African thinkers writing on, 287
    Callicott on, 285–286
    in Central Africa, 24–26
    and environmental justice, 43–47
    Maathai on creation of, 46–47
    meaning of, 15
    postcolonial influences on, in Central Africa, 46. *See also* Land ethics

Environmental sustainability
    and local control, 63, 241–243
    and mobility and "sense of place," 229–230
    understanding interconnectedness of social/cosmological worlds for, 144
Epulu, 1, 16, 18
    CEFRECOF scientist at, on population growth, 233
    conservation efforts at, 151
    experimental agroforestry garden at, 269(photo)
    focus group at, 54, 228
    problems of capitalist market discussed in focus group at, 236–237
    RFO, human-environment relations and, 259–260. *See also* Réserve de Faune à Okapis
"Ethics of consumption," 288
Evangelical Covenant Church of America, 16

Fairhead, James
    on effect of dualism on nature conservation/land management, 262–263
    on local peoples' role in deforestation, 260
Farmers in Ituri
    effects of RFO policy on, 205–208, 216–217
    relations between Mbuti forest-dwellers and, 196
Farming, 88
    CECRECOF limits on, 197, 198
    connection to nature through, 77
    impact of RFO on, 161
    and periodicity, 90–92
Female single heads of household, 100, 112
Finally Comes the Poet (Brueggemann), 23
Fines, 155, 171, 182, 216
Fish and fishing, 61, 77, 88
    ancestors' ideas about protecting, 69, 71
    boundaries on, 115
    poisoning of, 267
    restraint and limits on, 72–73
    rotational harvesting of, 92–94
    seining, 71
    techniques of Mbuja, 138
    tenure, 71
    traps, 96
Focus groups, 52–56
    discussion guide for, 54–55
Food, 90, 175–176, 178, 210. *See also* Diet; Fish and fishing; Gardens; Hunting
Forests, 128, 167–168
    African metaphysics and conservation of, 246
    ancestors' use/replenishment of, 69, 71, 139, 162
    complexity of, 157
    desire for money and impact on, 134–135
    "enticing" of, 201, 266

Index

as gardens, 215
hunting periodicity and protection of, 92
initiation camps within, 131–133
land use and human impact on, 115–116
meaning of, for local people, 215
rotational agriculture and regeneration of, 93
species harvested from, 88. *See also* Ituri Forest; Rainforests
Fulani, Tata
Bodangabo: story of changes in homeland, 130–131
consequences of aggrandizement for, 136–137
on fish traps, 96
on *Gaza*, 131–132
on leopard totem, 232
on microhydro and Mami wata, 10–13
on poisons in water, 267
on social/environmental healing, 140
on speaking for farmers, 145

Gabi (Kongo farmer)
on harvesting manioc, 68
on warfare, 104–105
Gaia Hypothesis, 35
Galeano, Eduardo
on narrative and interpretation, 56–57
Garbage pits
role of, in household economies/ecologies, 266–267
Gardens and gardening
animal damage/ravaging of, 152, 154, 178, 180, 209–211, 216, 264
boundaries on, 115
hunting integrated with, 92
individual-community relations over, 110–111, 114
kitchen, 266–267
and land grabbing, 115–116
and periodicity, 91
sanctions against in, 94–95. *See also* Farming
Gaza initiation rites
Christianity's role in ending of, 232
puberty rites of Ubangian ethnic groups, 131–133
revitalizing, 140
Geertz, Clifford
on narrative and interpretation, 56
Gilman Investment Company, 149, 151
Girls' initiations, 133
Global capitalism, 45
Global warming, 3
Gnosticism, 252(n2), 262, 277(n4)
Gold mining, 152, 159, 196, 182, 197

Greed, 125
Greenbelt Movement (Kenya), 46
Guha, Ramachandra
on American environmental movement, 147
Guinea
local peoples' reforestation in, 260
Guns, 89, 99, 139
impact of, on animal population, 97, 100, 136, 213–214
Gyekye, Kwame, 25
on individual-community relationship, 109
on reverence for life, 272

Harvesting
benign technologies for, 96–97
rotational, of game and fish, 92–94
Health and health care
devolution of, 152
garden-ravaging and effect on, 216
in Loko, 65, 66, 67
RFO hunting restrictions and impact on local people's, 175–176, 177, 178
Hebrews (ancient)
meaning of covenant for, 32–33, 34
Heri (Mbo elder), 183, 171
on balance between animals and humans, 169–170
on celebrating the forest, 167–168
on Mbo hunting restrictions, 164
on negative impact of RFO hunting restrictions on local people, 175–176
Honey gathering/honey season, 88, 90, 200, 244, 245
Hope, 47–48, 58, 288
Horton, Robin, 273
on African theoretical systems, 25
Hospitality, 61, 183, 184
ancestors' advice on, 123, 124–125
and forest, 167–168
Humans
Mbo view of interdependence between animals and, 169–170
Hunger
garden-ravaging animals related to, 210
reserve hunting restrictions and, 213
Hunt, Nancy Rose
on colonial birthrate policies, 105–106
Hunting, 77, 88, 92, 99, 115, 215
colonial laws on, 77
commercial, 89–90
gardens/farming correlated with, 90, 92
and local control, 242
meaning of procuring meat through, 177

nets, 187, 191–192, 203
and periodicity, 90–91, 180, 244
restraint and limits on, 72–73, 74
RFO policy on, 155, 187

ICCN. *See* Institut Congolais pour la Conservation de la Nature
"Idealism in Practice: Comprehending the Incomprehensible" (Midgeley), 143
IES. *See* Institute for Environmental Studies
Immigrant farmers, 2–3
"Indigenous" churches
and "Africanization" of Christianity, 141
environmental sustainability and practices in, 119
sankofa within, 142
Indigenous farmers, 200, 205–208, 216–217
Indigenous Technical Knowledge (or Indigenous Knowledge Systems), 6–7, 8, 13
Individual-community relationship
"both/and" perspective applied to, 254, 255, 256, 257
labor governed by, 114
living with neighbors, 106–107
Individuals or community
versus individuals and community, 256–257
Inheritance, and land, 115
Initiation ceremonies/rites, 45, 46, 133–135, 143
Christianity's role in ending of, 232
destruction of, under colonialism, 129
and local control, 243
reemergence of, 141
Institut Congolais pour la Conservation de la Nature, xvi, 151
Institute for Ecology, Justice, and Faith, 23, 30
Institute for Environmental Studies, 2
Institut Médical Evangélique Docteur Carlson, 66
Institut Supérieur de Pedagogie, 245
International Monetary Fund, 46
Interpretation and narrative, 56–57, 63
Interviews
versus focus groups, 53
Invention of Africa, The (Mudimbe), 39, 40
ISP. *See* Institut Supérieur de Pedagogie
ITK. *See* Indigenous Technical Knowledge
Ituri Forest, 1, 16, 190–191, 270
background on, 148–149, 151
community-reserve relations in, 156, 158
as garden for Mbuti, 191
and iconic visions of tropical rainforest, 264
major east-west road through, 2(photo)
okapi (rainforest giraffe) in, 149(photo)
within Orientale Region (Northeastern Congo), 150(map)
pit-sawn mahogany lumber produced in, 236(photo)
primary forest and plantain gardens in, 275(photo)
women planting peanuts in, 102(photo)
Ituri River, 148
Ivory market, 235

Jackson, Michael, 286
on fieldwork, 66
on lived experience, 50, 51, 60, 62
on pragmatism and research methodology, 49
James, Williams
notion of radical empiricism, 51

Kabe, Mutuza
on African communalistic vision, 255–256
Kagu, Mama (Ngbaka farmer)
on marriage, 101
on population growth, 100
Kinship
personhood, environmental responsibility and, 119
Kitchen gardens, 266–267
Krueger, Richard
on focus groups, 53
Kuranko people (Sierre Leone)
Jackson's work with, 50
Kwela, Professor
on ecological benefits within African metaphysics, 245–246
and equilibrium between African/Western approaches, 251

Labor
among Mbuti, 156
communal groups for, 114, 256, 258
global capitalism, profit maximization and, 45
work hours extended, 135, 136
Land, 115
in communal trust, 261
economic/political mismanagement of, 135
meaning of, 7, 10. *See also* Forests; Gardens and gardening
Land ethics, 5, 9, 15–16, 18, 21–59, 286
and actual practices of the people, 83, 84
ecological science and nature-culture relationship, 34–38
ecology, values, and value transformation, 22–23
environmental ethics and environmental justice, 43–47
environmental ethics in Central Africa, 24–26
environmental ethics in global perspective, 23–24

Index   315

  local people's voices, 38–43
  premodern world views and nature-culture
    relationship, 30–34
  premodern world views and postmodern
    ecological science, 29–30
  repair and reconstruction: signs of hope,
    47–48
  research model on: circles of investigation, 289,
    290(fig.)
  universalism and pluralism: "one/many problem,"
    26–29
Land grabbing, 115–116
Land tenure, 3, 256, 257
  in Central Africa, 109–110
  privatization/individualization of, 45
  and women, 112–113
Leach, Melissa
  on effect of dualism on nature conservation/land
    management, 262–263
  on local peoples' role in deforestation, 260
Leaders, hierarchical organization of, 222(n2)
Lega people, Mwindo Epic of, 286
Leopards, 117–118, 164, 165, 182, 232
Leopold, Aldo
  and features of "land ethic," 286
Libeka (Ubangian farmer)
  on ancestors and protection of forests/fishing, 69,
    71
Likilemba (communal labor groups), 114, 256, 258
Limits. See Restraints and limits
Lived experience, 48, 62
  Jackson on, 50, 51
  methods of interacting and participating in,
    49–52
  and values transformation, 57
Local people, 3, 4, 151
  challenge of listening to, 41
  common ground between project staff and,
    241–250
  differences in perspective between project staff
    and, 227
  environmental education, project and,
    238–239
  and local control, 241–243
  marginalization of, in natural resource
    management, 264
  narratives of, 57, 63
  reciprocity between nature and, 248–250
  reforestation by, in Kissidougou Prefecture,
    Guinea, 260
  social/environmental changes coming from,
    139–140
  voices of, and research objective, 48

  ways of resolving conflicts between RFO and,
    217, 218–219
  what is Africa?, listening to voices of, 38–43
Logging, 9, 196, 197, 234
Loko (northern Congo), 15–16, 18, 103, 233
  CEUM, community relations and, 254–256
  expertise from, to benefit RFO, 269
  focus group at, on role of poverty and economic
    injustice, 233–234
  focus group on role of Christian evangelism in,
    231
  history of, 65–66
  savanna-forest ecotone near, 64(photo)
  voices of farmers of, 63–68
Loko Health Zone, 67
Longo, Père, 153–154
Lopez, Barry
  on Navajo, 284
  on storytelling and transformation,
    23

Maathai, Wangari
  on environmental ethic and sustainable societies,
    46–47
Mahogany logging/lumber
  local villager uses chain saw for, 235(photo)
  production of pit-sawn, in Ituri, 236(photo)
Making a living
  and conservation, 87–88
  diversification of, 98
  from forest, 162
  and integration and periodicity, 90–92. See also
    Money
Malabo, Mama (Mbandja farmer)
  on hunting periodicity, 91
  on land tenure, 112–113
  on marriage, 101–102
Malengo (CEFRECOF researcher)
  on Catholic church and degrading of ancestral
    values, 232
Malnutrition, 183
Mambasa Zone, 148
Mami wata, 19(n6), 136
  microhydro and, 10–13, 275
Mamuya (Loko community development organizer)
  on role of State, 238
Mangabeys, 180, 181, 190, 216, 264
Manioc greens, 110
  and dietary changes due to hunting restrictions,
    175–176, 177, 178, 215
  harvesting, 68, 72
Mapoli (Bali elder), 183
  on overhunting, 180

on prohibitions against killing okapi, 181–182
on RFO and peace in the forest, 178, 179, 180
Marriage/marital status
    impact of, on population, 100–101
    women's opinions about, 101–102
Marxist societies
    community emphasized by, 106, 107
Masalito (Epulu Mbuti guide), 190
    on complexity of forest, 157
    on Mbuti hunting restrictions, 191–192
    on opposition to sharing forest with *bakpala*, 195–196
Masina, 201
    on Mbuti and the forest, 203, 204
Masolo, D. A.
    assessment of Vansina by, 284–285
    on Mudimbe, 41, 42
Matiyé (Budu farmer), 183
    on RFO and restrictions on local villagers, 159, 160–163
Maurice (Songye farmer), 183
    on benefits of RFO, 179
    on marketing garden produce, 222(n4)
    on overhunting, 180
    on reserve policies, 182
Maximization of production, 95–96
Mboya, Paulin
    on need for local people's support for reserve, 218–219, 220, 221
Mbuja people, 69
    fishing techniques of, 138
    hunting technologies used by, 90
Mbuti (foragers in Ituri), 3, 149, 156, 202–205
    hunting and gathering by, 60
    net hunting by, 177, 180
    periodicity among, 244–245
    perspectives of, on forest and RFO, 190–197, 199
    relations between farmers in Ituri and, 196
    reserve conservation initiatives and survival of, 207–208
    and reserve prohibitions against *bakpala*, 199–200
    RFO policies and, 155, 205–208, 216–217
McFague, Sallie
    on dangers of new universalism, 26
    feminist critique of science by, 43
Meat
    CEFRECOF limits on marketing of, 198, 199
    central role of, 177
    effects of reserve policy and limited availability of, 175, 176–178
    and Mbuti-villager exchange system, 179–180, 182
Metanoia, 6
Metaphysical realm
    social realm and natural realm intertwined with, 86
Methodology and research, 48–58
    circles of investigation, 290(fig.)
    conversation, lived experience, and radical empiricism, 48–52
    focus groups, 52–56
    narrative and interpretation, 56–57
Microhydro
    and Mami wata, 10–13, 136, 275
Midgeley, Mary
    on traditional ideals, 143
Migrants, Sharecroppers, Mountaineers (Coles), 62
Mining, 196
    impact of capitalism on, 130
    neocolonial operations, 9
Missionaries
    and destruction of ancestral values, 231
    destruction of initiations under, 132, 133
Mobility, and "sense of place," 229–230
Mobutu Sese Seko, President, xiii
    abuse of wealth by, 80–81, 134
    humanitarian aid, deteriorating infrastructure under, 65
    land tenure changes under, 115
    public health crisis under, 65
Moderators, focus groups, 53
Moke (Mbuti elder), 1, 201
    on Mbuti's relationship to forest, 202–205
Money
    and departure from ancestral ways, 185
    desire for, and impact on forest, 134–135
    desire for, and moral/social breakdown, 84
    rejection of ancestral ways and desire for, 129, 133–134
Monkeys, 135, 180–181, 190, 209
Mosolo, Abbé, 246
    on African metaphysics and environment, 247–248
    critique of Western exploitation by, 236–237
    on environmental awareness among students, 239–240
    on reciprocity between nature and local people, 248–249
    on Westernization and education in Africa, 226
Moyo
    Sindima on symbol of, 259

Mpanya, Mutombo, 9
    on African land ethics, 86
Mudimbe, V. Y., 39, 40, 284
    Masolo on, 41, 42
    on oral tradition methodologies, 285, 286
    on "what is Africa?," 40
Muir, John, 31
Musafiri (Bira elder)
    on manioc greens, 177
Mushrooms gathering/harvesting, 88, 90
Music and singing, 203
Mutualités, 256
Mwindo Epic, of Lega people, 286
Myths, 284
    Lévi-Strauss on, 283
    transformation through, 23

Nani (Pagabeti farmer/hunter)
    on economics of overhunting, 135–136
    on hunting periodicity, 91
Narratives
    and interpretation, 56–57
    letting them speak for themselves, 285, 286
National parks
    and dualism, 77, 79
Native Americans, 25, 31
Natural realm
    interconnectedness between social realm, cosmological realm and, 16, 86
Natural resources, 45, 134
    ancestors' laws governing, 124–125
    distributing weight of human impact on, 89
    restraint and limits on use of, 72–73, 74, 83, 84, 90–92
Nature, 37, 77, 118, 259, 273
    African understandings of, 9
    alienation from, 74–75, 77
    human communities integrated with, 78, 82–83, 84
    interconnectedness of society, cosmology and, 125–126, 136, 144
    is it normative?, 29–30, 35–36
    reciprocity between local people and, 248–250
Nature-culture relationship, 15
    dualism between, 263
    premodern world views and, 30–34
Navajo, 284
Ndjaosiko (Budu farmer's wife), 183
    on RFO and restrictions on local villagers, 160–163
Negritude, 8, 19n4, 40
Neocolonialism, 9, 47, 134, 239
Net hunting, 180

"New ecology," 260
New York Zoological Society, 16, 151
Ngambe (Ngbaka cultural organization), 142
Ngbaka cosmology
    place of *Seto* in, 121–124, 125
NGOs. *See* Nongovernmental organizations
Nguluma
    disappearance of, 130–131
Nkomo (Lokele fisherman/elder)
    on need for understanding and agreement between RFO and locals, 188–189
    on reasons for decrease in fish, 184–187
Nkrumah, Kwame, 8, 40, 41
Nongovernmental organizations, 109
Northeastern Congo
    Orientale Region: Ituri Forest and surrounding area, 150(map)
Northwestern Congo
    Équateur Region: Ubangi and Mongala Subregions, 70(map)
Ntembe na Mbeli
    communal gardening by, 114
Nutrition, 215
    and reserve policy, 175, 177–178. *See also* Diet; Food
Nyerere, Julius, 8, 40, 41
Nylon nets, and fish depopulation, 184
NYZS. *See* New York Zoological Society
Nzambe (Christian)
    entrance of, into Ngbaka cosmology, 122, 123, 124, 126

Ojibwa world views, 285
Okapi (rainforest giraffe), 1, 148, 149(photo), 187, 189, 249
    prohibitions against killing of, 164, 166, 173, 174, 181–182, 192
    and potential economic benefits of, for local people, 270
    protection of, 166
    taboos related to, 244. *See also* Réserve de Faune à Okapis
Okapi Capture Station (Epulu), 149
Omari, C. K., 9
One/many problem, 21–22, 26–29
    Callicott on, 29
    and quest for environmental ethics, 15
Onyewuenyi, Innocent
    on morality, 108
Optimization of production, 95–96
Oral tradition, 50, 58, 61
    Central African, 284–287
    Seto within, 121–124

Orientale Region (Northeastern Congo)
  Ituri Forest and surrounding area in, 150(map)
Overholt, Thomas
  on narratives, 283
  on Ojibwa world view, 285
Overhunting, 126, 135, 136, 180–181
  and benefits of RFO policies, 195

Paluku (Loko project staff member)
  on local control, 243
Parish development committees, 254, 256
Participant observation
  versus focus groups, 53
Paths in the Rainforests (Vansina), 44
Paul Carlson Foundation, 65
PDCs. *See* Parish development committees
Peanut planting, 97, 100, 102(photo)
Periodicity, 90–92, 180, 244, 261
Personhood (*umunthu*), 142
  environmental responsibility, kinship, and, 119
  and natural world, 117, 118
Pesticides, 267
  and fish depopulation, 138, 185
Pigs, 180, 181, 209, 216, 264
Pindwa (Bogofo farmer/hunter/leader), 67, 68, 90
  on child discipline and respect for nature, 80, 81
  on hunting periodicity, 91–92
  on individual *versus* community-based development, 256–257
  on losing contact with ancestral teachings, 69
Poachers, 139, 154
Poisons, 71, 99
  agricultural, 89
  and fish depopulation, 138, 185
  in water, 267
Population growth
  controls on, 99–106
  role of, in environmental deterioration, 233
  and warfare, 105
Postmodern ecology, 35
  and premodern world views, 29–30, 36, 38
Postmodern science
  and premodern teachings, 57–58
Poverty, 48, 143, 153, 233–235, 248, 249
Premodern world views
  and nature-culture relationship, 30–34
  and postmodern ecology, 29–30, 36, 38
Pristineness of nature, 263–264, 274, 277(n5)
Project staff
  on ancestral traditions, 227–229
  common ground between local people and, 241–250

differences in perspectives between local people and, 227
local people, environmental education and, 238–239. *See also* University educated staff
Projet Okapi, 20(n10)
Protestant Christianity
  ancestral traditions diminished by, 231

Qualitative and quantitative methods
  focus groups advantages over, 53

Radical empiricism, 48, 51
Radio CANDIP, 246
Rainforest giraffe. *See* Okapi
Rain forests, 64
  biodiversity juxtaposed with poverty in, 153
  community-individual balance in, 107
  iconic visions of, 264
  Nguluma within, 130. *See also* Forests; Ituri Forest
RDCs. *See* Regional development committees
Reforestation, 137–138
Regional development committees, 254, 256
Réserve de Faune à Okapis (Okapi Wildlife Reserve), 16, 18, 79, 150(map), 151, 246, 268–269
  Badengaido and living within, 158–159
  Central African holistic understanding applied to, 254
  effects of official policy of Mbuti-villager relations and forest use, 205–208, 216–217
  Epulu, human-environment relations and, 259–260
  hunting policy of, 155, 187
  informational placard about, at Bronx Zoo, 220(photo), 221(photo)
  local people's views on current conservation policies of, 160–163, 167–168, 169, 170, 171, 216–217
  managing for "misplaced pristineness" within, 264
  Mapoli on benefits of, 178, 179
  Mbuti-villager relations and policies of, 199–201
  need for respecting local people and support of, 174, 217, 218–219, 220, 249, 261–262
  patterns of power between local people and, 156–157
  protein deficiencies and hunting restrictions by, 178
  restrictions on garden-ravaging animals by, 209–211, 212. *See also* Ituri Forest
Reserve guards, 222(n3)
  abuse of power and authority by, 216, 218
  fear of reprisals from, 171, 176, 178, 206, 208
Resource management, 261

Index

benign technologies for, 96–97
conservation of, in making a living, 87–88
and household economy diversification, 88–90
integration and periodicity of, 90–92
population growth controls, 99–106
and production for consumption rather than markets, 97–99
rotational harvesting of game and fish, 92–94
sanctions against waste, 94–95
Respect
for local people, and support for reserve, 218–219, 220
between local people and university-educated, 251
Restraints and limits, 51, 72–73, 74, 84, 115, 192
"Restricted" warfare, 105
RFO. *See* Réserve de Faune à Okapis
Right relationship, 33, 52, 83, 84
Rituals, 23, 284
Roads, 153, 158–159, 200
Rothenberg, Deb, 63, 65, 66, 67, 80, 130, 144
Ruether, Rosemary Radford
on connecting with marginalized peoples, 48
on need for new ecological spirituality, 24

Sacredness of life, 247–248, 259, 272, 277
Sankofa, 142, 144, 240, 241
Sartre, Jean-Paul, 19(n4)
Savanna-forest ecotone, near Loko, 64(photo)
Schools and schooling
and cutback in time for initiations, 133
effect of depopulation on, 152
technical, 153, 154
Science
feminist critique of, 43
Sedentariness taboos, 229, 230
Selling and buying
ancestral attitude toward, 130
changes in style and scale, 98
Senghor, Léopold, 8, 19(n4)
"Sense of place," and mobility, 229–230
Serequeberhan, Tsenay, 25, 253
on destructiveness of colonial legacy, 45
Seto
place of, in Ngbaka cosmology, 121–124, 125
Settlement patterns
and Mbuti-villager relations, 200
rotational, 93, 229
and sedentariness, 230
Sharing ethic, 126, 204
Shipton, Parker
on African land tenure systems, 110
on symbolic thought and ecological practices, 73

SIFORZAL, 234
Simbotwe, M. P.
on conservation partnership with African traditional values, 174
on resentment over wildlife restrictions, 171
Sindima, Harvey J., 9
on African concept of creation, 272
on community in Africa, 108
on oneness of nature and person, 259
on value transformation, 142
on vision, 253
Skolimowski, Henryk
on conceptual alienation, 75
Sleeping sickness, 44, 65
Social realm
interconnectedness between natural realm, cosmological realm, and, 16, 86
Society
interconnectedness of nature, cosmology and, 125–126, 136, 144
Soil depletion, 93, 230
Songolo, Étienne (CEFRECOF researcher)
on communication between locals and project staff, 249–250
Soulé, Michael, 36
State, the
ecological damage by, 138–139
prohibitions against killing animals by, 192–193
role of, in protection/degradation of environment, 237–238
top-down control of local population through, 267, 268
Station de Capture des Okapis, 211
Stories and storytelling, 23, 50, 284
Symbols, 284
transformation through, 23

Taboos, 229, 244
Tambwe (Ubangian farmer)
on departure of ancestors and broken chain with natural world, 72
on replenishment of forests and water, 71
on solving social/environmental problems, 139
Taxes, 99, 152, 237. *See also* Fines
Termite gathering/harvesting, 88, 90
Things Fall Apart (Achebe), 34
Timber harvesting
pit-sawing of mahogany lumber, 236(photo)
Toma, Père, 153, 154
Totems
and conversion to Christianity, 232
ecological benefits of, 244

Traditions
    fluidity of, 193
    Vansina on, 44–45
Trapping, 88, 96
    cultivation correlated with, 91
    farming correlated with, 90
Tree plantings
    in Bodangabo, 137
    Eucharists, 119, 266
Tropical rainforests
    iconic visions of, 264. *See also* Rainforests
Tuan, Yi-Fu, 27, 32
    on appreciation of nature, 76
Turnbull, Colin
    on Père Longo, 154
    view on villagers by, 196, 200

Ubangi, the
    sankofa within, 142
Ubangian ethnic groups
    Gaza: role and demise of within, 131–133
"Unity in diversity," 29
Universalism and pluralism
    Engel on dialectic between, 27
    false dichotomy between, 38
    and "one/many problem," 21–22, 26–29
    relationship between, 25, 26
Université d'Aequatoria, 80–81
University-educated staff
    on ancestral traditions, 228
    perspective on insights about local villagers by, 225–226
    on role of Christian evangelism, 231–232. *See also* Project staff
University of Chicago, Ecology Group at, 35
University of Wisconsin-Madison, Central African oral traditions
    collections of, 286

Values, 3, 22–23
    transformation of, 5, 57, 142
Vansina, Jan, 26, 39
    on Atlantic trade, 129
    on holistic resource use, 90
    on human-environment relationship, 271
    on individual-community relations, 108
    Masolo's assessment of, 284–285
    quoted from *Paths in the Rainforests*, 44
    on role of garbage pits, 266
    on traditions, 44–45
    on types of warfare, 105

Warfare, 103, 105
Wasitdo (Ngbaka farmer)
    on diversifying livelihoods, 88
Water pollution, 77, 138–139, 185–186
Water spirits. *See* Mami wata
WCS. *See* Wildlife Conservation Society
Wealth
    in Central African societies, 131
    cosmological costs with, 95–96
    disparities of, under capitalism, 134
    distribution of, in community, 261
    rejection of ancestral ways and desire for, 133–134
Western environmental theory/ethics
    lessons from Central African human-environment relations for, 270–275
Western influences
    narrative critiques of, by project staff/academics, 229–237
Whitehead, Alfred North, 51
    on first hand knowledge, 49
    and misplaced pristineness, 263
Wilderness
    American notions of, 263, 264
    Cronon on idealizing of, 274
Wildlife Conservation Society, The, 16, 151, 220
Wiredu, Kwasi
    on stewardship of land, 83
Women
    farmers, 199–200
    and land tenure, 112–113
    peanut planting in the Ituri, 102(photo)
    female heads of household, 100, 112
World Bank, 46, 234
World Wildlife Fund, 151
Worster, Donald, 29
    on nature's patterns and ecological ethics, 36–37

Yambo (Ngombe farmer)
    on concern for future generations, 74
Youth
    ancestral ways rejected by, 81, 129
    in Bilenge ya Mwinda movement, 141
    influence of Western culture on, 46, 134
    reinvigorating ancestral traditions for, 142–143

Zaire, Democratic Republic of, xiii, 132. *See also* Congo, Democratic Republic of
Zoos, 74–75, 76, 77, 149